THE PHYSICS OF FLOCKING

In creatures ranging from birds to fish to wildebeest, we observe the collective and coherent motion of large numbers of organisms, known as "flocking." John Toner, one of the founders of the field of active matter, uses the hydrodynamic theory of flocking to explain why a crowd of people can all walk, but not point, in the same direction. Assuming a basic undergraduate-level understanding of statistical mechanics, the text introduces readers to dry active matter and describes the current status of this rapidly developing field. Through the application of powerful techniques from theoretical condensed matter physics, such as hydrodynamic theories, the gradient expansion, and the renormalization group, readers are given the knowledge and tools to explore and understand this exciting field of research.

This book will be valuable to graduate students and researchers in physics, mathematics, and biology with an interest in the hydrodynamic theory of flocking.

JOHN TONER is a theoretical physicist and Professor Emeritus at the University of Oregon, with a primary research focus in condensed matter physics. He predicted that soap is the best soundproofing material, and that quasicrystals are the hardest. He was awarded the American Physical Society's 2020 Onsager Prize in Statistical Physics for his pioneering work on flocking. He has also been the recipient of Simons and Gutzwiller Fellowships, and is a Fellow of the American Physical Society.

THE PHYSICS OF FLOCKING

Birth, Death, and Flight in Active Matter

JOHN TONER

University of Oregon

Shaftesbury Road, Cambridge CB2 8EA, United Kingdom

One Liberty Plaza, 20th Floor, New York, NY 10006, USA

477 Williamstown Road, Port Melbourne, VIC 3207, Australia

314–321, 3rd Floor, Plot 3, Splendor Forum, Jasola District Centre,
New Delhi – 110025, India

103 Penang Road, #05–06/07, Visioncrest Commercial, Singapore 238467

Cambridge University Press is part of Cambridge University Press & Assessment,
a department of the University of Cambridge.

We share the University's mission to contribute to society through the pursuit of
education, learning and research at the highest international levels of excellence.

www.cambridge.org
Information on this title: www.cambridge.org/9781108834568

DOI: 10.1017/9781108993623

When citing this work, please include a reference to the DOI 10.1017/9781108993623

First published 2024

A catalogue record for this publication is available from the British Library

A Cataloging-in-Publication data record for this book is available from the Library of Congress

ISBN 978-1-108-83456-8 Hardback

To Kim, for suggesting that I write this book, and for so many other things that this dedication would be longer than the book if I listed them all.

Contents

Preface

When I was first learning physics (back in the late Pleistocene), I often got the feeling that I'd been born too late. I thought that if I'd only been born in, say, 1642, *I*, rather than Newton, could have discovered the principles of mechanics. After all, I was *understanding* the principles of Newtonian mechanics as I was being taught them. So surely, I thought in my youthful arrogance and naivety, I could have *invented* it. Even at that early age, though, I recognized this as a vain self-delusion: We can all *learn* concepts that only someone *much* smarter than ourselves could *invent*.

Nonetheless, this experience repeated itself when I first learned fluid mechanics a few years later. As before, I thought, had I only been born at the time Navier and Stokes were, I could have derived the equations of fluid mechanics that now bear their names. At least, I could have done so (I thought) had I known the general principles of *hydrodynamics*, which is a way of thinking about macroscopic systems in general – i.e., not limited to simple fluids – that was developed long after Navier and Stokes did their pioneering work.

Again, of course, I dismissed this notion as pure (indeed, delusional) vanity.

But amazingly, in 1994, I actually got something like a chance to put this idea to the test. It happened when I went to a talk by Tamas Vicsek about his work on "flocking": the collective, coherent motion of large numbers of self-propelled "agents" (e.g., birds). Vicsek had developed a simulation of this phenomenon so clever, simple, and powerful that it is still extensively studied today, and which bears his name: the "Vicsek algorithm." Vicsek also pointed out in his talk that his simulation showed a phenomenon that should have shocked his audience, which consisted largely of statistical physicists: long-ranged order (in this case, as manifested by coherent motion of an arbitrarily large "flock" in *two* dimensions, even in the presence of noise). This, as Vicsek correctly noted, *looked* like a violation of one of the most cherished (because it's one of the few) theorems of statistical physics: the Mermin–Wagner–Hohenberg theorem, which states (loosely) that the phenomenon I just described – long-ranged order in two dimensions in the presence of nonzero noise – is impossible.

At least two of the statistical physicists in the audience – me, and Yuhai Tu – *were* shocked by this. Indeed, we were *so* shocked that we attempted to explain it that very afternoon. Not only did we succeed, but we did so by doing for flocks what Navier and Stokes had done for simple fluids: developing a "universal" theory that would describe *all* flocks.

That we could do so – and furthermore, do so in one afternoon and evening – was not because we're as brilliant as Navier and Stokes (although Yuhai might be close!), but because the general principles of hydrodynamics are so powerful that, once learned, they enable anyone who's learned them to formulate the hydrodynamic theory of *any* new system. In short, to go back to Newton again, we were standing on the shoulders of giants; in this case, the giants who developed our current understanding of hydrodynamics.

It is my modest purpose in this book to give readers precisely such an understanding, so that they too will be able to formulate hydrodynamic theories of heretofore unconsidered systems. I will do so using flocking systems – or, to use the current buzzwords, "dry active matter" – as an example.

In the process, I will also teach the reader the extremely powerful idea of the "renormalization group" – more specifically, the *dynamical* renormalization group (DRG). The ideas of the renormalization group – in particular, universality – are, in my opinion, the true basis of all hydrodynamic theories. Indeed, I'd go further, and concur with Paul Goldbart, who once said to me that "The renormalization group is not just a technique for calculating a few exponents. It's what explains why we can do physics at all."

I also hope the book will prove useful to those who already know these techniques, but want to get up to date on the current state of our understanding of these systems. This is particularly important because there are a number of incorrect papers about dry active systems out there (some of them written by me!), and this book provides me with an opportunity to set the record straight.

So that's what this book attempts to do. What it does *not* attempt to do is provide a comprehensive overview of everything that's happened in the field of "Active Matter" since its birth 28 years ago. There are review articles that have bravely tackled that task [1, 2], but this book does not attempt such breadth. Instead, I will focus exclusively on what are called "dry active matter" systems, which, crudely speaking, are systems in which a flock moves without momentum conservation. While appropriate for, e.g., ants moving over the ground, for which momentum can be lost due to friction with the ground, this does *not* describe the most familiar example of flocking: a flock of birds. That's because, as birds fly through the air, the total momentum of birds plus air is conserved. We know this has important consequences for bird motion – in particular, it leads to many instabilities, quite a few of which have been seen experimentally.

I've chosen to focus on "dry" systems for a number of reasons. First and foremost, they're simpler. Secondly, because "wet" (momentum conserving) systems tend to be unstable,[1] ordered states do not occur in them. And for a dyed-in-the-wool condensed matter theorist like me, ordered states are the most fascinating ones. Thirdly, when treating an unstable system, there's a limit to how much can be done analytically. Specifically, one can demonstrate the instability of the ordered state analytically. But then identifying what state such systems do ultimately reach, and what the properties of those states are, is often impossible to determine analytically. One is then forced to resort to numerical simulations, which, to put it mildly, are not my forte. Finally, I focus on dry systems because they're the ones about which I know the most.

The book is organized as follows.

In Chapter 1, I introduce the concept of flocking (or, for those who love jargon, "dry active matter"). In that chapter, I describe the Vicsek algorithm, and discuss the "hook" that got Yuhai and me interested in it all those years ago: the shocking fact that the algorithm exhibits long-ranged order in two dimensions, in apparent contradiction to the Mermin–Wagner–Hohenberg theorem.

In Chapter 2, I present a "dynamical" derivation of the Mermin–Wagner–Hohenberg theorem that none of those three giants would recognize. I also review the more conventional derivation. In doing so I also introduce the powerful concept of the gradient expansion, which is one of the essential tools of hydrodynamics in general (indeed, of *physics* in general), and of its application to flocking in particular.

Chapter 3 explains the dynamical renormalization group (DRG), illustrating it with the simplest (in some ways) example that I know: the Kardar–Parisi–Zhang (KPZ) equation. In addition to providing a simple illustration of the DRG, the KPZ equation will also make a later appearance in Chapter 8, where I'll show that one variant on the flocking problem in two dimensions can be mapped onto it.

In Chapter 4 I'll build on the ideas developed in the previous two chapters to derive the hydrodynamic equations of motion for flocking developed by Yuhai Tu and me. I'll also show that a linear analysis of those equations simply recovers the Mermin–Wagner–Hohenberg theorem; that is, it implies that flocks can *not* order in two dimensions.

Fortunately, as I'll show in Chapter 5 this linear analysis can be shown, using the DRG, to be wrong. The DRG further shows that the *nonlinear* terms in the hydrodynamic equations of motion modify the long-distance behavior of flocks in a way that makes them able to order in two dimensions. *Un*fortunately, it proves

[1] Oddly, the instabilities occur not for *high* Reynolds number, but for *low* Reynolds number. See [3, 4, 5, 6].

to be impractical to actually carry out the full DRG for this problem. Indeed, I estimate that, had we started such a calculation back in 1994, when Yuhai and I first formulated this problem, and worked on it to the exclusion of all other work since then, we *might* be finishing the calculation just about now. I think both of us are very glad we spent the intervening years doing other things. Some of those are described in the following chapters.

The first of those other things is discussed in Chapter 6: *incompressible* flocks, specifically in $d > 2$ spatial dimensions (which in practice obviously means $d = 3$). For this problem, we ("we" in this context being Chiu Fan Lee, Leiming Chen, and me) can not only get *exact* scaling laws, but we can do so just by counting on our fingers.

Indeed, so simple is this calculation that it proves possible to actually derive the results using a simple (at least, simple in hindsight once one knows the answer) hand-waving argument, which I present in Chapter 7.

Our old friend the KPZ equation returns in Chapter 8, in which I review the demonstration by the same gang of three (Chiu Fan, Leiming, and me) that *two*-dimensional incompressible flocks are actually described, albeit in a rather peculiar way, by the *one*-dimensional KPZ equation.

Finally, in Chapter 9, I'll discuss what I like to call "Malthusian" flocks; that is, flocks with birth and death. These prove to be treatable using the DRG. Unlike incompressible flocks, however, hand-waving is not sufficient for this problem: All of the machinery of the DRG must be brought to bear, in a rather formidable, but tractable calculation, done by the gang of three again, which yields very good approximate values for the scaling exponents describing these systems.

Chapters 1, 2, 4, and 5 appeared, in a somewhat different form, in reference [7].

Acknowledgments

I thank Yuhai Tu, Leiming Chen, and Chiu Fan Lee for their years of fruitful collaboration with me on these problems, and for graciously allowing me to use some jointly written material and many figures generated by them in this book. I also thank Tamas Vicsek for introducing us (and the world!) to this problem, Jim Sethna and Karen Dahmen for pointing out the existence of the λ_2 and λ_3 terms, and Pawel Romanczuk for pointing out the geometrical interpretation of the field $h(\mathbf{r})$ as the displacement of the flow lines.

The Max Planck Institute for the Physics of Complex Systems, Dresden, Germany, has been extremely supportive of me over the years, in particular through the Martin Gutzwiller Fellowship.

The Department of Bioengineering at Imperial College, London, The Higgs Centre for Theoretical Physics at the University of Edinburgh, and the Lorentz Center of Leiden University, also deserve thanks for their hospitality while this work was underway.

Finally, but most importantly, I thank Kim for everything.

1
Introduction and Motivation: Are Birds Smarter Than Nerds?

Everyone has seen "flocking," by which I mean the collective, coherent motion of large numbers of organisms [8, 9, 10, 11]. Flocks of birds and schools of fish, and herds of wildebeest, are all familiar sights (although the latter possibly only in nature documentaries). Perhaps nowadays it is most commonly seen in the simulations used for digital cinematic special effects [8, 9, 10, 11]; these have led to the only Oscar ever given for a physics project!

In the past couple of decades, many synthetic systems of self-propelled particles have been fabricated [12, 13] that also exhibit flocking. In addition to providing important experimental realizations of this phenomenon, these experiments make clear that flocking does *not* depend on intelligent decision making by the flockers, but, rather, can arise spontaneously from simple short-ranged interactions.

I will hereafter refer to all such collective motions – flocks, swarms, herds, collections of synthetic self-propelled objects, etc. – as "flocking"; for convenience, I will also refer to the "flockers" as "birds," or, alternatively, "boids."

Note that flocking can occur over an enormous range of length scales: from kilometers (herds of wildebeest) to microns (e.g., the microorganism *Dictyostelium discoideum* [14, 15, 16, 17]).

Remarkably, despite the familiarity and widespread nature of the phenomenon, it is only in the past three decades that many of the universal features of flocks have been identified and understood. It is my goal in this book to explain how we've come to understand one particular type of "flocking": namely, "polar ordered dry active fluids," which I'll define soon. In the process, I hope to introduce those of you unfamiliar with it to the "hydrodynamic" approach, which is a powerful technique that can be applied to any large-scale collective phenomenon.

1.1 An Example of Flocking: the Vicsek Model

To my knowledge, the first physicist to think about flocking – certainly the physicist who kicked off the modern field of active matter – was Thomas Vicsek [18, 19].

He was, as far as I know, the first to recognize that flocks fall into the broad category of nonequilibrium dynamical systems with many degrees of freedom that has, over the past few decades, been studied using powerful techniques originally developed for equilibrium condensed matter and statistical physics (e.g., scaling, the renormalization group, etc.). In particular, Vicsek noted an analogy between flocking and ferromagnetism: The velocity vector of the individual birds is like the magnetic spin of an iron atom in a ferromagnet. The usual "moving phase" of a flock, in which all the birds, on average, are moving in the same direction, is then the analog of the "ferromagnetic" phase of iron, in which all the spins, on average, point in the same direction. Another way to say this is that the development of a nonzero mean center of mass velocity $\langle \mathbf{v} \rangle$ for the flock as a whole therefore requires spontaneous breaking of a continuous symmetry (namely, rotational), precisely as the development of a nonzero magnetization $\mathbf{M} \equiv \langle \mathbf{S} \rangle$ of the spin in a ferromagnet breaks the continuous[1] spin rotational symmetry of the Heisenberg magnet.

Because $\langle \mathbf{v} \rangle$ is only nonzero in the ordered state, it is an "order parameter" for flocking [20].

To make this analogy complete obviously requires that the birds, like the spin in a ferromagnet, live in a rotation invariant environment; that is, that the spins have nothing external that tells them in which direction to point, and the birds have nothing external that tells them which way to fly.

To study this phenomenon (the spontaneous breaking of rotation invariance by collective motion – which is what I will mean henceforth by the term "flocking"), Vicsek formulated his deservedly famous algorithm. I will now describe this algorithm in detail.

The model incorporates the following general features.

(1) A large number (a "flock") of point particles ("boids"[2]) each move over time through a space of dimension d $(= 2, 3, \ldots)$, *attempting* at all times to "follow" (i.e., move in the same direction as) their neighbors.
(2) The interactions are purely short ranged: Each "boid" responds only to its neighbors, defined as those "boids" within some fixed, finite distance R_0, which is assumed to be independent of L, the linear size of the "flock."
(3) The "following" is not perfect: The "boids" make errors at all times, which are modeled as a stochastic noise. This noise is assumed to have only short-ranged spatio-temporal correlations.

[1] Of course, in a real crystalline ferrogmagnet, crystal symmetry breaking fields make the rotational symmetry of the spins discrete, rather than continuous, since there are only a discrete set of orientations for the spin preferred by the lattice. In flocks, there are no such symmetry breaking fields, so the rotational symmetry *is* continuous, as it is in the idealized $0(n)$ Heisenberg model of a ferromagnet. Everything I say hereafter about ferromagnetic systems implicitly refers to this fully rotationally invariant $0(n)$ model.

[2] I will frequently use the term "boid" (a short form for "birdoid"), coined by C. Reynolds [8].

(4) The underlying model has complete rotational symmetry: The flock is equally likely, a priori, to move in any direction.

Any model that incorporates these general features should belong to the same "universality class," in the sense that term is used in critical phenomena and condensed matter physics. That is, all such systems should be described by the same simple, universal scaling laws at large distances and times. To see this "universality," of course, we need a large flock: The universality becomes exact in the "thermodynamic" limit; i.e., as $N \to \infty$, where N is the number of "boids" in the flock. The specific model proposed and simulated numerically by Vicsek is the following.

In Vicsek's discrete time model, a number of birds labeled by i move in a two-dimensional plane with positions $\{\mathbf{r}_i(t)\}$, with time t being a discrete (integer) variable. At each integer time, all of the birds simultaneously choose the direction they will move on the next time step (taken to be of duration $\Delta t = 1$) by averaging the directions of motion of all of those birds within a circle of radius R_0 (in the most convenient units of length $R_0 = 1$) on the previous time step (i.e., updating is simultaneous). The distance R_0 is assumed to be $\ll L$, the size of the flock. The direction the bird actually moves on the next time step differs from the previously described direction by a random angle $\eta_i(t)$, with zero mean and standard deviation Δ. The distribution of $\eta_i(t)$ is identical for all birds, time independent, *and* uncorrelated between different birds and different time steps. Each bird then, on the next time step, moves in the direction so chosen a distance $v_0 \Delta t$, where the speed v_0 is the same for all birds.

To summarize, the rule for bird motion in $d = 2$ is

$$\theta_i(t+1) = \left\langle \theta_j(t) \right\rangle_n + f_i(t), \tag{1.1.1}$$

$$\mathbf{r}_i\,(t+1) = \mathbf{r}_i(t) + v_0\,(\cos\theta(t+1), \sin\theta(t+1))\,, \tag{1.1.2}$$

$$\langle f_i(t) \rangle = 0, \tag{1.1.3}$$

$$\left\langle f_i(t) f_j(t') \right\rangle = 2D\delta_{ij}\delta_{tt'}, \tag{1.1.4}$$

where the symbol $\langle\rangle_n$ denotes an average over "neighbors," which are defined as the set of birds j satisfying

$$\left|\mathbf{r}_j(t) - \mathbf{r}_i(t)\right| < R_0\,. \tag{1.1.5}$$

Here $\langle\rangle$ without the subscript n denote averages over the random distribution of the noises $f_i(t)$, and $\theta_i(t)$ is the angle of the direction of motion of the ith bird (relative to some fixed reference axis) on the time step that ends at t.

The quantity $\langle\theta_j(t)\rangle_n$ is defined via

$$\langle\theta_j(t)\rangle_n \equiv \arctan\left(\frac{\langle\sin(\theta_j(t))\rangle_n}{\langle\cos(\theta_j(t))\rangle_n}\right), \tag{1.1.6}$$

where

$$\langle\sin(\theta_j(t))\rangle_n = \frac{\sum_{j\in n}\sin(\theta_j(t))}{N_n},$$
$$\langle\cos(\theta_j(t))\rangle_n = \frac{\sum_{j\in n}\cos(\theta_j(t))}{N_n} \tag{1.1.7}$$

with $\sum_{j\in n}$ denoting a sum over all neighbors (that is, all birds) satisfying (1.1.5), and N_n the number of birds satisfying (1.1.5).

This definition is equivalent to saying that the direction each bird moves between time t and time $t+1$ would, in the absence of noise, be the direction of the average of the velocity *vectors* $\mathbf{v}_j(t)$ of its neighbors at time t.

The reason for this convoluted definition of the average $\langle\theta_j(t)\rangle_n$ is that using the more obvious definition

$$\langle\theta_j(t)\rangle_{n\text{wrong}} = \frac{\sum_{j\in n}\theta_j(t)}{N_n} \tag{1.1.8}$$

has pathologies associated with the fact that θ_j, like all angles, is defined only modulo 2π. To have an unambiguous definition of θ_j, therefore, one must introduce a "cut"; that is, define θ_j to always lie within some range of width 2π.

For example, one could choose to define θ_j to always lie in the interval $0 < \theta_j < 2\pi$, with $\theta_j = 0$ defined to point to the east. But then consider a situation in which a bird had two neighbors, one heading one degree due south of east, the other heading one degree due north of east. With our convention, we'd define $\theta_1 = 1°$, and $\theta_2 = 359°$. Thus, on our next step, if we used the rule (1.1.8), our bird would head off at 180°; i.e., almost exactly *opposite* the direction of its neighbors. This is clearly *not* following your neighbors!

You might think you could fix this problem by putting the "cut" due west; that is, by defining θ_j to always lie in the interval $-\pi < \theta_j < \pi$. However, you can easily convince yourself that, while this fixes the problem just described when your neighbors are heading almost due east, it gives you the same problem if they're heading almost due west. Indeed, in general, one will always have problems with the rule (1.1.8) if your neighbors are moving close to, but on opposite sides of, the direction in which you choose to put the cut.

You can easily convince yourself that the rule (1.1.7) has no such problems.

The flock evolves through the iteration of this rule. Note that the "neighbors" of a given bird may change on each time step, since birds do not, in general, move in exactly the same direction as their neighbors.

I have been rather precise and detailed in explaining this algorithm. However, we actually believe that most of the details of this algorithm do not matter for the scaling properties of the flock. Only a few features (all of which the Vicsek algorithm possesses) *do* matter for those scaling laws. These features are: activity, conservation laws, symmetries, short-ranged interactions, and noisiness. We now elaborate on these.

(1) Activity: A large number (a "flock") of point particles ("boids") each move over time through a space of dimension d ($= 2, 3, \ldots$), *attempting* at all times to "follow" (i.e., move in the same direction as) their neighbors. This motion is due to some form of self-propulsion; in Vicsek's algorithm, the rule is that the speed of each creature is constant. Departures from this rule are not important, provided that the boids prefer to be in a state of motion, rather than at rest. This is what is meant by the word "active" in "polar ordered dry active fluids."
This self-propulsion requires an energy source; it also requires that the system be out of equilibrium. Dead birds don't flock!
(2) Conservation laws: The underlying model does *not* conserve momentum; the total momentum of the flock can change. Indeed, it does so every time a creature turns. We imagine this violation of momentum conservation can happen because the creatures move either over a fixed surface, in two dimensions, or through some fixed matrix (e.g., a gel) in three dimensions, with which they interact frictionally. This surface or matrix therefore acts like a momentum "source" or "sink."
This lack of momentum conservation is what is meant by the term "dry" in "polar ordered dry active fluids." Note that many active systems – e.g., many active nematics – are "wet," by which we mean momentum is conserved. Note, incidentally, that real birds (and not only water birds!) are "wet" in this sense, since the sum of their momentum and the momentum of the air through which they fly is conserved. This changes the dynamics considerably. The problem of wet flocks can still be treated by a hydrodynamic approach [3, 4, 5, 6], but the hydrodynamic model is different because of momentum conservation. I will not discuss that case further here.
There *is* one conservation law in the Vicsek algorithm, however: The number of birds is conserved. That is, birds are not being born or dying "on the wing." You laugh, but there are many biological situations – bacteria swarms, and tissue development to name just two – in which this is not a good approximation: Bacteria or cells are being born and dying on the time scale of the motion.

The hydrodynamics of this case is quite interesting [21, 22, 23], and will be discussed in Chapter 9.

(3) Symmetry: The underlying model has complete rotational symmetry; the flock is equally likely, a priori, to move in any direction. I will here consider models that do *not* have Galilean invariance: that is, they have a preferred Galilean frame. This frame is the one in which the background medium over or through which the boids move is stationary. In the Vicsek algorithm, this is the unique frame in which the *speeds* (i.e., the *magnitudes* of the velocity vectors, but *not* the velocity vectors themselves) are the same (and given by v_0).

(4) The interactions are purely short ranged: In Vicsek's model, each "boid" only responds to its neighbors. In Vicsek's model, these are defined as those "boids" within some fixed, finite distance R_0, which is assumed to be independent of L, the linear size of the "flock." Hence, in the limit of flock size going to infinity – i.e., the "thermodynamic limit" – the range of interaction is much smaller than the size of the flock. Variants on this rule – for example, interactions whose strength falls off exponentially with distance – can also be considered short ranged.

(5) The "following" is not perfect: The "boids" make errors at all times, which are modeled as a stochastic noise. This noise is assumed to have only short-ranged spatio-temporal correlations. Its role in this problem is very similar to the role of temperature in equilibrium systems: It tends to disorder the flock. As you'll see, one of the most interesting questions in this problem is whether the ordered state can survive this noise.

In addition to these symmetries of the equations of motion, which reflect the underlying symmetries of the physical situation under consideration, it is also necessary to treat correctly the symmetries of the *state* of the system under consideration. These may be different from those of the underlying system, precisely because the system may spontaneously break one or more of the underlying symmetries of the equations of motion. Indeed, this is precisely what happens in the ordered state of a ferromagnet: The underlying rotation invariance of the system as a whole is broken by the system in its steady state, in which a unique direction is picked out – namely, the direction of the spontaneous magnetization.

As should be apparent from our earlier discussion, this is also what happens in a spontaneously moving flock. Indeed, the symmetry that is broken – rotational – and the manner in which it is broken – namely, the development of a nonzero expectation value for some vector (the spin $\mathbf{S}$ in the ferromagnetic case; the velocity $\mathbf{v}$ in the flock) – are precisely the same in both cases.[3]

[3] The isotropic Heisenberg model of magnetism is invariant under uniform rotation of all the spins, without a corresponding rotation of the lattice on which they live. A flock, like an ordinary collection of interacting

The fact that it is a unique *vector* that is singled out, rather than merely a unique *axis*, is the meaning of the word "polar" in "polar ordered dry active fluids."

Many different "phases,"[4] in this sense of the word, of a system with a given underlying symmetry are possible. Indeed, I have already described two such phases of flocks: the "ferromagnetic" or moving flock, and the "disordered," "paramagnetic," or stationary flock.

In equilibrium statistical mechanics, this is precisely how we classify different phases of matter: by the underlying symmetries that they break. Crystalline solids, for example, differ from fluids (liquid and gases) by breaking both translational and orientational symmetry. Less familiar to those outside the discipline of soft condensed matter physics are the host of mesophases known as liquid crystals, in some of which (e.g., nematics [24]) only orientational symmetry is broken, while in others, (e.g., smectic [24], which we'll revisit in Chapter 8) translational symmetry is only broken in *some* directions, not all.

It seems clear that, at least in principle, every phase known in condensed matter systems could also be found in flocks. In this book, I'm going to focus on just one phase: the "polar ordered dry active fluid phase," in which rotational symmetry is completely broken by the development of a nonzero average flock speed $\langle \mathbf{v} \rangle$, but all of the other symmetries of the dynamics (e.g., translation invariance) are preserved. The word "fluid" in "polar ordered active fluids" is what tells us that these are systems in which translational invariance is *not* broken.

The first, and to my mind still the biggest, surprise in the entire field of active matter is that a "polar ordered dry active fluid phase" is even possible in two dimensions. The reason Yuhai and I (and Vicsek) found this so surprising is the well-known "Mermin–Wagner–Hohenberg theorem" [25, 26, 27] of equilibrium statistical mechanics. This theorem states that in a thermal equilibrium model at nonzero temperature with short-ranged interactions, it is impossible to spontaneously break a continuous symmetry. This implies in particular that the equilibrium or "pointer" version of Vicsek's algorithm described earlier, in which the birds carry a vector $\mathbf{v}_i$ whose direction is updated according to Vicsek's algorithm, but in which the birds do not actually move, can never develop a true long-ranged ordered state in which all the $\mathbf{v}_i$ point, on average, in the same direction (more precisely, in which $\langle \mathbf{v} \rangle \equiv \frac{\Sigma_i \mathbf{v}_i}{N} \neq \vec{0}$), since such a state breaks a continuous symmetry, namely rotation invariance.

Yet the *moving* flock evidently has no difficulty in doing so; as Vicsek's simulation shows, even two-dimensional flocks with rotationally invariant dynamics,

molecules, such as those which form liquid crystals, is invariant only under spatial rotations, which rotate both the position and the velocity vectors of the creatures of the flock.

[4] By "phases" in systems far from equilibrium, I simply mean nonequilibrium steady states of a given symmetry.

short-ranged interactions, and noise – i.e., seemingly all of the ingredients of the Mermin–Wagner–Hohenberg theorem – *do* move with a nonzero macroscopic velocity, which requires $\langle \mathbf{v} \rangle \neq \mathbf{0}$, which, in turn, breaks rotation invariance, in seeming violation of the theorem.

There are a pair of gedanken experiments that make the very paradoxical and surprising nature of this result more obvious. Both experiments start by putting a million people on a flat, featureless plane in the fog. The featurelessness of the plane, and the fog, ensure rotation invariance (since they leave the people with no external indication of a preferred direction), while the fog has the further role of ensuring that each person can see only a few of her nearest neighbors.

The first experiment now consists of asking everyone to try to point in the same direction. The result is that the people cannot all point in the same direction, no matter how good a job they do at aligning with their nearest neighbors (unless, of course, the alignment is perfect). If they make the slightest errors, those will accumulate over distance, so that, even though a given person may point in roughly the same direction as others not too far away from her, widely separated people will inevitably be pointing in wildly different directions.

The second gedanken experiment consists of slightly modifying the instructions given to these million folks: Now ask them to all *walk* in the same direction.

Amazingly, if this instruction is given to the same people, in the same fog, with the same errors, they *can* all walk in the same direction. Moving, apparently, is fundamentally different from pointing.

Why? There is a very simple explanation for this *apparent* "violation" of the Mermin–Wagner–Hohenberg theorem: One of the essential premises of the Mermin–Wagner–Hohenberg theorem does *not* apply to movers, namely, they are *not* systems in thermal equilibrium. The nonequilibrium aspect arises from the motion: You can't move forever in a medium with friction unless you're alive. And, if you're alive, you're not in thermal equilibrium (that's why we say "cold and dead").

Clearly, motion must be what stabilizes the order in $d = 2$: As described above, the motion is the *only* difference between the pointing and moving gedanken experiments just described.

But *how* does motion get around the Mermin–Wagner–Hohenberg theorem? And, more generally, how best to understand the large-scale, long-time dynamics of a very large, moving flock?

The answer to these questions can be found in the field of hydrodynamics. I will apply that body of knowledge to flocks in Chapter 4. But first, I'll explain why the Mermin–Wagner–Hohenberg theorem is true for pointers, which will give us some insight into why it's *not* true for walkers.

2

Dynamical Derivation (and a More Conventional One) of the Mermin–Wagner–Hohenberg Theorem; or, Why We Can't All Point the Same Way

You will not, with good reason, see anything like the following derivation in any textbook on statistical mechanics. The usual derivation involves the powerful tools of equilibrium statistical mechanics: Boltzmann weights, Hamiltonians, and the like. Since the Mermin–Wagner–Hohenberg theorem was derived for equilibrium systems, for which all these tools are available, it would be completely nuts (to use the technical term) not to take advantage of those powerful tools. Indeed, at the end of this chapter, I'll review that "standard" approach.

However, *none* of those very powerful tools is available for nonequilibrium systems like flocks. It's therefore useful, I think, to attempt the seemingly crazy stunt of deriving the Mermin–Wagner–Hohenberg theorem in a purely dynamical way that can be generalized to nonequilibrium systems. In this way I hope to elucidate exactly what it is about moving that is fundamentally different from pointing, and in particular, how that difference makes long-ranged order literally infinitely more robust in two dimensions in a moving system than a pointing one.

2.1 Describing "Pointers" by a Noisy Diffusion Equation

So let's think about those million pointers on the featureless plane in the fog. Consider in particular the angle $\theta_i(t)$ between the direction $\hat{\mathbf{n}}$ at which a given pointer labeled by i is pointing at time t and some fixed reference direction, as illustrated in Figure 2.1.1.

A "Vicsek-like" algorithm for pointers which try to align with their neighbors is the following updating rule for θ_i:

$$\theta_i(t+1) = \langle \theta_j(t) \rangle_n + f_i(t), \tag{2.1.1}$$

where the average over neighbors, and the neighbors themselves, are defined as described earlier for the Vicsek model.

Fig. 2.1.1 Illustration of the "Vicsek-like" algorithm for pointers. The angle θ_i between the ith pointer and some common reference direction is updated on each step to be the average over its "neighbors," defined to be those that lie within a distance R_0 of it. Reproduced from [7] by permission of Oxford University Press.

The extra term f_i is a random noise that takes into account the fact that the pointers will inevitably make mistakes in aligning with their neighbors. We'll assume this has zero mean (that is, the pointers are no more likely to err to the left than to the right), and variance $2D$, and that it is uncorrelated between pointers (i,j), and between successive time steps. That is,

$$\langle f_i(t)\rangle = 0, \tag{2.1.2}$$

$$\left\langle f_i(t) f_j(t')\right\rangle = 2D\delta_{ij}\delta_{tt'}, \tag{2.1.3}$$

where $\langle\rangle$ without the subscript n denote averages over the random distribution of the noises $f_i(t)$. Here the noise strength D will play the role of temperature, in the sense that larger D will lead to more fluctuations, and hence, presumably, less order.

The flock evolves through the iteration of this rule. Note that the "neighbors" of a given pointer do not change on each time step. To foreshadow where I'm ultimately going here, this is not true for movers, which can change their neighbors due to the differences in the motion of different movers within the flock. This is the fundamental difference between pointers and movers that makes the movers capable of aligning in two dimensions, while the pointers cannot.

But let's not get ahead of ourselves here. Returning to the pointers problem, I note, as first pointed out by Vicsek himself, that this model is exactly a simple, relaxational dynamical model for an equilibrium ferromagnet. That is, if we interpret each unit vector $\mathbf{n}_i$ that gives the direction the ith pointer is pointing as a "spin" carried by that pointer, and update each of them according to the above rule, then the model is easily shown to be an equilibrium ferromagnet, which will relax to the Boltzmann distribution for an equilibrium Heisenberg model (albeit with the "spin" living not on a periodic lattice, as they usually do in most models and in real ferromagnets, but, rather, on a random set of points). In the absence of noise (i.e., for $D = 0$), this algorithm will, unsurprisingly, lead to a "ferromagnetic" state, characterized by a nonzero "magnetization":

$$\mathbf{M} \equiv \langle \mathbf{n} \rangle \equiv \frac{1}{N} \sum_{i=1}^{N} \mathbf{n}_i, \tag{2.1.4}$$

where in this expression the $\langle \rangle$ mean an average over all the pointers. I'll assume throughout these notes that this average is equal to an average over the noise; in equilibrium physics, this is sometimes called the assumption of "ergodicity."

At zero noise, we would expect to, and do, eventually reach a state in which $|\mathbf{M}| = 1$; i.e., perfect alignment of all the pointers. The big question is: What happens when there is noise (i.e., when $D \neq 0$)?

To answer this, begin by noting that the dynamical rule (2.1.1) is actually a disguised version of a noisy diffusion equation. To see this, recall that one of the numerical algorithms for solving Laplace's equation $\nabla^2\theta = 0$ is to replace the value of the field θ at each point with the average of its neighbors. Thus, in the absence of noise, the dynamics (2.1.1) will eventually relax the field θ to a state in which $\nabla^2\theta = 0$, which implies that the rate of change of θ (again, in the absence of noise) is itself proportional to $\nabla^2\theta$ (since it vanishes when $\nabla^2\theta = 0$). Indeed, one can very simply derive this result as follows.

Consider for simplicity (although it is not necessary) a two-dimensional collection of pointers arranged on a square grid of lattice constant a, as illustrated in Figure 2.1.2.

The pointer at position $\mathbf{r}_i = (x_i, y_i)$, where my x- and y-axes are aligned with the square grid, has four neighbors, one to its right at $\mathbf{r}_1 = (x_i + a, y_i)$, a second to its left at $\mathbf{r}_2 = (x_i - a, y_i)$, a third above at $\mathbf{r}_3 = (x_i, y_i + a)$, and the fourth below at $\mathbf{r}_4 = (x_i, y_i - a)$. Thus, the dynamical rule (2.1.1) can be rewritten:

$$\begin{aligned}\theta_i(x_i, y_i, t+1) - \theta_i(x_i, y_i, t) = &\frac{1}{4}[\theta(x_i + a, y_i, t) + \theta(x_i - a, y_i, t) + \theta(x_i, y_i + a, t) \\ &+ \theta(x_i, y_i - a, t)] - \theta_i(x_i, y_i, t) + f_i(t),\end{aligned} \tag{2.1.5}$$

where I have subtracted the value $\theta_i(t)$ of θ_i on the previous time step from both sides, so as to make the left-hand side look like a discrete representation of a time derivative. In the process, I have made the right-hand side into a discrete version of the Laplacian. To see this, just reorganize the right-hand side as follows:

$$\begin{aligned}\theta_i(x_i, y_i, t+1) - \theta_i(x_i, y_i, t) = &\frac{1}{4}[\{\theta(x_i + a, y_i, t) - 2\theta_i(x_i, y_i, t) + \theta(x_i - a, y_i, t)\} \\ &+ \{\theta(x_i, y_i + a, t) - 2\theta_i(x_i, y_i, t)) + \theta(x_i, y_i - a, t)\}] \\ &+ f_i(t).\end{aligned} \tag{2.1.6}$$

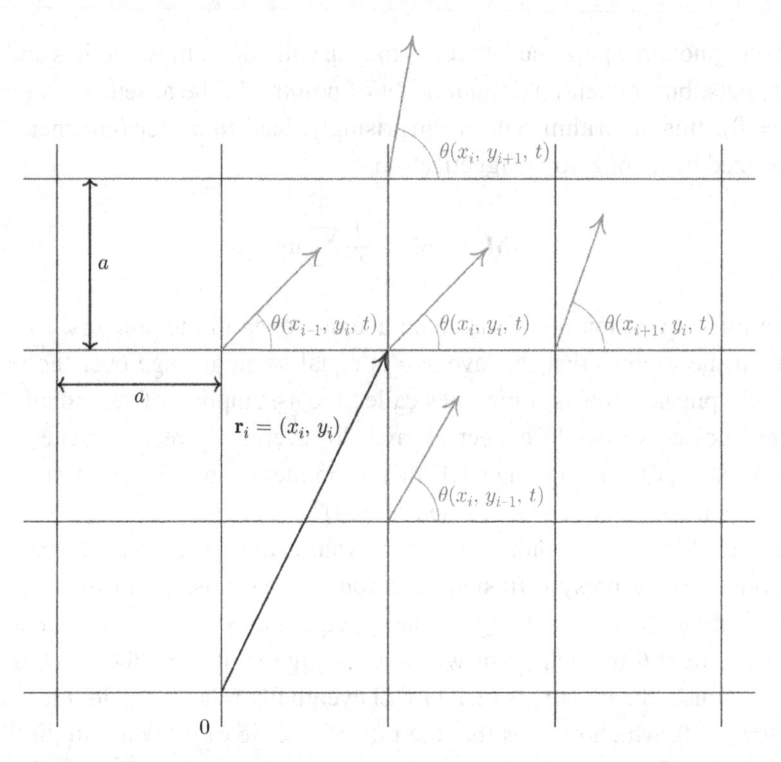

Fig. 2.1.2 A grid arrangement of pointers, to illustrate the diffusive nature of the angle dynamics, as explained in the text. Reproduced from [7] by permission of Oxford University Press.

Now note that, just as the left-hand side can be approximated as the time derivative of θ if θ varies slowly in time – that is, we can write $\theta_i(x_i, y_i, t+1) - \theta_i(x_i, y_i, t) \approx \partial_t\theta$ – the term in the first parenthesis on the right-hand side can be approximated by the second derivative of θ with respect to x: $\theta(x_i + a, y_i, t) - 2\theta_i(x_i, y_i, t) + \theta(x_i - a, y_i, t) \approx a^2\partial_x^2\theta$, provided that θ varies slowly with position. Likewise, the second term can be approximated by the second derivative of θ with respect to y: $\theta(x_i, y_i + a, t) - 2\theta_i(x_i, y_i, t)) + \theta(x_i, y_i - a, t) \approx a^2\partial_y^2\theta$. Hence, equation (2.1.6) can be approximated as

$$\partial_t\theta = \frac{a^2}{4}\left[\partial_x^2\theta + \partial_y^2\theta\right] + f = \nu\nabla^2\theta + f\,, \tag{2.1.7}$$

where I've defined the "diffusion constant" $\nu \equiv \frac{a^2}{4}$.

Note that I could also have derived this result purely on symmetry grounds: $\partial_t\theta$ must be a scalar made out of θ itself and its derivatives. By rotation invariance, it must vanish if θ is spatially uniform. By the isotropy of space, it must be an isotropic, scalar operator. The only thing you can make that does this to second order in gradients of θ (and linear order in θ) is $\nabla^2\theta$.

In writing this noisy diffusion equation (2.1.7), we have gone over to a continuum description. To complete this description, we must also specify the correlations of $f(\mathbf{r}, t)$ in continuum form as well. That form is

$$\langle f_i(\mathbf{r}, t) f_j(\mathbf{r}', t') \rangle = 2D\delta_{ij}\delta^d(\mathbf{r} - \mathbf{r}')\delta(t - t') \,. \tag{2.1.8}$$

The noisy diffusion equation (2.1.7) is also sometimes called the Edwards–Wilkinson equation [28].

2.2 Using the Noisy Diffusion Equation to Derive the Mermin–Wagner–Hohenberg Theorem; or: Mermin and Wagner Would Turn Over in Their Graves, If They Weren't Still Alive

So what are the consequences of the fact that θ obeys a diffusion equation? There are two that are important for our discussion:

(1) θ is slow, and
(2) θ is conserved (in the absence of noise).

To be more precise about point (1), the form of the diffusion equation implies that an initially localized departure of $\theta(\mathbf{r}, t = 0)$ from spatial homogeneity spreads very slowly. One can read this off by power counting from the form of the diffusion equation: A time derivative of θ can be estimated as roughly θ over a time t, while the Laplacian of θ can be estimated as θ divided by a distance r squared. Equating these gives

$$r^2 \propto t \,, \tag{2.2.1}$$

or, equivalently,

$$r \propto \sqrt{t} \,. \tag{2.2.2}$$

The exact solution (in the absence of noise) of the diffusion equation in d-spatial dimensions for an initially localized θ, which is

$$\theta(\mathbf{r}, t) = \theta_0 \exp\left(-\frac{r^2}{4\nu t}\right) / (4\pi \nu t)^{\frac{d}{2}} \,, \tag{2.2.3}$$

where $\theta_0 \equiv \int \mathrm{d}^d r\, \theta(\mathbf{r}, t)$ is a constant, clearly obeys this scaling law.

This is very slow; indeed, anything moving at any constant speed, however small, will eventually outrun diffusive spreading, since $t \gg \sqrt{t}$ as $t \to \infty$. This is why you stir your coffee after adding milk to it: Even a slow stir leads to far faster mixing than diffusion. We'll see later that the reason flocks can order in two dimensions is essentially that, by their motion, they stir themselves.

Turning now to point (2), we can see that θ is conserved in the absence of noise by setting $f = 0$ and integrating both sides of the diffusion equation (2.1.7) over all $\mathbf{r}$. This gives

$$\frac{\mathrm{d}}{\mathrm{d}t}\int \mathrm{d}^d r\, \theta(\mathbf{r}, t) = \nu \int \mathrm{d}^d r\, \nabla^2 \theta(\mathbf{r}, t) = \nu \int \mathrm{d}^d r\, \nabla \cdot \nabla \theta(\mathbf{r}, t) = \nu \int_S \mathrm{d}\mathbf{A} \cdot \nabla \theta(\mathbf{r}, t) \,, \tag{2.2.4}$$

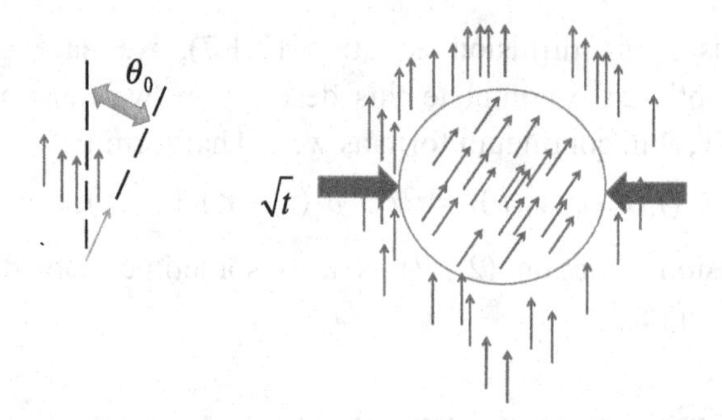

Fig. 2.2.1 Evolution of a single error in the pointer problem. The original error θ_0 gets shared evenly after a time t among all of the pointers within a distance $\propto \sqrt{t}$ of the original pointer. Reproduced from [7] by permission of Oxford University Press.

where in the last equality I've used the divergence theorem, with S being the surface bounding my volume of integration. This shows that the integral of θ over any region of space can change only if there is a current (proportional to $\nabla\theta$) through the surface of that region. Basically, θ acts like the milk in your coffee: The total quantity of it is conserved; diffusion can only redistribute it in space. And it can only do *that* very slowly; i.e., the spatial spread $r(t)$ after a time t only grows like $\sqrt{t}$.

The consequence of these two observations is that fluctuations decay very slowly in the pointer system. To illustrate this dramatically, consider a pointer system with no noise, and with an almost perfectly ordered initial condition: Only the pointer at the center is pointing in a different direction from any of the others, and he is pointing at an angle θ_0 to the right of the direction the others are pointing. (See Figure 2.2.1). Defining $\theta = 0$ as the direction in which the others are pointing, we have $\int d^d r\, \theta(\mathbf{r}, t = 0) = \theta_0$.

What will this collection look like after time t, if there is no noise? Well, by point (1), the initial error will now be spread out over all the pointers within a distance $r(t) \propto \sqrt{t}$. By point (2), the sum of the deviations of all of these pointers (including the original error-making pointer) from the original direction of most of them must still be θ_0, since θ is conserved. So the original fluctuation of θ_0 must now be distributed over all of those pointers within that distance $r(t) \propto \sqrt{t}$. Hence, we can crudely estimate the angular deviation after a time t by assuming (as proves to be the case) that this initial error is spread roughly uniformly over all $N(t)$ of the pointers within this distance $r(t)$. That number $N(t)$ is easy to estimate; it's just the density times the volume (or hypervolume, if we're considering $d \neq 3$) of the region of radius $r(t) \propto \sqrt{t}$. Assuming the density is roughly constant, at least over a sufficiently large region (as indeed it is for a random set of points; fluctuations in the density of a random set of points over a volume V scale as $\frac{1}{\sqrt{V}} \to 0$ for $V \to \infty$), it's clear that

$$N(t) \propto [r(t)]^d \propto t^{d/2}, \tag{2.2.5}$$

where I've used $r(t) \propto \sqrt{t}$.

Since the original total error of θ_0 is now divided among all $N(t)$ of these pointers, the typical fluctuation $\theta(t)$ of each of them, including the original error-maker, is now

$$\theta(t) \approx \frac{\theta_0}{N(t)} \propto \frac{\theta_0}{t^{d/2}}. \tag{2.2.6}$$

Note that the exact solution (2.2.3) to the diffusion equation agrees with the result: For $r \lesssim \sqrt{\nu t}$, the exponential in (2.2.3) is $\mathcal{O}(1)$, so the exponential itself is also $O(1)$. Hence, in that region of radius $\sim \sqrt{\nu t}$, $\theta(t) \propto \frac{\theta_0}{t^{d/2}}$, as (2.2.6) asserts.

I want to call your attention to two things about this result.

(1) The decay is extremely slow; specifically, it is a power law in time. Hence, it is asymptotically slower than *any* exponential decay. This is a consequence of the conservation law for θ, which is in turn a consequence of the underlying rotation invariance. This means θ is a *Goldstone mode* of the system, a concept that may already be familiar to some readers. I'll discuss this more in Section 2.3.

(2) The power law of this decay is dimension dependent, with slower decay in lower dimensions. This is a general and recurring theme in statistical and condensed matter physics: Fluctuations decay more slowly, and, hence, are more important, in lower dimensions. Ultimately, this is why the Mermin–Wagner–Hohenberg theorem applies to low-dimensional systems – specifically, $d \leq 2$ – but not higher-dimensional ones.

With this result (2.2.6) for a single initial error in hand, let's now go back and consider our original model with noise. Now the situation is even worse: While any initial errors are very slowly decaying according to (2.2.6), more errors are constantly being made. The question now becomes, can the slow decay of (2.2.6) keep up with the accumulation of new errors? Given the dimension dependence of (2.2.6), you won't be surprised to learn that the answer to this question is also dimension dependent: The errors can be kept under control for spatial dimensions $d > 2$, but not for $d \leq 2$. This is the Mermin–Wagner–Hohenberg theorem [25, 26, 27].

To see this, consider the "blob" of pointers that can have exchanged information diffusively with some central pointer after a time t. As noted earlier, this blob will have radius $r(t) \propto \sqrt{t}$, or, equivalently, given a radius r of the blob, the time required for all parts of that blob to be able to communicate with each other is

$$t \propto r^2. \tag{2.2.7}$$

This blob will contain $N(t) \propto r^d$ pointers. How many errors will these pointers collectively have made? Well, each of them will have made

$$\#\,\text{of errors/pointer} \propto t \propto r^2 ; \tag{2.2.8}$$

hence, the full collection of $N(t) \propto r^d$ of them will have made

$$\text{total}\,\#\,\text{of errors} \propto N(t)t \propto r^{d+2} . \tag{2.2.9}$$

Since the typical magnitude of the sum of a number of independent random variables with zero mean is proportional to the square root of that number, we have

$$\sqrt{\langle \theta^2 \rangle} \approx \frac{\sqrt{\text{total}\,\#\,\text{of errors}}}{N(t)} \propto \frac{r^{\frac{d+2}{2}}}{r^d} \propto r^{1-d/2} , \tag{2.2.10}$$

which diverges as r (or, equivalently, time t) goes to infinity for $d < 2$. As often happens, the vanishing of this exponent $1 - d/2$ in (2.2.10) in $d = 2$ indicates *not* a constant, but a logarithm: In fact, in Section 2.3 I'll show that a slightly more careful version of the reasoning used here, applied in exactly $d = 2$, implies that

$$\sqrt{\langle \theta^2 \rangle} \propto \sqrt{\ln(r)} \to \infty \tag{2.2.11}$$

in $d = 2$.

So we've shown by this purely dynamical argument that, for $d \leq 2$, fluctuations diverge in the limit of an infinitely large system. Hence, there can be no long-ranged order in our system of pointers for those spatial dimensions. This is the Mermin–Wagner–Hohenberg theorem, derived in a very unorthodox dynamical way. In Section 2.4, I'll show that modifying this argument to take into account motion shows that movers *can* order in $d = 2$. But first, I'll derive the Mermin–Wagner–Hohenberg theorem more formally and systematically using hydrodynamics.

2.3 More Formal Solution of the Dynamical Equation (and How You Get That Log in $d = 2$)

We can now solve the noisy diffusion equation (2.1.7) by looking for plane wave normal modes, or, to put it more technically, by Fourier transforming. While Fourier transformation is a familiar trick for solving both ordinary and partial differential equations, I'll briefly review it here in some detail, both for completeness, and to make my normalization conventions (which are common, but not universal) clear.

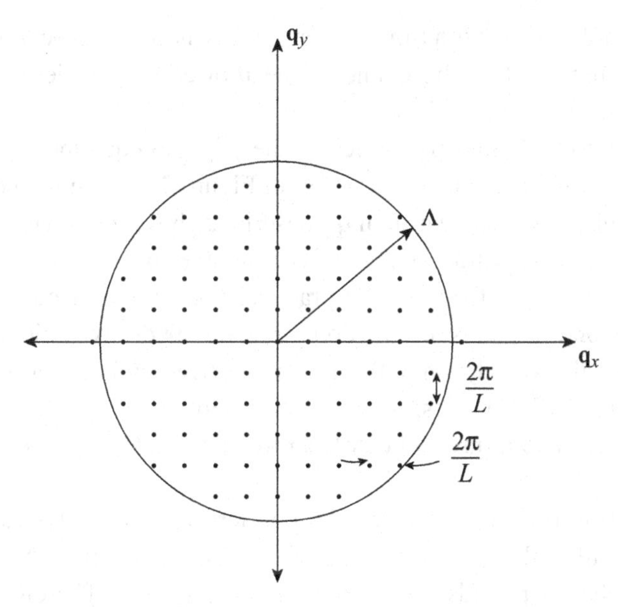

Fig. 2.3.1 Illustration of the set of **q**'s appearing in the sums for Fourier transforms. These form a grid of spacing $\frac{2\pi}{L}$, where L is the linear extent of the system in *real* space. We also impose an ultraviolet cutoff Λ on those sums, as illustrated.

I replace the fluctuating field $\theta(\mathbf{r}, t)$ with a new set of dynamical variables $\theta(\mathbf{q}, \Omega)$, which I'll call the "Fourier transformed fields" defined via

$$\theta(\mathbf{r}, t) = \frac{1}{\sqrt{VT}} \sum_{\mathbf{q},\Omega} \theta(\mathbf{q}, \Omega) e^{i(\mathbf{q}\cdot\mathbf{r} - \Omega t)} . \tag{2.3.1}$$

This definition requires some explanation. I am imagining my system is confined in a volume (or hypervolume, if $d \neq 3$) V. Specifically, I consider a hypercube with edge length L; hence, $V = L^d$. I also assume periodic boundary conditions at the walls of this hypercube, so that $\theta(\mathbf{r} + L\hat{x}_i, t) = \theta(\mathbf{r}, t)$ for all directions $i = (1, 2, \ldots, d)$. As usual for Fourier transforms, this quantizes the allowed values of **q**: The ith component of every **q** in the sum over **q** in (2.3.1) must be an integral multiple of $\frac{2\pi}{L}$. That is, the set of allowed **q**'s is given by

$$\mathbf{q} = \frac{2\pi}{L}\mathbf{n}, \tag{2.3.2}$$

where **n** is a vector all of whose components are integers.

Thus, as illustrated in Figure 2.3.1, the set of allowed **q**'s over which the sum in (2.3.1) extends forms a hypercubic grid, whose nearest neighbor spacings are all $\frac{2\pi}{L}$, which gets very small (that is, the set of points gets very dense) in the limit in which the system dimension $L \to \infty$.

I'll play a similar game with time: The system is assumed to be temporally periodic as well, with period T. This quantizes the allowed frequencies Ω to be integral multiples of $\frac{2\pi}{T}$.

One other important concept implicit in the $\sum_{\mathbf{q},\Omega}$ in equation (2.3.1) is that of the *ultraviolet cutoff*. That is, as illustrated in Figure 2.3.1, we include in the sum only those members of the fine grid of $\mathbf{q}$'s described above those with wavenumbers $q \equiv |\mathbf{q}| < \Lambda$, where the ultraviolet cutoff Λ is another parameter of our model.

There are two reasons for this. The practical reason is to avoid divergences at large $\mathbf{q}$. The second reason is that we don't trust *any* hydrodynamic theory at large $\mathbf{q}$ anyway, since all hydrodynamic theories are derived by keeping only terms that are important at small distances, or large wavevector.

Note that there is *no* ultraviolet cutoff on frequencies Ω, because we don't need one.

It is also useful to be able to invert the Fourier transform. This can be done by noting that, for all of the $\mathbf{q}$'s in our fine grid (2.3.2), the function $\mathrm{e}^{\mathrm{i}\mathbf{q}\cdot\mathbf{r}}$ goes through an integer number of periods in the volume of the system. Therefore, unless that integer is zero *for all Cartesian components of the vector* $\mathbf{q}$, we have

$$\int_V \mathrm{d}^d r\, \mathrm{e}^{\mathrm{i}\mathbf{q}\cdot\mathbf{r}} = 0, \quad \mathbf{q} \neq \mathbf{0}, \tag{2.3.3}$$

where the subscript V denotes that the region of integration is our hypercube of volume V. Obviously, if $\mathbf{q} = \mathbf{0}$, $\mathrm{e}^{\mathrm{i}\mathbf{q}\cdot\mathbf{r}} = 1$, and we trivially have

$$\int_V \mathrm{d}^d r\, \mathrm{e}^{\mathrm{i}\mathbf{q}\cdot\mathbf{r}} = V, \quad \mathbf{q} = \mathbf{0}. \tag{2.3.4}$$

We can summarize (2.3.3) and (2.3.4) using the handy "Kronecker delta"

$$\int_V \mathrm{d}^d r\, \mathrm{e}^{\mathrm{i}\mathbf{q}\cdot\mathbf{r}} = \delta^K_{\mathbf{q}} V, \tag{2.3.5}$$

where by definition

$$\delta^K_{\mathbf{q}} = \begin{cases} 1, & \mathbf{q} = \mathbf{0}, \\ 0, & \mathbf{q} \neq \mathbf{0}. \end{cases} \tag{2.3.6}$$

Similarly,

$$\int_0^T \mathrm{d}t\, \mathrm{e}^{\mathrm{i}\Omega t} = \delta^K_{\Omega} T. \tag{2.3.7}$$

Using this result, I can now obtain an expression for the Fourier transformed field $\theta(\mathbf{q}', \Omega')$ for some particular $\mathbf{q}'$ and Ω' in our set of allowed $\mathbf{q}$'s and Ω's in terms of the real-space field $\theta(\mathbf{r}, t)$ by multiplying both sides of the definition by

$e^{-i(\mathbf{q}'\cdot\mathbf{r}-\Omega' t)}$, integrating both sides over all $\mathbf{r}$ in the hypercube, and over time from 0 to T, and using the "orthogonality relations" (2.3.5) and (2.3.7). This gives

$$\int_V d^d r \int_0^T dt\, \theta(\mathbf{r},t) e^{-i(\mathbf{q}'\cdot\mathbf{r}-\Omega' t)} = \sqrt{VT} \sum_{\mathbf{q},\Omega} \theta(\mathbf{q},\Omega) \delta^K_{\mathbf{q}-\mathbf{q}'} \delta^K_{\Omega-\Omega'} . \tag{2.3.8}$$

Since the Kronecker deltas in this expression are nonzero only when $\mathbf{q} = \mathbf{q}'$ and $\Omega = \Omega'$, only that term in the $\sum_{\mathbf{q},\Omega}$ survives. The right-hand side of (2.3.8) therefore involves only $\theta(\mathbf{q}',\Omega')$:

$$\int_V d^d r \int_0^T dt\, \theta(\mathbf{r},t) e^{-i(\mathbf{q}'\cdot\mathbf{r}-\Omega' t)} = \sqrt{VT}\theta(\mathbf{q}',\Omega') , \tag{2.3.9}$$

which is trivially solved for $\theta(\mathbf{q}',\Omega')$:

$$\theta(\mathbf{q}',\Omega') = \frac{1}{\sqrt{VT}} \int_V d^d r \int_0^T dt\, \theta(\mathbf{r},t) e^{-i(\mathbf{q}'\cdot\mathbf{r}-\Omega' t)} . \tag{2.3.10}$$

Since this works for *any* $\mathbf{q}'$ and Ω', I can drop the primes in this expression and write

$$\theta(\mathbf{q},\Omega) = \frac{1}{\sqrt{VT}} \int_V d^d r \int_0^T dt\, \theta(\mathbf{r},t) e^{-i(\mathbf{q}\cdot\mathbf{r}-\Omega t)} . \tag{2.3.11}$$

Now let's apply this to the noisy diffusion equation. In doing so, I will Fourier transform the random force $f(\mathbf{r},t)$ in exactly the same way that I did the angle field $\theta(\mathbf{r},t)$.

Inserting the Fourier decomposition (2.3.1) into the equation of motion (2.1.7), and then multiplying both sides by $e^{-i(\mathbf{q}\cdot\mathbf{r}-\Omega t)}$, integrating both sides over all $\mathbf{r}$ in the hypercube, and over time from 0 to T, and using the "orthogonality relations" (2.3.5) and (2.3.7), I get

$$-i\Omega\theta(\mathbf{q},\Omega) = -\nu q^2 \theta(\mathbf{q},\Omega) + f(\mathbf{q},\Omega) . \tag{2.3.12}$$

In deriving this expression, I have used the facts that

$$\partial_t e^{-i(\mathbf{q}\cdot\mathbf{r}-\Omega t)} = -i\Omega e^{-i(\mathbf{q}\cdot\mathbf{r}-\Omega t)}, \quad \nabla^2 e^{-i(\mathbf{q}\cdot\mathbf{r}-\Omega t)} = -q^2 e^{-i(\mathbf{q}\cdot\mathbf{r}-\Omega t)} . \tag{2.3.13}$$

The correlations of the Fourier transformed noise can easily be obtained by inverting the Fourier transform. That is, since

$$f(\mathbf{q},\Omega) = \frac{1}{\sqrt{VT}} \int_V d^d r f(\mathbf{r},t) e^{-i(\mathbf{q}\cdot\mathbf{r}-\Omega t)} , \tag{2.3.14}$$

it follows that

$$\langle f(\mathbf{q},\Omega) f(\mathbf{q}',\Omega')\rangle = \frac{1}{VT} \int_V d^d r \int_0^T dt \int_V d^d r' \int_0^T dt'\, \langle f(\mathbf{r},t) f(\mathbf{r}',t')\rangle e^{-i(\mathbf{q}\cdot\mathbf{r}-\Omega t)} e^{-i(\mathbf{q}'\cdot\mathbf{r}'-\Omega' t')} . \tag{2.3.15}$$

Using our postulated correlations (2.1.8) for the noise in this expression gives

$$
\begin{aligned}
&\langle f(\mathbf{q},\Omega) f(\mathbf{q}',\Omega')\rangle \\
&= \frac{1}{VT}\int_V \mathrm{d}^d r \int_0^T \mathrm{d}t \int_V \mathrm{d}^d r' \int_0^T \mathrm{d}t'\, 2D\delta^d(\mathbf{r}-\mathbf{r}')\delta(t-t')\mathrm{e}^{-\mathrm{i}(\mathbf{q}\cdot\mathbf{r}-\Omega t)}\mathrm{e}^{-\mathrm{i}(\mathbf{q}'\cdot\mathbf{r}'-\Omega' t')} \\
&= \frac{1}{VT}\int_V \mathrm{d}^d r \int_0^T \mathrm{d}t\, 2D\mathrm{e}^{-\mathrm{i}((\mathbf{q}+\mathbf{q}')\cdot\mathbf{r}-(\Omega+\Omega')t)} \\
&= 2D\delta^K_{\mathbf{q}+\mathbf{q}'}\delta^K_{\Omega+\Omega'}.
\end{aligned}
\tag{2.3.16}
$$

Note that the Fourier transformed equation of motion (2.3.12) has achieved of goal of decoupling: Every Fourier mode $\theta(\mathbf{q},\Omega)$ has its own equation of motion (2.3.12), with its own noise $f(\mathbf{q},\Omega)$, and both the equations, and the noises, are decoupled from all of the other $\theta(\mathbf{q}',\Omega')$ at different $\mathbf{q}$'s and Ω's. Furthermore, these decoupled equations are themselves just simple linear equations, which we can trivially solve for the fields $\theta(\mathbf{q},\Omega)$ in terms of the noises $f(\mathbf{q},\Omega)$:

$$
\theta(\mathbf{q},\Omega) = G(\mathbf{q},\Omega) f(\mathbf{q},\Omega),
\tag{2.3.17}
$$

where I've defined the "propagator"

$$
G(\mathbf{q},\Omega) \equiv \frac{1}{-\mathrm{i}\Omega + \nu q^2}.
\tag{2.3.18}
$$

Autocorrelating the solution (2.3.17) with itself, and using the result (2.3.16) for the autocorrelations of the noise f, gives

$$
\begin{aligned}
\langle\theta(\mathbf{q},\Omega)\theta(\mathbf{q}',\Omega')\rangle &= G(\mathbf{q},\Omega)G(\mathbf{q}',\Omega')\langle f(\mathbf{q},\Omega) f(\mathbf{q}',\Omega')\rangle \\
&= 2DG(\mathbf{q},\Omega)G(\mathbf{q}',\Omega')\delta^K_{\mathbf{q}+\mathbf{q}'}\delta^K_{\Omega+\Omega'} \\
&= 2DG(\mathbf{q},\Omega)G(-\mathbf{q},-\Omega)\delta^K_{\mathbf{q}+\mathbf{q}'}\delta^K_{\Omega+\Omega'} \equiv C(\mathbf{q},\Omega)\delta^K_{\mathbf{q}+\mathbf{q}'}\delta^K_{\Omega+\Omega'},
\end{aligned}
\tag{2.3.19}
$$

where I've defined the "correlation function"

$$
\begin{aligned}
C(\mathbf{q},\Omega) \equiv \langle\theta(\mathbf{q},\Omega)\theta(-\mathbf{q},-\Omega)\rangle &= \langle|\theta(\mathbf{q},\Omega)|^2\rangle \\
&= 2DG(\mathbf{q},\Omega)G(-\mathbf{q},-\Omega) = \frac{2D}{\Omega^2+\nu^2 q^4}.
\end{aligned}
\tag{2.3.20}
$$

The second equality (i.e., the fact that $\langle\theta(\mathbf{q},\Omega)\theta(-\mathbf{q},-\Omega)\rangle = \langle|\theta(\mathbf{q},\Omega)|^2\rangle$), follows from the fact, easily verified from the equation (2.3.11) for the inverse Fourier transform, that the Fourier transform of a real-valued field like $\theta(\mathbf{r},t)$ obeys

$$
\theta(-\mathbf{q},-\Omega) = \theta^*(\mathbf{q},\Omega),
\tag{2.3.21}
$$

where $\theta^*(\mathbf{q},\Omega)$ is the complex conjugate of $\theta(\mathbf{q},\Omega)$.

With this result in hand, we can calculate the equal-time correlations of the spatial Fourier transform $\theta(\mathbf{q}, t)$, defined by Fourier transforming the real-space field $\theta(\mathbf{r}, t)$ only in space, but not in time. That is, I define

$$\theta(\mathbf{r}, t) = \frac{1}{\sqrt{V}} \sum_{\mathbf{q}} \theta(\mathbf{q}, t) e^{i\mathbf{q}\cdot\mathbf{r}} , \tag{2.3.22}$$

from which I can, as I did above for the spatio-temporal Fourier transform, derive the inverse Fourier transform

$$\theta(\mathbf{q}, t) = \frac{1}{\sqrt{V}} \int_V d^d r\, \theta(\mathbf{r}, t) e^{-i\mathbf{q}\cdot\mathbf{r}} . \tag{2.3.23}$$

We can get the equal time correlations of this spatial Fourier transform by integrating the full spatio-temporal correlation (2.3.19) over frequency. I begin by using the inverse Fourier transform (2.3.23) to write

$$\langle \theta(\mathbf{q}, t)\theta(\mathbf{q}', t)\rangle = \frac{1}{V} \int_V d^d r \int_V d^d r' \langle \theta(\mathbf{r}, t)(\mathbf{r}', t)\rangle e^{i(\mathbf{q}\cdot\mathbf{r}+\mathbf{q}'\cdot\mathbf{r}')} . \tag{2.3.24}$$

Now rewriting the real-space fields $\theta(\mathbf{r}, t)$ and $\theta(\mathbf{r}', t)$ in terms of the full spatio-temporal Fourier transforms $\theta(\mathbf{q}, \Omega)$ using (2.3.1) gives

$$\langle \theta(\mathbf{q}, t)\theta(\mathbf{q}', t)\rangle$$
$$= \frac{1}{V^2 T} \int_V d^d r \int_V d^d r' \sum_{\mathbf{p},\Omega} \sum_{\mathbf{p}',\Omega'} \langle \theta(\mathbf{p}, \Omega)\theta(\mathbf{p}', \Omega')\rangle e^{i((\mathbf{q}+\mathbf{p})\cdot\mathbf{r}+(\mathbf{q}'+\mathbf{p}')\cdot\mathbf{r}')} e^{-i(\Omega+\Omega')t} . \tag{2.3.25}$$

Now using my results (2.3.19) and (2.3.20) for the correlations of the spatio-temporally Fourier transformed fields $\theta(\mathbf{q}, \Omega)$ and $\theta(\mathbf{q}', \Omega')$, I get

$$\langle \theta(\mathbf{q}, t)\theta(\mathbf{q}', t)\rangle$$
$$= \frac{1}{V^2 T} \int_V d^d r \int_V d^d r' \sum_{\mathbf{p},\Omega} \sum_{\mathbf{p}',\Omega'} C(\mathbf{p}, \Omega)\delta^K_{\mathbf{p}+\mathbf{p}'} \delta^K_{\Omega+\Omega'} e^{i((\mathbf{q}+\mathbf{p})\cdot\mathbf{r}+(\mathbf{q}'+\mathbf{p}')\cdot\mathbf{r}')} e^{-i(\Omega+\Omega')t} . \tag{2.3.26}$$

Now we can once again use the magic of the Kronecker deltas to reduce the sums over $\mathbf{p}'$ and Ω' to a single term:

$$\langle \theta(\mathbf{q}, t)\theta(\mathbf{q}', t)\rangle = \frac{1}{V^2 T} \int_V d^d r \int_V d^d r' \sum_{\mathbf{p},\Omega} C(\mathbf{p}, \Omega)\delta^K_{\mathbf{q}+\mathbf{q}'} \delta^K_{\Omega+\Omega'} e^{i((\mathbf{q}+\mathbf{p})\cdot\mathbf{r}+(\mathbf{q}'-\mathbf{p})\cdot\mathbf{r}')} . \tag{2.3.27}$$

Note that all time dependence has disappeared because the $\delta^K_{\Omega+\Omega'}$ forces $\Omega + \Omega' = 0$, which kills off the only time dependence in (2.3.26), which is in the $e^{-i(\Omega+\Omega')t}$ factor. This makes sense, since we're studying a (statistically) steady state, so equal-time correlations should not depend on the time t at which they're evaluated.

Now we can evaluate the integrals over $\mathbf{r}$ and $\mathbf{r}'$ using the orthogonality relations (2.3.5); this gives

$$\langle\theta(\mathbf{q},t)\theta(\mathbf{q}',t)\rangle = \frac{1}{T}\sum_{\mathbf{p},\Omega} C(\mathbf{p},\Omega)\delta^K_{\mathbf{q}+\mathbf{p}}\delta^K_{\mathbf{q}'-\mathbf{p}}\,. \tag{2.3.28}$$

Again, the Kronecker deltas allow only the term $\mathbf{p} = -\mathbf{q}$ to survive in the sum over $\mathbf{p}$, and even that term only survives if $\mathbf{q}' = -\mathbf{q}$. We therefore get

$$\langle\theta(\mathbf{q},t)\theta(\mathbf{q}',t)\rangle = \frac{1}{T}\sum_{\Omega} C(\mathbf{q},\Omega)\delta^K_{\mathbf{q}+\mathbf{q}'}\,. \tag{2.3.29}$$

(Note that in writing this equation I have used the fact that $C(-\mathbf{q},\Omega) = C(\mathbf{q},\Omega)$.)

Recall that the sum over Ω is over Ω's given by

$$\Omega = \frac{2\pi n}{T}\,, \tag{2.3.30}$$

where n is an integer. Thus, we can think of the sum over Ω in (2.3.29) as a sum over an integer n running from $-\infty$ to ∞. Thus

$$\langle\theta(\mathbf{q},t)\theta(\mathbf{q}',t)\rangle = \frac{1}{T}\sum_{n=-\infty}^{\infty} C(\mathbf{p},\Omega)\delta^K_{\mathbf{q}+\mathbf{q}'}\,. \tag{2.3.31}$$

Since we wish to consider our period T to be extremely large, the steps in this sum are very small. Hence, we can approximate the sum by an integral, which gives

$$\langle\theta(\mathbf{q},t)\theta(\mathbf{q}',t)\rangle = \frac{1}{T}\int_{-\infty}^{\infty} \mathrm{d}n\, C(\mathbf{q},\Omega)\delta^K_{\mathbf{q}+\mathbf{q}'}\,. \tag{2.3.32}$$

Changing the variable of integration from n to Ω using the relation (2.3.30) between Ω and n, we see that the period T drops out our final result (which is a good thing, since it was introduced as a somewhat arbitrary contrivance!), and we have

$$\langle\theta(\mathbf{q},t)\theta(\mathbf{q}',t)\rangle = C_{\mathrm{ET}}(\mathbf{q})\delta^K_{\mathbf{q}+\mathbf{q}'}\,, \tag{2.3.33}$$

where I've defined the equal-time correlation function

$$C_{\mathrm{ET}}(\mathbf{q}) \equiv \int_{-\infty}^{\infty} \frac{\mathrm{d}\Omega}{2\pi}\, C(\mathbf{q},\Omega)\,. \tag{2.3.34}$$

The results (2.3.33) and (2.3.34) are completely general; all that changes from model to model is $C(\mathbf{q},\Omega)$.

Using (2.3.34) for our XY model, and using our result (2.3.19) and (2.3.20) for $C(\mathbf{q}, \Omega)$ for that model, gives

$$C_{\rm ET}(\mathbf{q}) = \int \frac{d\omega}{2\pi} \frac{2D}{(\omega^2 + \nu^2 q^4)} = \frac{D}{\nu q^2}. \tag{2.3.35}$$

Note that this correlation function diverges like $1/q^2$ as $q \to 0$. We will see in Section 2.4 that this is exactly the result of a standard equilibrium statistical mechanics analysis of this problem.

It is this divergence at small $\mathbf{q}$ that leads to the Mermin–Wagner–Hohenberg theorem in this approach. To see this, let's calculate the *real-space* fluctuations $\langle(\theta(\mathbf{r}, t))^2\rangle$ of the angle field $\theta(\mathbf{r}, t)$. Clearly, if these diverge as the system size goes to infinity for any temperature T (not to be confused with the period T that I introduced earlier; here T is proportional to D, as we'll see later), a sufficiently large system must always be disordered. Contrariwise, if this is *finite*, and proportional to temperature, then long-ranged order is *guaranteed* at sufficiently low temperature.

I'll now show that the first case – divergent fluctuations at *all* nonzero temperatures in the limit of infinite system size – holds for spatial dimensions $d \leq 2$. This proves the Mermin–Wagner–Hohenberg theorem.

I'll begin by relating the equal-time real-space fluctuations $\langle(\theta(\mathbf{r}, t))^2\rangle$ to the equal-time Fourier transformed correlation functions (2.3.35) just calculated. Going back again to the connection (2.3.22) between the real and Fourier space fields, I have

$$\begin{aligned}\langle(\theta(\mathbf{r}, t))^2\rangle &= \frac{1}{V}\sum_{\mathbf{q}}\sum_{\mathbf{q}'}\langle\theta(\mathbf{q}, t)\theta(\mathbf{q}', t)\rangle e^{i(\mathbf{q}+\mathbf{q}')\cdot\mathbf{r}} = \frac{1}{V}\sum_{\mathbf{q}}\sum_{\mathbf{q}'} C_{\rm ET}(\mathbf{q}) e^{i(\mathbf{q}+\mathbf{q}')\cdot\mathbf{r}}\delta^K_{\mathbf{q}+\mathbf{q}'} \\ &= \frac{1}{V}\sum_{\mathbf{q}} C_{\rm ET}(\mathbf{q}) = \int \frac{d^d q}{(2\pi)^d} C_{\rm ET}(\mathbf{q}).\end{aligned} \tag{2.3.36}$$

This relation is completely general for any translationally invariant model (translation invariance is required to obtain the Kronecker delta $\delta^K_{\mathbf{q}+\mathbf{q}'}$ in the second equality). Now specializing to our XY model by using our expression (2.3.35) for the Fourier transformed equal time correlation function, I get

$$\langle(\theta(\mathbf{r}, t))^2\rangle = \frac{D}{\nu} I(\Lambda, L), \tag{2.3.37}$$

where I've defined

$$I(\Lambda, L) \equiv \int_{\frac{2\pi}{L}<p<\Lambda} \frac{d^d p}{(2\pi)^d} \frac{1}{p^2}. \tag{2.3.38}$$

The lower limit $\frac{2\pi}{L}$ on this integral appears because I have excluded the $\mathbf{q} = \mathbf{0}$ term from the sum on $\mathbf{q}$, since that term just corresponds to uniform rotations of all the spins. Such uniform rotations clearly do not affect the presence or absence of long-range order. Once those terms are excluded, the smallest $\mathbf{q}$ in the sum is clearly $\frac{2\pi}{L}$. It is straightforward to show that, for $d \neq 2$,

$$I(\Lambda, L) = \left(\frac{K_d}{d-2}\right)\left(\Lambda^{d-2} - \left(\frac{L}{2\pi}\right)^{2-d}\right), \tag{2.3.39}$$

where I've defined

$$K_d \equiv \frac{S_d}{(2\pi)^d}, \tag{2.3.40}$$

with S_d the surface hyperarea of a d-dimensional sphere of unit radius.

Clearly, $d = 2$ is a special case. One can either take the tricky limit $d \to 2$ of (2.3.39) with a little help from the Marquis de l'Hôpital, or simply reevaluate the integral exactly in $d = 2$. Either way, one gets

$$\int_{\frac{2\pi}{L}<p<\Lambda} \frac{d^2p}{(2\pi)^2}\frac{1}{p^2} = \frac{1}{2\pi}\ln\left(\frac{\Lambda L}{2\pi}\right). \tag{2.3.41}$$

To summarize, we've found

$$\langle(\theta(\mathbf{r},t))^2\rangle \propto \begin{cases} L^{2-d}, & d < 2, \\ \ln\left(\frac{\Lambda L}{2\pi}\right), & d = 2, \\ \text{finite (independent of} L), & d > 2. \end{cases} \tag{2.3.42}$$

Note that (2.3.42) goes to ∞ as $L \to \infty$ for $d \leq 2$. This implies the Mermin–Wagner–Hohenberg theorem: The real-space fluctuations in $\theta(\mathbf{r}, t)$, as given by equation (2.3.42), diverge as system size $L \to \infty$ for all spatial dimensions $d \leq 2$. Note also that the *way* in which they diverge – that is, like L^{2-d} – agrees with the result of my earlier hand-waving argument of Section 2.2. Note also that the more formal approach I've taken in this section enables us to show that the borderline case $d = 2$ also has divergent fluctuations, a conclusion that was actually beyond the simple hand-waving argument. Sometimes, knowing a little math *is* useful!

2.4 Connection to the Textbook Derivation of the Mermin–Wagner–Hohenberg Theorem, and Introduction to the Gradient Expansion

While the argument just presented might seem less likely to provoke Mermin and Wagner into turning over in their (in fact unoccupied) graves, it is still *not* the argument used by Mermin, Wagner, and Hohenberg. Their argument relied completely on equilibrium statistical mechanics. In particular, it used concepts like Boltzmann weights, partition functions, etc., none of which involves any discussion of time dependence *at all*. In this section, I'll review that equilibrium argument for completeness, and show how it connects to the dynamical arguments presented above.

This analysis will also give me the chance to introduce a few concepts that prove to be hugely important throughout condensed matter physics, and, in particular, in studying the topics I'll discuss in this book. These concepts are: locality, symmetry arguments, and gradient expansions. I'll refer to the approach that uses these ideas as "the hydrodynamic approach." I'll begin by illustrating this for the pointer problem.

To treat the pointer problem as an equilibrium statistical mechanics problem, all we need to do is formulate a Hamiltonian $H(\{\theta(\mathbf{r},t)\})$ for the continuous angle field $\theta(\mathbf{r},t)$ that we introduced above. The curly brackets $\{\}$ in the argument of H are meant to suggest that H is a function*al* of the entire configuration $\theta(\mathbf{r},t)$. That is, H can depend on spatial *derivatives* of $\theta(\mathbf{r},t)$, as well as directly on $\theta(\mathbf{r},t)$ itself. This is important, because it turns out that symmetry *only* allows terms involving spatial derivatives, as we'll see in a moment.

I'll assume that H depends only on the instantaneous spatial configuration of $\theta(\mathbf{r},t)$, not on its time derivatives. This essentially drops time out of the problem, so I'll stop writing the time argument in my $\theta(\mathbf{r},t)$ henceforth (until I return at the end of this section to a discussion of equilibrium dynamics).

Now let's start using the aforementioned important concepts.

First, locality: This implies that H should be an integral over space of some *local* energy density $h(\mathbf{r})$, which itself depends only on the values of $\theta(\mathbf{r})$ and its spatial derivatives at the point $\mathbf{r}$.

Second, symmetries: We assume that the Hamiltonian, like the dynamics, is *globally rotation invariant*; this means that the Hamiltonian must be unchanged if we rotate *all* of the spins by the same angle θ_0. That is,

$$H(\{\theta(\mathbf{r})\}) = H(\{\theta(\mathbf{r}) + \theta_0\}). \tag{2.4.1}$$

This condition imposes severe constraints on the energy density h: It can only depend on spatial *derivatives* of the field θ, since the uniform rotation θ_0 in (2.4.1) drops out of all such derivatives. Therefore we have

$$h = h(\partial_i\theta, \partial_i\partial_j\theta, \ldots), \tag{2.4.2}$$

where $h(\ldots)$ is some function that remains to be determined.

Now comes the idea of the *gradient expansion*: Since we expect, and will show a posteriori, that the field $\theta(\mathbf{r})$ will usually be slowly varying in space, we can expand the function $h(\ldots)$ *in powers of the gradients*. That is, we'll keep only terms in the expansion of h that have the *smallest possible number of spatial derivatives*.

But what *is* the smallest possible number of spatial derivatives? Well, we've already established that it can't be zero; rotation invariance forbids any such term (except for irrelevant constants – that is, terms altogether independent of $\theta(\mathbf{r})$). What about terms involving one spatial derivative? Well, any such term can be written as a total divergence: For example, if $h = C\partial_x\theta$, then $h = C\nabla\cdot\mathbf{J}$ with $\mathbf{J} = \theta\hat{\mathbf{x}}$. The change in the Hamiltonian is then just the integral of a total divergence, which can always be reduced to a surface term. If we impose periodic boundary conditions, any such surface term will vanish.

So we need to go to terms that involve two derivatives. There are only two such types of terms: terms involving one power of a second derivative of θ, or terms quadratic in spatial derivatives of θ. The former can easily be shown to be total divergences; for example, $\partial_x \partial_y \theta = \nabla \cdot \mathbf{J}$ with $J = \partial_x \theta \hat{\mathbf{y}}$. So we can drop those terms, and are therefore left with only terms quadratic in first spatial derivatives of θ. Equivalently, this means we can take the energy density h to depend quadratically on the *gradient* $\nabla\theta$ of θ; that is,

$$h = h(\nabla\theta), \tag{2.4.3}$$

where the functional dependence is quadratic.

Now we can use another symmetry of the system. We've already used what I'll call rotation invariance in *spin* space – i.e., the symmetry (2.4.1). But we haven't yet used rotation invariance in *real* space; that is, the fact that all directions in real space are equivalent, since we placed our pointers at random positions in the space.

This means that the energy density (2.4.3) should not depend on the *direction* of the real-space vector $\nabla\theta$. More precisely, two local configurations whose $\nabla\theta$'s are related by a simple rotation should have the same local energy density; that is, if the *magnitudes* of the two $\nabla\theta$'s are the same, their local energy densities h should be the same. Another way to say this is that $h(\nabla\theta)$ should only be a function of the magnitude $|\nabla\theta|$ of $\nabla\theta$.

So, we know h should depend only on $|\nabla\theta|$, and that it should be a quadratic function of the components of $\nabla\theta$. What quantity satisfies both of these conditions? There's only one, and that's $|\nabla\theta|^2$. Hence, h must have the form

$$h = \frac{K}{2}|\nabla\theta(\mathbf{r})|^2 \tag{2.4.4}$$

for some constant K, which we call the "spin wave stiffness."

This implies that

$$H_{\mathrm{XY}} = \frac{K}{2}\int \mathrm{d}^d r |\nabla\theta(\mathbf{r})|^2. \tag{2.4.5}$$

Note that, while I derived this form assuming that the pointers were positioned at random,[1] so that we had full real-space rotation invariance, I can also argue for

[1] For random arrangements of pointers, one should also allow some randomness in the value of K at different points in space. That is, one should replace (2.4.5) by

$$H_{\mathrm{XY}} = \frac{1}{2}\int \mathrm{d}^d r K(\mathbf{r})|\nabla\theta(\mathbf{r})|^2 \tag{2.4.6}$$

with the local spin wave stiffness having some "quenched" (i.e., frozen) randomness. However, such randomness, if sufficiently weak, can be shown, by extending the sort of renormalization group techniques we've used here to include quenched disorder, to be "irrelevant"; that is, to leave the long-ranged behavior of the form (2.4.5) for pointers on a lattice, as long as the lattice has sufficiently high symmetry. To see this, note that I can write an *arbitrary* quadratic function of the components of $\nabla\theta$ as

$$h = M_{ij}(\partial_i\theta)(\partial_j\theta) \tag{2.4.7}$$

for some matrix symmetric matrix M_{ij} (I can make it symmetric because $(\partial_i\theta)(\partial_j\theta)$ is obviously symmetric under interchange of i and j). But I can always rotate my real-space coordinates to the orthonormal frame formed by the eigenvectors of the matrix M_{ij} (which is orthonormal *because* the matrix is symmetric). In that coordinate system, we can write

$$h = \sum_{i=1}^{d} \lambda_i(\partial_i\theta)^2\,, \tag{2.4.8}$$

where $i = (1, 2, \ldots, d)$ denotes the d eigendirections of the matrix M_{ij}, with corresponding eigenvalues λ_i.

Now consider, for example, a square lattice in $d=2$, a cubic lattice in $d=3$, or, most generally, a d-dimensional hypercubic lattice. Clearly, by symmetry, if the edges of the squares, cubes, or hypercubes are parallel to the $\hat{\mathbf{x}}_i$ axes, the eigendirections of the matrix M_{ij} *must* also be along those axes (assuming that the interactions between nearest pointers separated in the $\hat{\mathbf{x}}_i$-direction are the same as those in the $\hat{\mathbf{x}}_j$-direction, for all directions i and j, which is what I mean by a hypercubic lattice). Thus, $1 = x$ and $2 = y$ in (2.4.8). Furthermore, in this highly symmetric lattice, the eigenvalues $\lambda_1 = \lambda_x = \lambda_y = \lambda_2$. Defining this common eigenvalue to be $\frac{K}{2}$, I can rewrite (2.4.8) as

$$h = \frac{K}{2}\left[(\partial_x\theta)^2 + (\partial_y\theta)^2\right] = \frac{K}{2}|\nabla\theta(\mathbf{r})|^2\,, \tag{2.4.9}$$

thereby recovering (2.4.4).

I will leave the slightly more complicated exercise of proving that (2.4.4) also holds for two-dimensional *hexagonal* lattices to the reader.

Note that, even if we are dealing with a low-symmetry lattice, the expression (2.4.8) holds for *any* lattice. Hence, defining $\lambda_i \equiv \frac{K_i}{2}$, we can write

$$H_{\mathrm{XY}} = \frac{1}{2}\int \mathrm{d}^d r \sum_{i=1}^{d} K_i(\partial_i\theta)^2\,, \tag{2.4.10}$$

whose behavior is just a mildly anisotropic version of the special (but not terribly special) case (2.4.4) which applies to all hypercubic lattices, and

the system unchanged. (This is true in the ordered phase, but not necessarily at the transition.) I'll therefore ignore it here.

to random arrangements of pointers. Furthermore, with a simple rescaling of coordinates

$$r_i \equiv \sqrt{\left(\frac{K_i}{\tilde{K}}\right)} r_i', \tag{2.4.11}$$

where I've defined $\tilde{K}$ to be the geometric mean of all the K_i; that is,

$$\tilde{K} \equiv \left(\prod_{i=1}^{d} K_i\right)^{1/d}, \tag{2.4.12}$$

so (2.4.10) becomes (2.4.5) with $K = \tilde{K}$ and $\mathbf{r}$ replaced with $\mathbf{r}'$.

How do we now solve the statistical mechanics problem defined by this XY Hamiltonian (2.4.4)? As in the dynamical approach of Section 2.3, we do so by looking for plane wave normal modes, or, to put it more technically, by Fourier transforming. Using the purely spatial Fourier transform (2.3.22), it is easy to see that

$$\nabla\theta(\mathbf{r}) = \frac{1}{\sqrt{V}} \sum_{\mathbf{q}} \mathrm{i}\mathbf{q}\theta(\mathbf{q})\mathrm{e}^{\mathrm{i}\mathbf{q}\cdot\mathbf{r}}. \tag{2.4.13}$$

Inserting this into the XY Hamiltonian (2.4.5) gives:

$$H_{\mathrm{XY}} = -\frac{1}{V} \sum_{\mathbf{q}} \sum_{\mathbf{q}'} \mathbf{q}\cdot\mathbf{q}'\theta(\mathbf{q})\theta(\mathbf{q}') \int_V \mathrm{d}^d r\, \mathrm{e}^{\mathrm{i}(\mathbf{q}+\mathbf{q}')\cdot\mathbf{r}}. \tag{2.4.14}$$

Using the orthogonality relation (2.3.5) gives us a Kronecker delta which forces $\mathbf{q}' = -\mathbf{q}$. Keeping only that term in the $\sum'_{\mathbf{q}}$ gives us our final form for the Hamiltonian:

$$H_{\mathrm{XY}} = \sum_{\mathbf{q}} \frac{Kq^2}{2} |\theta(\mathbf{q})|^2. \tag{2.4.15}$$

We now need only do equilibrium statistical mechanics on this Hamiltonian. In fact, we will only need one result from equilibrium statistical mechanics, namely the *equipartition theorem*, which states that, if a variable appears in a Hamiltonian as a single quadratic term, the thermal average of that single term is $\frac{k_B T}{2}$, where T is the temperature, and k_B is Boltzmann's constant.

Applying this theorem to our Hamiltonian (2.4.15) gives

$$\left\langle \frac{Kq^2}{2} |\theta(\mathbf{q})|^2 \right\rangle = \frac{k_B T}{2}, \tag{2.4.16}$$

which is trivially solved to give the mean squared fluctuations of the Fourier transformed θ:

$$\langle|\theta(\mathbf{q})|^2\rangle = \frac{k_B T}{Kq^2}\,. \tag{2.4.17}$$

Note that this result (2.4.17) is identical to the result of the purely dynamical analysis of Section 2.3 if

$$\frac{k_B T}{K} = \frac{D}{\nu}\,. \tag{2.4.18}$$

This is not a coincidence, as can be seen by putting time back into this equilibrium picture. We can do this by writing down the simplest equilibrium dynamical model for the time evolution of the angle field $\theta(\mathbf{r},t)$, which is called the "time-dependent Ginzburg–Landau model" (which I'll hereafter abbreviate as "TDGL" model). This model is purely relaxational, and can be derived directly from the Hamiltonian. It is

$$\partial_t\theta = -\Gamma\frac{\delta H}{\delta\theta} + f \tag{2.4.19}$$

with the noise f again being zero mean and Gaussian, with variance given by (2.1.3). Evaluating the functional derivative in (2.4.19) gives

$$\partial_t\theta = \Gamma K\nabla^2\theta + f, \tag{2.4.20}$$

which the alert reader will recognize as identical to our earlier equation of motion (2.1.7) if we identify

$$\nu = \Gamma K\,. \tag{2.4.21}$$

Note that with this identification, our earlier condition (2.4.18) for recovering the equilibrium result implies that

$$D = \Gamma k_B T\,. \tag{2.4.22}$$

This condition is an example of an "Einstein relation." It is also called the "fluctuation–dissipation theorem." It turns out that, quite generally, *any* model of the purely relaxational form (2.4.19) will relax to an equilibrium Boltzmann distribution with the Hamiltonian H at temperature T provided that the relation (2.4.22) is satisfied, *regardless* of the form of the Hamiltonian H.

So that's how our dynamical model of Section 2.3 connects to the equilibrium results, and, therefore, to the Mermin–Wagner–Hohenberg theorem.

In Chapter 3, I'll turn to a model that can *not* be written as in the purely relaxational form (2.4.19), so that conventional equilibrium statistical mechanics is no help. The model I'll discuss is the KPZ equation.

3

The Dynamical Renormalization Group; or, Why We Can Do Physics, Illustrated by the KPZ Equation

The demonstration just given of the solution to the "pointers" problem may have given you an unwarranted overconfidence in our ability to solve the "flockers" problem. Our success in solving the pointers problem was ultimately attributable to the fact that the equation of motion for that system is *linear*. Unfortunately, it turns out, as we'll see in Chapter 4, that the important differences between flockers and pointers are all manifest in *nonlinear* terms in the equation of motion for the latter.

To assess the effect of those new nonlinear terms, we'll have to use the dynamical renormalization group (RG) [29]. Readers interested in a general discussion of the RG approach (albeit focused on critical phenomena) are referred to the book by Ma [30]. The best treatment I know of the dynamical RG for noncritical systems, which is what I'll be using throughout this book, is [29].

Before applying this technique to flocks, I'll illustrate it with the simplest non-trivial example known to me: the Kardar–Parisi–Zhang (KPZ) equation [31]. This model also has the virtue in the present context of reducing to the dynamical model (2.1.7) that we've just solved, if one ignores nonlinearities.

This equation was first proposed by the eponymous trio as a model for a growing interface, as illustrated in Figure 3.0.1. For a growing interface in our real, three-dimensional world, it describes the interface by a height field $h(\mathbf{r}, t)$, where $\mathbf{r}$ is a two-component vector giving positions in a two-dimensional reference plane parallel to the (spatially) averaged orientation of the growing interface, and t is time. The field $h(\mathbf{r}, t)$ simply gives the height at time t of the point on the interface that is directly "above" (that is, separated along the direction perpendicular to the reference plane) the point $\mathbf{r}$ in the reference plane.

We often consider generalizations of the KPZ equation to other spatial dimensions. For reference, I will refer to the physical case just described (i.e., an interface in a three-dimensional world) as the "two-dimensional" KPZ equation. More generally, when I talk about the "d-dimensional" KPZ equation, I mean the case in which the argument $\mathbf{r}$ of the height field $h(\mathbf{r}, t)$ has d components. Other authors

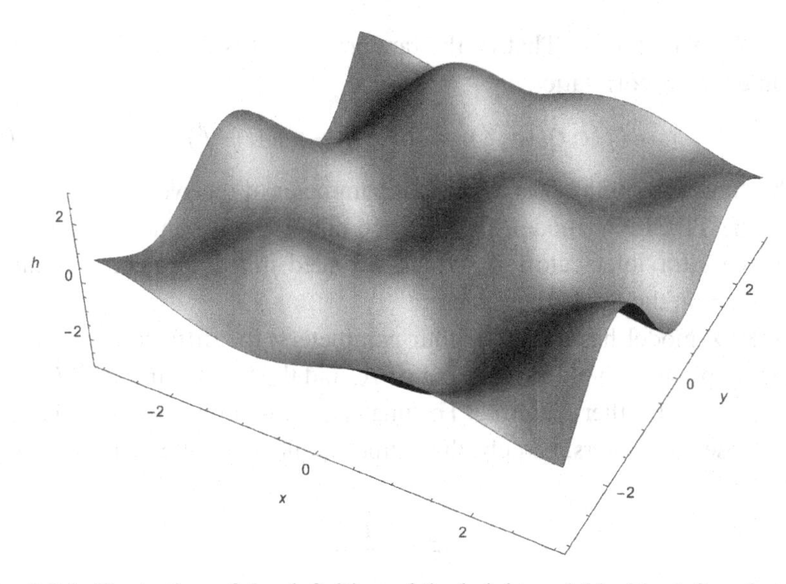

Fig. 3.0.1 Illustration of the definition of the height variable $h(\mathbf{r},t)$ for what I'll call a two-dimensional KPZ equation. Here $\mathbf{r} = (x,y)$.

will use the phrase "$d+1$-dimensional KPZ equation" to describe this system, just to make it clear that time t is *not* being counted as one of the d dimensions.

This model, although derived for a growing interface, actually proves to describe an *enormous* number of systems including, as we shall see in Chapter 8, "incompressible" two-dimensional flocks (which, as we'll see there, map on to the *one*-dimensional KPZ equation). Many other examples are also known, quite a few of which are discussed by Kardar, Parisi, and Zhang (K, P, and Z) in their first papers [31].

I should emphasize here that *every* result I will present in this chapter, and the arguments which lead to them, are all taken shamelessly from the work of KPZ.

I will defer discussion of the symmetry arguments used by K, P, and Z to justify their proposed equation to Section 3.9 at the end of this chapter. I also refer any reader who is interested in that derivation (or, indeed, in any other aspect of this problem) to the excellent original paper [31]. For now I will simply quote their result, which is identical to the "TDGLXY" model I discussed in Chapter 2, except for one additional nonlinear term:

$$\partial_t h(\mathbf{r},t) = \nu\nabla^2 h(\mathbf{r},t) + \frac{\lambda}{2}|\nabla h(\mathbf{r},t)|^2 + f(\mathbf{r},t), \tag{3.0.1}$$

where the diffusion constant ν and the nonlinear coupling λ are constant parameters of the model, and the $\mathbf{f}$ term is a random driving force exactly like the white noise of the "TDGLXY" model. That is, the random force **f** is assumed to be Gaussian with "white" noise correlations:

$$\langle f(\mathbf{r},t)f(\mathbf{r}',t')\rangle = 2D\delta^d(\mathbf{r}-\mathbf{r}')\delta(t-t'), \tag{3.0.2}$$

where the "noise strength" D is a constant parameter of the system.

As for the TDGLXY model, there is one additional parameter hidden in this problem: the ultraviolet cutoff Λ, which we need in this problem for the same reasons we did in the flocking problem.

So the KPZ model has a total of four parameters: the diffusion constant ν, the nonlinear coupling λ, the ultraviolet cutoff Λ, and the "noise strength" D.

This may sound rather daunting. Fortunately, it turns out that only one combination of these parameters, namely, the "dimensionless coupling strength" g, given by

$$g = \frac{S_d}{(2\pi)^d}\frac{\lambda^2 D\Lambda^{d-2}}{\nu^3}, \tag{3.0.3}$$

actually matters. Here, S_d is the surface area (or, strictly speaking, "hyperarea") of a d-dimensional unit sphere. That is, $S_3 = 4\pi$, $S_2 = 2\pi$, and so on. This constant, $O(1)$ factor is introduced only to simplify some of our later expressions.

The physical significance of this dimensionless coupling constant g is that it provides a measure of how important nonlinear effects are in this system. That is why it is proportional to a positive power of the coefficient λ of the nonlinear term in (3.0.1); clearly, if λ gets bigger, nonlinear effects will become more important. It is proportional to the noise strength D because the larger the noise, the larger the typical amplitude of the field h, and, hence, the more important the nonlinearity $|\nabla h(\mathbf{r},t)|^2$ relative to the linear terms in (3.0.1). And finally, the nonlinearity becomes *less* important as the diffusion constant ν grows, because a bigger diffusion constant makes the fluctuations decay more rapidly, and, hence, reduces the typical magnitude of h.

That g is, indeed, the only important dimensionless parameter in the problem can be shown by rescaling. First, I'll rescale lengths to make the ultraviolet cutoff equal to 1. This entails replacing the spatial coordinate **r** with a new, dimensionless coordinate **R** related to **r** via

$$\mathbf{r} = \frac{\mathbf{R}}{\Lambda}. \tag{3.0.4}$$

Now I will rescale time t and the field h to make the diffusion constant and the noise correlation equal to 1. (In fact, I'll make the noise correlation equal to 2, but that's an unimportant detail.) That is, I'll define a dimensionless time variable τ and a dimensionless height variable H via

$$t \equiv \alpha\tau, \quad h \equiv \beta H, \tag{3.0.5}$$

where the rescaling factors α and β will be chosen to make the coefficient analogous to ν for the rescaled variables, and the analog of the noise strength D in the renormalized variables, both equal 1.

Making the changes of variables (3.0.4) and (3.0.5) in (3.0.1) gives

$$\frac{\beta}{\alpha}\partial_\tau H(\mathbf{R},\tau) = \beta\Lambda^2\nu\nabla'^2 H(\mathbf{R},\tau) + \frac{\lambda}{2}\Lambda^2\beta^2|\nabla' H(\mathbf{R},\tau)|^2 + f(\mathbf{R},\tau)\,, \tag{3.0.6}$$

where the primes on the gradients denote gradients with respect to the new variable $\mathbf{R}$, which can easily be simplified to

$$\partial_\tau H(\mathbf{R},\tau) = \alpha\Lambda^2\nu\nabla'^2 H(\mathbf{R},\tau) + \frac{\lambda}{2}\alpha\beta\Lambda^2|\nabla' H(\mathbf{R},\tau)|^2 + \frac{\alpha}{\beta}f(\mathbf{R},\tau)\,. \tag{3.0.7}$$

Defining

$$F \equiv \frac{\alpha}{\beta}f, \tag{3.0.8}$$

we can write this as

$$\partial_\tau H(\mathbf{R},\tau) = \alpha\Lambda^2\nu\nabla'^2 H(\mathbf{R},\tau) + \frac{\lambda}{2}\alpha\beta\Lambda^2|\nabla' H(\mathbf{R},\tau)|^2 + F(\mathbf{R},\tau)\,. \tag{3.0.9}$$

Using (3.0.2) and (3.0.8), it is straightforward to show that the correlations of F are given by

$$\begin{aligned}\langle f(\mathbf{R},\tau)F(\mathbf{R}',\tau')\rangle &= \frac{\alpha^2}{\beta^2}\langle f(\mathbf{r},t)f(\mathbf{r}',t')\rangle \\ &= 2\frac{\alpha^2}{\beta^2}D\delta^d(\mathbf{r}-\mathbf{r}')\delta(t-t') \\ &= 2\frac{\alpha^2}{\beta^2}D\delta^d\left(\frac{\mathbf{R}-\mathbf{R}'}{\Lambda}\right)\delta(\alpha(\tau-\tau')) \\ &= 2\frac{\alpha\Lambda^d}{\beta^2}D\delta^d(\mathbf{R}-\mathbf{R}')\delta(\tau-\tau'),\end{aligned} \tag{3.0.10}$$

where in the last equality I have used the well-known property of delta functions that

$$\delta(kx) = \frac{1}{k}\delta(x) \tag{3.0.11}$$

for any positive constant k, and have been careful to remember that $\delta^d(\mathbf{R}-\mathbf{R}')$ is shorthand notation for a product of d delta functions (one for each Cartesian component of $\mathbf{R}$ and $\mathbf{R}'$), each of which contributes a factor of Λ as a result of the rescaling (3.0.4).

Now, as I said earlier, I will choose α to make the coefficient of $\nabla'^2 H(\mathbf{R},\tau)$ in (3.0.9) equal to 1. This clearly implies

$$\alpha = \frac{1}{\nu\Lambda^2}\,. \tag{3.0.12}$$

Next, I will choose β so that the "dimensionless noise strength" – that is, the coefficient of $\delta^d(\mathbf{R}-\mathbf{R}')\delta(\tau-\tau')$ in (3.0.10) – is equal to 2. This obviously gives

$$\beta = \sqrt{\alpha D \Lambda^d} = \sqrt{\frac{D\Lambda^{d-2}}{\nu}}\,. \tag{3.0.13}$$

With these choices of rescaling (3.0.4), (3.0.5), (3.0.12), and (3.0.13), the rescaled KPZ equation becomes

$$\partial_\tau H(\mathbf{R},\tau) = \nabla'^2 H(\mathbf{R},\tau) + \frac{\sqrt{g}}{2}|\nabla' H(\mathbf{R},\tau)|^2 + F(\mathbf{R},\tau)\,, \tag{3.0.14}$$

with ultraviolet cutoff equal to 1,

$$\langle f(\mathbf{R},\tau)F(\mathbf{R}',\tau')\rangle = 2\delta^d(\mathbf{R}-\mathbf{R}')\delta(\tau-\tau')\,, \tag{3.0.15}$$

and

$$\sqrt{g} = \lambda\alpha\beta\Lambda^2 = \frac{\lambda\Lambda^2}{\nu\Lambda^2}\sqrt{\frac{D\Lambda^{d-2}}{\nu}} = \lambda\sqrt{\frac{D\Lambda^{d-2}}{\nu^3}}\,. \tag{3.0.16}$$

The last expression clearly implies

$$g = \frac{\lambda^2 D\Lambda^{d-2}}{\nu^3}\,, \tag{3.0.17}$$

which is, of course, exactly the g we defined earlier in (3.0.3), up to the arbitrary numerical, parameter independent prefactor $\frac{S_d}{(2\pi)^d}$.[1]

This demonstrates that g is the only meaningful parameter in the KPZ equation, in the sense that any two KPZ equations with the same value of g will have exactly the same correlations, up to a trivial rescaling of position $\mathbf{r}$, height h, and time t given by equations (3.0.4), (3.0.5), (3.0.12), and (3.0.13).

Having established that the dimensionless number g is the parameter that determines the importance of nonlinearities in this problem, one might think that, were we lucky enough to have a system in which g was small, we could ignore nonlinearities altogether, and use a linear theory – like the one we used to study the XY model previously – to solve the problem. Unfortunately, this proves to be true *only* in spatial dimensions $d > 2$. For $d \leq 2$, nonlinearities *always* change the long distance scaling of the problem, no matter how weak they are (i.e., no matter how small g is).

We'd now like to understand the behavior of this fluctuating, dynamic surface. In particular, what is the scale of the height fluctuations? This, of course, we can estimate as the root mean squared average of h; i.e., $\sqrt{\langle|h(\mathbf{r},t)\rangle|^2}$. We can also ask for the two-point correlation function

$$C_h(\mathbf{r},t) \equiv \langle (h(\mathbf{R},T) - h(\mathbf{R}+\mathbf{r},T+t))^2\rangle\,, \tag{3.0.18}$$

[1] Note that [31] erroneously misses the factor of Λ^{d-2} that I find here.

which, since it depends on time t, as well as the separation $\mathbf{r}$ of the two points being correlated, gives us information about the evolution of the surface with time.

We'll begin by looking at the linear theory.

3.1 Linear Theory

As a first step towards understanding any new dynamical problem, it's best to begin by linearizing the equation of motion. The reason this is helpful is that, for linear equations, we need only find the normal modes, which almost always prove to be plane waves, as we saw in our treatment of the "pointers" problem earlier. Since these normal modes are decoupled, we can study the dynamics of each of them independently, thereby effectively reducing a stochastic *partial* differential equation to a stochastic *ordinary* differential equation. In fact, by Fourier transforming in time as well as space, we'll be able to reduce the problem to a set of independent linear stochastic equations, which are then quite easy to solve for the correlation functions.

This approach works for almost *any* hydrodynamic theory. That is, it "works" in the sense that it is possible using this approach to calculate any property of the system in which one is interested. Whether or not the results of the linear theory are *correct*, on the other hand, requires further analysis. Even that analysis, however, still takes the linearized theory as an input, so linearization is always a good place to start.

Let's therefore linearize the KPZ equation. This obviously amounts to nothing more than dropping the λ term in that equation, which leaves us with

$$\partial_t h(\mathbf{r}, t) = \nu \nabla^2 h(\mathbf{r}, t) + f(\mathbf{r}, t) , \tag{3.1.1}$$

with the statistics of the noise f still being given by (3.0.2).

This linearized equation of motion is just a noisy diffusion equation, which is also sometimes called the Edwards–Wilkinson equation [28]. Note also that it is *exactly* the same as the dynamical version of the XY model (the "time-dependent Ginzburg–Landau model, or "TDGL" model) that I discussed in Chapter 2. Hence, we can simply transcribe all of the results we obtained in Section 2.4 here, just by replacing θ with h everywhere. Thus we have the solution for the spatiotemporally Fourier transformed fields $h(\mathbf{q}, \Omega)$, defined just as we did for the θ field in Chapter 2:

$$h(\mathbf{r}, t) = \frac{1}{\sqrt{VT}} \sum_{\mathbf{q},\Omega} h(\mathbf{q}, \Omega) e^{i(\mathbf{q}\cdot\mathbf{r} - \Omega t)} \tag{3.1.2}$$

in terms of the spatiotemporally Fourier transformed noises $f(\mathbf{q}, \Omega)$:

$$h(\mathbf{q}, \Omega) = G(\mathbf{q}, \Omega) f(\mathbf{q}, \Omega) , \tag{3.1.3}$$

where the "propagator"

$$G(\mathbf{q}, \Omega) \equiv \frac{1}{-\mathrm{i}\Omega + \nu q^2} \tag{3.1.4}$$

is the same as that found earlier for the XY model.

Likewise,

$$\begin{aligned}\langle h(\mathbf{q}, \Omega) h(\mathbf{q}', \Omega')\rangle &= G(\mathbf{q}, \Omega) G(\mathbf{q}', \Omega')\langle f(\mathbf{q}, \Omega) f(\mathbf{q}', \Omega')\rangle \\ &= 2D G(\mathbf{q}, \Omega) G(\mathbf{q}', \Omega') \delta^K_{\mathbf{q}+\mathbf{q}'} \delta^K_{\Omega+\Omega'} \\ &= 2D G(\mathbf{q}, \Omega) G(-\mathbf{q}, -\Omega) \delta^K_{\mathbf{q}+\mathbf{q}'} \delta^K_{\Omega+\Omega'} \equiv C(\mathbf{q}, \Omega) \delta^K_{\mathbf{q}+\mathbf{q}'} \delta^K_{\Omega+\Omega'} ,\end{aligned} \tag{3.1.5}$$

with

$$C(\mathbf{q}, \Omega) = \frac{2D}{(\Omega^2 + \nu^2 q^4)} , \tag{3.1.6}$$

which, again as for the XY model, leads to divergent equal-time correlations:

$$C_{\mathrm{ET}}(\mathbf{q}) = \int \frac{\mathrm{d}\Omega}{2\pi} \frac{2D}{(\Omega^2 + \nu^2 q^4)} = \frac{D}{\nu q^2} . \tag{3.1.7}$$

As for the XY model, here too for the linearized KPZ equation, this correlation function diverges like $1/q^2$ as $q \to 0$. This means that the surface will, according to the linearized theory, be algebraically "rough" in spatial dimensions $d < 2$, logarithmically rough in $d = 2$, and "smooth" in $d > 2$. Specifically,

$$\langle (h(\mathbf{r}, t))^2 \rangle \propto \begin{cases} L^{2-d}, & d < 2, \\ \ln\left(\frac{\Lambda L}{2\pi}\right), & d = 2, \\ \text{finite (independent of } L), & d > 2. \end{cases} \tag{3.1.8}$$

I can also calculate spatio-temporal correlation functions of $h(\mathbf{r}, t)$. It's particularly instructive to look at the two-point correlation function

$$C_h(\mathbf{r}, t) \equiv \langle (h(\mathbf{R}, T) - h(\mathbf{R} + \mathbf{r}, T + t))^2 \rangle . \tag{3.1.9}$$

Performing the sort of manipulations with Fourier transforms that should by now be familiar, I find that this correlation function is quite generally related to the spatio-temporally Fourier transformed correlation function $C(\mathbf{q}, \Omega)$ via

$$C_h(\mathbf{r}, t) = 2 \int \frac{\mathrm{d}\Omega}{2\pi} \int \frac{\mathrm{d}^d q}{(2\pi)^d} C(\mathbf{q}, \Omega)[1 - \cos(\mathbf{q} \cdot \mathbf{r} - \Omega t)] . \tag{3.1.10}$$

By "quite generally" in the preceding sentence, I mean that this relation holds for "any" dynamical model. That is, it is simply a property of Fourier transforms.

What is specific to the KPZ equation, or, at least its linearized version, equation (3.1.1), is the result (3.1.6) for $C(\mathbf{q}, \Omega)$. Inserting that result into (3.1.10), I have

$$C_h(\mathbf{r}, t) = 4D \int \frac{\mathrm{d}\Omega}{2\pi} \int \frac{\mathrm{d}^d q}{(2\pi)^d} \left(\frac{1 - \cos(\mathbf{q} \cdot \mathbf{r} - \Omega t)}{\Omega^2 + \nu^2 q^4} \right) . \tag{3.1.11}$$

This integral can be, rather unenlighteningly, expressed in terms of special functions. It is much more illuminating, however, to simply tease out its scaling behavior with some simple changes of variable of integration. Specifically, let's define new variables of integration $\mathbf{Q}$ and ω via

$$\mathbf{q} \equiv \mathbf{Q}/r, \quad \Omega \equiv \nu\omega/r^2 . \tag{3.1.12}$$

This very straightforwardly gives (provided one remembers that the change of variables in $\mathbf{q}$ is actually d changes of variable, one for each Cartesian component of $\mathbf{q}$)

$$C_h(\mathbf{r}, t) = r^{2-d} F_s \left(t/r^2 \right) , \tag{3.1.13}$$

where I've defined the scaling function

$$F_s(x) \equiv \frac{4D}{\nu} \int \frac{\mathrm{d}\omega}{2\pi} \int \frac{\mathrm{d}^d Q}{(2\pi)^d} \left(\frac{1 - \cos(\mathbf{Q} \cdot \hat{\mathbf{r}} - \omega\nu x)}{\omega^2 + Q^4} \right) . \tag{3.1.14}$$

Note that (3.1.13) can also be written in the generic scaling form

$$C_h(\mathbf{r}, t) \equiv \langle (h(\mathbf{R}, T) - h(\mathbf{R} + \mathbf{r}, T + t))^2 \rangle = r^{2\chi} F_s(t/r^z) \tag{3.1.15}$$

with the "dynamic exponent" z and the "roughness exponent" χ given, in this linear theory, by

$$\chi = \chi_{\mathrm{lin}} = \frac{2-d}{2}, \quad z = z_{\mathrm{lin}} = 2 . \tag{3.1.16}$$

We will see later that, for the KPZ equation in spatial dimensions $d \leq 2$, the scaling form (3.1.15) continues to hold, but with the exponents z and χ taking on different values from those predicted by the linear theory. The scaling function F_s also changes. However, the exponents $z(d)$ and $\chi(d)$ remain *universal* for $d \leq 2$; that is, they are uniquely determined by the spatial dimension d of the system, and independent of all of the parameters ν, λ, and D in the KPZ equation. The scaling function is likewise universal, up to a nonuniversal prefactor, and a nonuniversal multiplicative factor in the argument of the scaling function. In the linear theory (which proves to be valid for $d > 2$, as we'll see later), these nonuniversal factors are $\frac{4D}{\nu}$ and ν, respectively.

The universal exponents χ and z appear everywhere. The roughness exponent χ gives the scaling of a typical height h with length scale L or r, while z gives the scaling of times t with length scales r or L. Note, for example, that in the linear theory, the same exponent $\chi = \frac{2-d}{2}$ governs *both* the scaling of the two-point correlation function (3.1.13) with the separation r of the two points, and that of the single-point mean squared height fluctuations (3.1.8) with the system size L, which is the relevant length scale in that case.

We can tease out the scaling behavior of the correlation function $C_h(\mathbf{r}, t)$ in various limits directly from the scaling form (3.1.15). First note that we expect, and its integral expression (3.1.14) confirms, that the scaling function $F_s(x)$ goes to a finite, nonzero constant as its argument $x \to 0$. Thus

$$C_h(\mathbf{r}, t) \propto r^{2\chi}, \quad r^z \gg |t| \,. \tag{3.1.17}$$

Obviously, the condition $r^z \gg t$ is, strictly speaking, meaningless, since the two sides have different dimensions. In the linear theory, the real condition is $r^2 \gg \nu|t|$, where I've used the linearized value $z = 2$, and the diffusion constant ν makes the units (meters squared) the same on both sides of the inequality. That diffusion constant is, of course, nonuniversal (indeed, it's one of the free parameters of the model). In $d \leq 2$, where the nonlinearities become important, the true condition for (3.1.17) to hold will be $\left(\frac{r}{\xi}\right)^z \gg \left(\frac{t}{\tau}\right)$, where the characteristic length ξ and the characteristic time scale τ are likewise nonuniversal. For now, to keep my expressions simple, I'll ignore this subtlety, and simply write the condition for (3.1.17) to hold – that is, for the spatio-temporal correlation function $C_h(\mathbf{r}, t)$ to become t independent, and become a simple power law in r – as $r^z \gg t$.

In the opposite, $r^z \ll |t|$, limit, we expect $C_h(\mathbf{r}, t)$ to become r independent (as it certainly must do for the equal-space correlations – i.e., when $r = 0$). Clearly, the only way this can happen is for the r dependence of the scaling function $F_s(x)$ in (3.1.15) to cancel off the $r^{2\chi}$ factor in front. This implies that $F_s(x)$ must scale like $r^{-2\chi}$ for small x. But since $F_s(x = \frac{t}{r^z})$ can only be a function of the scaling combination $x = \frac{t}{r^z}$, the only way it can also scale like $r^{-2\chi}$ for small x is for it to scale like $x^{\frac{2\chi}{z}}$ for small x. This is easily seen to imply

$$C_h(\mathbf{r}, t) \propto t^{\frac{2\chi}{z}}, \quad r^z \gg |t| \,. \tag{3.1.18}$$

I can summarize the results of this scaling analysis as follows:

$$C_h(\mathbf{r}, t) \equiv \langle (h(\mathbf{R}, T) - h(\mathbf{R} + \mathbf{r}, T + t))^2 \rangle = r^{2\chi} f(t/r^z) \propto \begin{cases} r^{2\chi}, & r^z \gg t \,, \\ |t|^{\frac{2\chi}{z}}, & r^z \ll t \,, \end{cases} \tag{3.1.19}$$

with

$$\chi_{\text{lin}} = \frac{2-d}{2}, \quad z_{\text{lin}} = 2 \,. \tag{3.1.20}$$

We will see in Section 3.2 that these values of the scaling exponents z and χ no longer hold for $d \leq 2$ due to the effects of nonlinearities. However, the scaling law (3.1.19) continues to apply, but with different values of the exponents $z(d)$ and $\chi(d)$. As mentioned earlier, these exponents remain *universal* for $d \leq 2$; that is, they are uniquely determined by the spatial dimension d of the system, and independent of all of the parameters ν, λ, and D in the KPZ equation. The scaling function is likewise universal, in the sense of being the same for all KPZ equations in the same spatial dimension d, up to a nonuniversal prefactor, and a nonuniversal multiplicative factor in the argument of the scaling function.

Of course, at this point, I have only proved that these claims of universality hold in the linear theory. To decide when the linear theory actually holds, and what happens when it does not, I obviously need to consider the effect of the nonlinear term $\lambda|\nabla h|^2$ in the KPZ equation (3.0.1). I'll turn to that question now.

3.2 The Effect of Nonlinearities

The linear theory just presented has, paradoxically, provided us with enough information to show that it breaks down for spatial dimensions $d \leq 2$. We can see this by power counting. I'll first present this argument in its crudest possible form, and then give progressively more sophisticated versions of the same argument, until, finally, we reach the renormalization group.

3.2.1 Simple Power Counting

The crudest argument is to simply take the ratio of the nonlinear term $\lambda|\nabla h|^2$ to the $\nabla^2 h$ term. Very crudely speaking, when we take this ratio, we can "cancel off" the two spatial derivatives in $\lambda|\nabla h|^2$ with the two in $\nabla^2 h$. We can also "cancel off" one of the two height fields h in $\lambda|\nabla h|^2$against the one in the $\nabla^2 h$ term. Thus, we are left with the (admittedly very crude) conclusion that the ratio of these two terms scales like the height field h itself.

Since the fluctuations in the height field diverge as the system size $L \to \infty$, according to the linear theory, like $h \propto L^\chi = L^{\frac{2-d}{2}}$ for $d < 2$, and like $\sqrt{\langle h^2(\mathbf{r},t)\rangle} \propto \sqrt{\ln\left(\frac{L}{a}\right)}$ for $d = 2$, this suggests that, for $d \leq 2$, the nonlinear term will *never* be negligible in a sufficiently large system. To say this another way, the linear theory must *always* fail, in a sufficiently large system, for $d \leq 2$.

This proves to be true, as can be confirmed by the following more systematic power counting argument.

The idea of this argument – often also called a "scaling argument" – is to determine how various quantities in the equation of motion scale with the length scale L under consideration.

Some of these scalings are fairly obvious. Clearly, r, being a length scale itself, scales like L. This implies that every spatial gradient scales like $1/L$.

Other scalings are a bit more subtle, but can be read off from the linear theory we've just done. For example, we found in the linear theory that time scales t scale like length scales r^z with $z = 2$. Hence, *time* derivatives scale like $L^{-z} = L^{-2}$.

Finally, we found that, in the linear theory, the height field h scaled like r^χ, with $\chi = \frac{2-d}{2}$.

Putting all of these facts together, we can estimate the scalings of the various terms in the KPZ equation as follows: The $\partial_t h$ term contains one height field h and one time derivative. Hence, it will scale like L^{-z} (from the time derivative) times L^χ (from the height field h), for an overall scaling

$$\partial_t h \sim L^{\chi - z} = L^{-1-\frac{d}{2}} . \tag{3.2.1}$$

Very similar analysis gives

$$\nabla^2 h \sim L^{\chi-2} = L^{-1-\frac{d}{2}} . \tag{3.2.2}$$

Note that both of these terms scale exactly the same way with L. This is unsurprising: In a sense, we obtained the value of the roughness exponent χ by balancing all of the linear terms in the KPZ equation. Indeed, we can obtain both scaling exponents χ and z simply by precisely this sort of power counting argument without going through the detailed calculations of Section 3.1, as I'll now show.

From the f–f correlation function (3.0.2), we see that $|\mathbf{f}|^2$ scales like $\delta^d(\mathbf{r})\delta(t)$. Since a delta function is a density in its argument, and being careful to note that $\delta^d(\mathbf{r})$ is a product of d delta functions – one for each Cartesian component of $\mathbf{r}$ – this implies that $|\mathbf{f}|^2$ scales like L^{-d-z}. This obviously implies that $\mathbf{f}$ itself scales like $L^{-\left(\frac{d+z}{2}\right)}$.

Thus, we can summarize the scaling of the three linear terms in the KPZ equation as follows:

$$\partial_t h \sim L^{\chi - z}, \quad \nabla^2 h \sim L^{\chi - 2}, \quad f \sim L^{-\left(\frac{d+z}{2}\right)} . \tag{3.2.3}$$

Requiring that all three of these scale the same way with L leads to two linear equations for the unknown exponents χ and z:

$$\chi - 2 = \chi - z, \quad \chi - 2 = -\left(\frac{d+z}{2}\right), \tag{3.2.4}$$

whose solutions are trivial and are found to be $z = 2$ and $\chi = \frac{2-d}{2}$. These are, of course, exactly the same scaling exponents we found through our far more detailed analysis of the linearized equation of motion in Section 3.1.

With these linearized exponents in hand, we can now use them to determine the importance of the $\lambda|\nabla h|^2$ term. Applying the same power counting we've just done for the linear terms, it's easy to see that this term scales as

$$|\nabla h|^2 \sim L^{2\chi-2} = L^{-d}, \tag{3.2.5}$$

where in the last equality I've used $\chi = \frac{2-d}{2}$.

This scaling exponent implies that the $\lambda|\nabla h|^2$ will become important at large length scales relative to the linear terms when its scaling exponent $(-d)$ becomes greater than or equal to that of the linear terms $(-1 - \frac{d}{2})$. That is, the nonlinearity becomes important if

$$-d \geq -1 - \frac{d}{2}, \tag{3.2.6}$$

that is, for $d \leq 2$.

To put the above analysis on a firmer footing, and to address the much harder question of how to deal with the nonlinearity when $d \leq 2$, I'll use the dynamical renormalization group (DRG), which I'll now describe.

3.2.2 Dynamical Renormalization Group 1: How?

The DRG starts by averaging the equations of motion of whatever dynamical model is under consideration over the short-wavelength fluctuations: i.e., those with support in the "shell" of Fourier space $b^{-1}\Lambda \leq |\mathbf{q}| \leq \Lambda$, where Λ is the "ultraviolet cutoff" I introduced earlier, and b is an arbitrary rescaling factor. We will do this using a perturbation theory approach, in which the nonlinearity is treated as a small perturbation to the linear theory. I'll describe this technique in excruciating detail later. Then, one rescales lengths, time, and fields so as to restore the ultraviolet cutoff to its original value. The rescaling of wavevectors that accomplishes this is, obviously, the change of variable

$$\mathbf{q} = b^{-1}\mathbf{q}'. \tag{3.2.7}$$

Because of the inverse relation between wavevectors and positions (which is why Fourier space $\mathbf{q}$ is often referred to as "reciprocal space"), this implies a rescaling of spatial coordinates that goes exactly the opposite way; i.e.,

$$\mathbf{r} = b\mathbf{r}'. \tag{3.2.8}$$

Once we start rescaling, why stop here? Instead, we'll also rescale time t, and the height field h, by the following simple rescalings:

$$t = b^z t', \quad h = b^\chi h', \tag{3.2.9}$$

where the "dynamical exponent" z and the "roughness exponent" χ are, at this point, arbitrary. However, the special values of z and χ that produce *fixed points* of the RG process – I'll define what I mean by "fixed points," and their significance, later – prove to be exactly those that appear in the scaling laws for the spatio-temporal correlations $C_h(\mathbf{r},t) \equiv \langle (h(\mathbf{R},T) - h(\mathbf{R}+\mathbf{r},T+t))^2 \rangle$ of the field $h(\mathbf{r},t)$. In particular, I'll show how the RG directly implies that $C_h(\mathbf{r},t)$ obeys the scaling law (3.2.21), but with exponents z and χ which are different, for $d < 2$, than those predicted by the linear theory, and which we can calculate directly from the RG (for $d > 2$, and $d = 1$) by simply finding the special values of z and d that produce RG fixed points.

The net result of the two step RG process described above is a new equation of motion – often called the "renormalized" equation of motion – written in terms of the rescaled variables. If one began with the most general equations of motion allowed by symmetry and conservation laws for the problem under consideration, then the *form* of these "renormalized" equations of motion must be the same as that of the original equation (since the same symmetries and conservation laws apply to it).

Therefore, the only difference between the renormalized equation of motion and the original one (sometimes called the "bare" equation of motion, a terminology that leads to many obvious bad puns!) must be in the values of the hydrodynamic parameters appearing in those equations. Therefore, the net result of the two-step RG process is the same equation of motion as the original, but with "renormalized" parameters. These renormalized parameters are, obviously, functions of the original parameters. The formulae relating the new, renormalized parameters ν_1, λ_1, and D_1 to the originals ν_0, λ_0, and D_0 are called the "recursion relations," and take the form

$$\nu_1 = f_\nu(\nu_0, \lambda_0, D_0, b), \quad \lambda_1 = f_\lambda(\nu_0, \lambda_0, D_0, b), \quad D_1 = f_D(\nu_0, \lambda_0, D_0, b)\,. \quad (3.2.10)$$

Since the form of the equations is unchanged after one complete RG cycle (averaging, then rescaling), the result of going through *another* RG cycle of averaging and rescaling must lead to the same formulae for the "doubly renormalized" renormalized parameters in terms of the "once renormalized" parameters as the formula that relates the "bare" parameters to the "once renormalized" parameters. That is, the recursion relations on each step of the RG are the same.

Therefore, once we've completed one step of our RG process, and gotten our recursion relations for the parameters, all we need to do to assess the effects of further iterations of the RG is to iterate those recursion relations. That is, the values of the parameters on successive steps of the RG obey

$$\nu_{n+1} = f_\nu(\nu_n, \lambda_n, D_n, b), \quad \lambda_{n+1} = f_\lambda(\nu_n, \lambda_n, D_n, b), \quad D_{n+1} = f_D(\nu_n, \lambda_n, D_n, b)\,. \quad (3.2.11)$$

We can then decide which parameters are important in the model ("relevant," in the jargon of the renormalization group) and which are "irrelevant" – i.e., unimportant at long wavelengths[2] – simply by using these recursion relations. There are some major qualifications to this statement, however, which I'll discuss below.

Parameters that flow to zero under repeated iteration of these recursion relations are called "irrelevant," and are usually unimportant for the long-wavelength scaling behavior of the system, and so can be dropped from the model. Parameters that remain nonzero upon renormalization are called "relevant," and must be kept in the model.

3.2.3 Dynamical Renormalization Group 2: Why?

> The eternal mystery of the world is its comprehensibility... The fact that it is comprehensible is a miracle.
>
> – Albert Einstein

> When people see other magic acts, they ask "How did he do that?" When they see ours, they ask "Why?"
>
> – Penn Jillette, of the magic act Penn and Teller

> The renormalization group is not just a technique for calculating a few exponents. It's what explains why we can do physics at all.
>
> – Paul Goldbart

Readers may be feeling like the audience at one of Penn and Teller's shows right now, so I'll address the question of why we do the RG at all.

The first part of the answer lies in the fact that, if we *can* do the RG, then we can directly determine the scaling laws, and the universal scaling exponents, from it. The second part of the answer is that the RG actually guides us in the construction of hydrodynamic models for *any* system – e.g., in the example we're discussing here, the formulation of the KPZ equation. More generally, as Paul Goldbart noted, it's nothing less than the answer to Einstein's famous question about why we can do physics.

I'll discuss the last point in Section 3.9. In this subsection, I'll illustrate how the RG can be used to obtain scaling laws. I'll do this first for the two-point correlation function introduced earlier:

$$\langle (h(\mathbf{R},T) - h(\mathbf{R}+\mathbf{r},T+t))^2 \rangle \equiv C_h(r,t;\nu_0,\lambda_0,D_0). \tag{3.2.12}$$

[2] In some problems, there are parameters that are "irrelevant" in the sense that they flow to zero under renormalization, but which cannot simply be set equal to zero in certain calculations because doing so leads to infinities. Such parameters are called "dangerously irrelevant variables." They can still be handled using the renormalization group, but rather more delicately (as their name suggests). We will not encounter any such variables in the problems I'll consider in this book. For a discussion of dangerously irrelevant variables, and how to deal with them, see, e.g., Ma [30].

I'm now considering this in the full, nonlinear equation of motion. Note that I have also replaced the argument $\mathbf{r}$ of this correlation function by its magnitude r; this is justified by the fact that our model is isotropic, so the correlation function must be as well. Hence, it can only depend on the magnitude r of the spatial separation $\mathbf{r}$ of the two points being correlated.

We can relate this directly to the same correlation function evaluated for the *renormalized* system by taking into account the rescalings (3.2.8), and (3.2.9) of position $\mathbf{r}$, time t, and fields h, with the result

$$C_h(r,t;\nu_0,\lambda_0,D_0) = b^{2\chi}C_h(b^{-1}r,b^{-z}t;\nu_1,\lambda_1,D_1)\,. \tag{3.2.13}$$

We can now continue iterating the RG; after n iterations, we'll have

$$C_h(r,t;\nu_0,\lambda_0,D_0) = b^{2n\chi}C_h(b^{-n}r,b^{-nz}t;\nu_n,\lambda_n,D_n)\,. \tag{3.2.14}$$

Now suppose we choose our rescaling exponents z and χ to produce a *fixed point*; that is, to make all of the parameters ν_n, λ_n, and D_n go to constant values ν_*, λ_*, and D_* as the number of iterations n of the RG goes to infinity. Whether or not this is possible will require actually performing the RG. Let's suppose we find that it *is* possible. Then the scaling relation (3.2.14) becomes

$$C_h(r,t;\nu_0,\lambda_0,D_0) = b^{2n\chi}C_h(b^{-n}r,b^{-nz}t;\nu_*,\lambda_*,D_*)\,. \tag{3.2.15}$$

Note that the scaling exponents z and χ in this expression are no longer arbitrary; rather, they are completely determined by the requirement that we obtain a fixed point. We can determine the magic values of z and χ that do this entirely from the recursion relations for the parameters ν_n, λ_n, and D_n. I will now show that these magic values are also the values of the scaling exponents appearing in the scaling law (3.2.21).

To see this, let's choose the number of iterations n of our renormalization group to make the spatial argument $b^{-n}r$ on the right-hand side of (3.2.15) equal to some *fixed* microscopic length a (e.g., $\frac{2\pi}{\Lambda}$, where Λ is the ultraviolet cutoff). That is, we choose $n = n_*(r)$ such that

$$b^{-n_*(r)}r = a\,, \tag{3.2.16}$$

which is easily solved to give

$$n_*(r) = \frac{\ln\left(\frac{r}{a}\right)}{\ln b}\,. \tag{3.2.17}$$

The alert reader will be concerned about the fact that (3.2.17) may not yield an integer solution. Ultimately, we will solve this problem by choosing the rescaling factor b (which the reader, being alert, will remember is also arbitrary) to be very close to one, so that choosing the nearest integer to the value of n_* given by (3.2.17) will make a negligible error. Indeed, I will ultimately choose

$$b = 1 + d\ell\,, \tag{3.2.18}$$

with $d\ell$ infinitesimal, in which limit the error vanishes.

There are a number of other good reasons to choose the rescaling factor b to be close to 1, as in (3.2.18). I'll discuss these as we come to them.

Evaluating (3.2.15) with $n = n^*$ gives

$$C_h(r,t;\nu_0,\lambda_0,D_0) = \left(\frac{r}{a}\right)^{2\chi} C_h\left(a, \left(\frac{a}{r}\right)^{z} t; \nu_*, \lambda_*, D_*\right) \equiv r^{2\chi} f(t/r^z)\,, \tag{3.2.19}$$

where I've defined the scaling function

$$f(u) \equiv a^{-2\chi} C_h(a, u; \nu_*, \lambda_*, D_*)\,. \tag{3.2.20}$$

Note that (3.2.19) is exactly the scaling form (3.2.21) that I argued for earlier.

Note that even *without* performing RG *explicitly*, the very idea that we *can*, even in principle, shows how we get scaling laws. Furthermore, we've now got a prescription for calculating the exponents in those scaling laws: Just find the values of z and χ that lead to fixed points of the RG. I'll hereafter refer to these values of the exponents as the "physical" exponents.

Note also that, as described earlier, this scaling form alone leads to the large r and large t limits of the correlation function:

$$C_h(\mathbf{r},t) \equiv \langle (h(\mathbf{R},T) - h(\mathbf{R}+\mathbf{r},T+t))^2 \rangle = r^{2\chi} f(t/r^z) \propto \begin{cases} r^{2\chi}, & r^z \gg t\,, \\ |t|^{\frac{2\chi}{z}}, & r^z \ll t\,. \end{cases} \tag{3.2.21}$$

Now let's apply this reasoning to the KPZ equation.

3.3 The RG Version of the Scaling Arguments; or How to Analyze a Nonlinear Stochastic, Partial Differential Equation by Counting On Your Fingers

The first step of the RG process – the averaging step described above – will lead to corrections to the diffusion constant ν and the noise strength D. One might a priori expect that it would also lead to corrections to the nonlinear coupling λ as well, but these prove to vanish, as I'll show later. For the moment, I won't use this fact – since I haven't proven it yet – and will allow for a correction to λ from this step as well.

To actually calculate these corrections *exactly* would be as difficult as solving the KPZ equation. Since that equation, in the presence of noise, has the fully trifecta of difficulty – it's nonlinear, it's stochastic, and it's a partial differential equation – you won't be surprised to learn that no one knows how to do this. Fortunately, it (often) proves to be sufficient to do the averaging in perturbation theory in the coupling λ. This perturbation theory is quite straightforward, as you'll see in a few pages. Quite amazingly, the RG allows us to leverage these perturbation theory results to make statements about the scaling of the KPZ equation even if λ is *not* small. I'll explain this cryptic comment later.

The changes that occur in the parameters due to the first step will be small if λ is small. More precisely, they'll be small if the dimensionless coupling constant g that I introduced earlier in equation (3.0.3) is small compared to one. This fact by itself is sufficient to enable us to argue that the critical dimension of the KPZ equation is two from the second, rescaling step of the RG, without actually calculating these corrections at all.

At the end of step 1 of the RG, therefore, we'll have a new KPZ equation of the same form as our original KPZ equation (3.0.1), but with the diffusion constant ν, the nonlinear coefficient λ, and the noise correlation D replaced by "intermediate" values ν_I, λ_I, and D_I (here the subscript "I" stands for "intermediate," since we have only done the first step of the two-step RG process). That is, the new, "intermediate" KPZ equation now reads

$$\partial_t h(\mathbf{r},t) = \nu_I \nabla^2 h(\mathbf{r},t) + \frac{\lambda_I}{2}|\nabla h(\mathbf{r},t)|^2 + f_I(\mathbf{r},t)\,, \tag{3.3.1}$$

where the correlation of the force f_I has been modified from (3.0.2) by the replacement of the original D with D_I given by the "intermediate" renormalized D obtained after the first step of the DRG. That is,

$$\langle f_I(\mathbf{r}_1',t_1')f_I(\mathbf{r}_2',t_2')\rangle = 2D_I\delta^d(\mathbf{r}_1-\mathbf{r}_2)\delta(t_1-t_2)\,. \tag{3.3.2}$$

As noted above, the corrections $\delta\nu$, $\delta\lambda$, and δD to the parameters arising on this first step, defined via

$$\nu_I \equiv \nu + \delta\nu, \quad \lambda_I \equiv \nu + \delta\lambda, \quad D_I \equiv D + \delta D, \tag{3.3.3}$$

will be small if λ is small. This obvious, and innocuous sounding, statement will actually enable me to show that the critical dimension for the KPZ equation is two *without* actually calculating those corrections $\delta\nu$, $\delta\lambda$, and δD at all, as you'll see in a few paragraphs.

Now we must perform the second step of the RG; that is, rescaling the spatial coordinates $\mathbf{r}$, the time t, and the height field h according to (3.2.8) and (3.2.9). Note that the rescalings of time t and spatial coordinates $\mathbf{r}$ imply the opposite rescaling of time derivatives and spatial derivatives; that is,

$$\nabla = b^{-1}\nabla', \quad \partial_t = b^{-z}\partial_{t'}, \tag{3.3.4}$$

where ∇' denotes a gradient with respect to $\mathbf{r}'$. Taking these, and the field rescaling (3.2.9), into account, it's quite straightforward to see that the KPZ equation, with the intermediate parameters defined above, rescales to become

$$b^{\chi-z}\partial_{t'}h'(\mathbf{r}',t') = b^{\chi-2}\nu_I\nabla'^2 h'(\mathbf{r}',t') + b^{2\chi-2}\frac{\lambda_I}{2}|\nabla h'(\mathbf{r}',t')|^2 + f_I(\mathbf{r}',t'). \tag{3.3.5}$$

We can cast this in the form of the original KPZ equation, in which $\partial_t h$ appears with coefficient 1, simply by dividing the entire equation by $b^{\chi-z}$. This gives

$$\begin{aligned}\partial_{t'}h'(\mathbf{r}',t') &= b^{z-2}\nu_I\nabla'^2 h'(\mathbf{r}',t') + b^{z+\chi-2}\frac{\lambda_I}{2}|\nabla h'(\mathbf{r}',t')|^2 + b^{z-\chi}f_I(\mathbf{r}',t') \\ &\equiv \nu_R\nabla'^2 h'(\mathbf{r}',t') + \frac{\lambda_R}{2}|\nabla h'(\mathbf{r}',t')|^2 + f_R(\mathbf{r}',t'),\end{aligned} \tag{3.3.6}$$

which looks exactly like the original KPZ equation, but with renormalized diffusion constant ν_R and nonlinear coupling λ_R given by

$$\nu_R = b^{z-2}\nu_I = b^{z-2}(\nu + \delta\nu), \tag{3.3.7}$$

and

$$\lambda_R = b^{z+\chi-2}\lambda_I = b^{z+\chi-2}(\lambda + \delta\lambda). \tag{3.3.8}$$

To complete our recursion relations, we need to find the recursion relation for the effective noise strength D_R. We can get this from the way the forces renormalize. To do so, we must take into account the fact that the forces appearing in the fully renormalized equation of motion are also affected by the rescaling. Specifically, we now have

$$f_R(\mathbf{r}',t') = b^{z-\chi}f_I(\mathbf{r}',t'). \tag{3.3.9}$$

The correlations of this field are given by

$$\langle f_R(\mathbf{r}'_1,t'_1)f(\mathbf{r}'_2,t'_2)\rangle = b^{2(z-\chi)}\langle f_I(\mathbf{r}'_1,t'_1)f_I(\mathbf{r}'_2,t'_2)\rangle. \tag{3.3.10}$$

The correlations of the intermediate force $f_I(\mathbf{r}'_1,t'_1)$ are given by

$$\langle f_I(\mathbf{r}'_1,t'_1)f_I(\mathbf{r}'_2,t'_2)\rangle = 2D_I\delta^d(\mathbf{r}_1-\mathbf{r}_2)\delta(t_1-t_2). \tag{3.3.11}$$

There is one very important subtlety hidden in (3.3.11): The space and time coordinates $\mathbf{r}_{1,2}$ and $t_{1,2}$ appearing in the delta functions are *unprimed*. This is *not* a typo: These are the coordinates *before* the rescaling $\mathbf{r}_{1,2} = b\mathbf{r}'_{1,2}$ and $t_{1,2} = b^z t'_{1,2}$ has been done (because, after all, we calculated the intermediate D_I from the f_I correlations on step one of the two-step RG process; i.e., *before* we did the rescaling.

To calculate the fully renormalized D (where by "fully," I mean after *both* integrating out the short wavelength degrees of freedom *and* the rescaling), I need to rewrite the delta functions in (3.3.11) in terms of the *rescaled* space and time coordinates $\mathbf{r}_{1,2}$ and $t_{1,2}$. This is readily done:

$$\delta^d(\mathbf{r}_1 - \mathbf{r}_2) = \delta^d(b(\mathbf{r}'_1 - \mathbf{r}'_2)) = b^{-d}\delta^d(\mathbf{r}'_1 - \mathbf{r}'_2), \tag{3.3.12}$$

where in the last equality I have again used the well-known property of delta functions that

$$\delta(kx) = \frac{1}{k}\delta(x) \tag{3.3.13}$$

for any positive constant k, and have been careful to remember that $\delta^d(\mathbf{r}_1 - \mathbf{r}_2)$ is shorthand notation for a product of d delta functions (one for each Cartesian component of $\mathbf{r}_{1,2}$), each of which contributes a factor of b^{-1} as a result of the rescaling. Note that this is the *only* point in the rescaling part of the RG process at which the spatial dimension d of the system has played any role. As we shall see in a moment, this is an important role indeed.

The delta function in time can be dealt with similarly:

$$\delta(t_1 - t_2) = \delta(b^z(t'_1 - t'_2)) = b^{-z}\delta(t'_1 - t'_2). \tag{3.3.14}$$

Using (3.3.12) and (3.3.14) in (3.3.11), I can rewrite the intermediate noise correlations entirely in term of the rescaled (primed) coordinates:

$$\langle f_I(\mathbf{r}'_1, t'_1) f_I(\mathbf{r}'_2, t'_2)\rangle = 2D_I b^{-z-d}\delta^d(\mathbf{r}'_1 - \mathbf{r}'_2)\delta(t'_1 - t'_2). \tag{3.3.15}$$

Now using this in our expression (3.3.11) for the fully renormalized noise correlations gives

$$\langle f_R(\mathbf{r}'_1, t'_1) f(\mathbf{r}'_2, t'_2)\rangle = b^{z-2\chi-d} 2D_I\delta^d(\mathbf{r}'_1 - \mathbf{r}'_2)\delta(t'_1 - t'_2) \equiv D_R\delta^d(\mathbf{r}'_1 - \mathbf{r}'_2)\delta(t'_1 - t'_2), \tag{3.3.16}$$

which implies that the fully renormalized noise strength D_R of the noise in the renormalized equation of motion (3.3.6) is given by

$$D_R = b^{z-2\chi-d}D_I = b^{z-2\chi-d}(D + \delta D). \tag{3.3.17}$$

We have now completed one step of the DRG. The results are summarized by the three expressions (3.3.7), (3.3.8), and (3.3.17) giving the renormalized diffusion constant ν_R, nonlinearity λ_R, and noise strength D_R in terms of the original ν, λ, and D. I'll gather all of those results here for easier reading:

$$\nu_R = b^{z-2}(\nu + \delta\nu), \quad \lambda_R = b^{z+\chi-2}(\lambda + \delta\lambda), \quad D_R = b^{z-2\chi-d}(D + \delta D). \tag{3.3.18}$$

Now we can ask: In this renormalized KPZ equation, are the nonlinear effects bigger or smaller than they were in the original equation? One might think that this question could be answered simply by asking whether the renormalized nonlinear coefficient λ_R is bigger or smaller than the original ("bare") coefficient λ. But this is not, in fact, the case. The reason for this is that, since the diffusion coefficient ν and the noise strength D *also* change as a result of the renormalization group process, the scale of the height fluctuations *also* changes upon renormalization, since, as we saw in our treatment of the linear theory (and as we could have guessed even *without* having done that detailed analysis) the height fluctuations depend on ν and D. This means that not only does λ in the term $\lambda|\nabla h|^2$ change upon renormalization; $|\nabla h|^2$ changes as well.

However, there *is* a way to avoid this complication: We can choose the heretofore arbitrary rescaling exponents χ and z to keep ν and D fixed (that is, unchanged as a result of the RG process). In RG jargon, we call this choosing the scaling exponents to "find a fixed point."

If we make this choice, then we *can* tell whether the nonlinearity has become more or less important upon RG simply by asking whether or not the nonlinear coupling λ has grown or shrunk upon renormalization.

So let's determine what this choice of scaling exponents is. We already have enough information to do so for the case of small λ, because then the "graphical" corrections $\delta\nu$, $\delta\lambda$, and δD are all small. In fact, all of those corrections prove to be proportional to λ^2 (or even smaller in the case of λ, for which $\delta\lambda = 0$). Even without doing the calculation of the graphical corrections, we know that they must be proportional to λ to some positive integer power, since they must vanish as $\lambda \to 0$, and should be analytic in λ for small λ.

Therefore, in the limit of small λ, if we choose z and χ to keep ν and D fixed, (3.3.18) then obviously implies

$$z - 2 = O(\lambda^2), \quad z - 2\chi - d = O(\lambda^2), \tag{3.3.19}$$

which are easily solved to give

$$z = 2 + O(\lambda^2), \quad \chi = \frac{2-d}{2} + O(\lambda^2). \tag{3.3.20}$$

Of course, these (ignoring the $O(\lambda^2)$ corrections) are exactly the values of the linearized exponents that we have previously found by several other arguments.

Using these results in our recursion relation (3.3.18) for λ gives

$$\lambda_R = b^{\frac{2-d}{2}}(\lambda + O(\lambda^2)). \tag{3.3.21}$$

This implies that if λ is sufficiently small to make the $O(\lambda^2)$ term in (3.3.21) negligible – and such a range of λ is guaranteed to exist, since that term will be much less than λ for sufficiently small λ – then the renormalized λ_R will be smaller than the bare λ for $d > 2$, and *bigger* than the bare λ for $d < 2$.

So, in the case $d > 2$, a sufficiently bare λ will get smaller upon renormalization. Upon repeated renormalizations, therefore, it will continue to get even smaller (since the $O(\lambda^2)$ term will become even *more* negligible as we continue renormalizing and λ gets smaller). Thus, for $d > 2$, we will find, at least for sufficiently small bare λ, that asymptotically, the nonlinearity is *irrelevant* (that is, flows to zero upon renormalization), and the long-distance behavior of the KPZ equation will be accurately described by the linear theory.

For $d < 2$, however, the opposite is the case: A very small initial λ will *grow* upon renormalization. Indeed, it can only *stop* growing if the graphical corrections cause it to do so. But this can only happen if the nonlinearity becomes important. Therefore, for $d < 2$, the nonlinearity is *always* important, no matter how small it is initially (i.e., how small its bare value is). In RG jargon, we say that λ is "relevant" for $d < 2$.

For $d = 2$, λ is "marginal," which simply means we don't know what it will do without further work.

So we can already make two very important conclusions about the long-distance, long-time behavior of the KPZ equation. First, for $d > 2$, at least for sufficiently small λ, the linear theory developed in Section 3.1 will suffice. Second, for $d \leq 2$, that linear theory will *never* suffice. That means we'd better learn how to deal with the nonlinearity, at least perturbatively. I'll discuss perturbation theory in Section 3.4. In Section 3.5, I'll show how to use it in the RG to calculate the corrections to the parameters that occur in the first step of the RG.

3.4 Perturbation Theory

Let's try to solve the nonlinear equation of motion (3.0.1) the way we solved the linear theory: by Fourier transforming. In the linear theory, this worked, because different Fourier modes decoupled. Here, it doesn't, because of the damned nonlinear term $\frac{\lambda}{2}|\nabla h(\mathbf{r}, t)|^2$.

In fact, upon Fourier transforming, and using the orthogonality relations (2.3.5) and (2.3.7), we can rewrite the KPZ equation in Fourier space as

$$G^{-1}(\mathbf{q}, \Omega)h(\mathbf{q}, \Omega) = \frac{1}{\sqrt{V}}\left(\frac{\lambda}{2}\right)\sum_{\mathbf{p}}\sum_{\omega}[\mathbf{p}\cdot(\mathbf{p}-\mathbf{q})]h(\mathbf{p}, \omega)h(\mathbf{q}-\mathbf{p}, \Omega-\omega)+f(\mathbf{q}, \Omega), \tag{3.4.1}$$

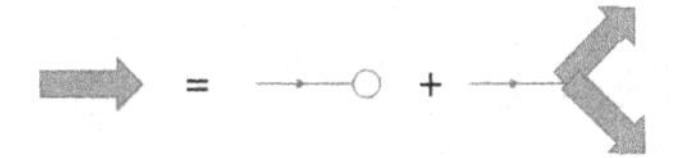

Fig. 3.4.1 Feynman diagram representation of the formal solution (3.4.3) of the equation of motion. Thick arrows represent the field h. The thin lines with a small arrow represent the propagator $G_0 = (-i\omega + \nu q^2)^{-1}$. The circle represents the noise. The three-pronged graph represents the λ nonlinearity in the equation of motion.

where

$$G^{-1}(\mathbf{q}, \Omega) = -i\Omega + \nu q^2 \tag{3.4.2}$$

is simply the inverse of the propagator (3.1.4) that we found in the linearized theory. We can see explicitly from (3.4.1) that the nonlinear λ term couples $h(\mathbf{q}, \Omega)$ to the h's at all other wavevectors $\mathbf{p}$ and frequencies ω.

So how do we proceed? The first step is formally to solve (3.4.1), which we can do simply by multiplying both sides by $G(\mathbf{q}, \Omega)$. This gives

$$h(\mathbf{q}, \Omega) = G(\mathbf{q}, \Omega) f(\mathbf{q}, \Omega) + \frac{G(\mathbf{q}, \Omega)}{\sqrt{V}} \lambda \sum_{\mathbf{p}} \sum_{\omega} [\mathbf{p} \cdot (\mathbf{p} - \mathbf{q})] h(\mathbf{p}, \omega) h(\mathbf{q} - \mathbf{p}, \Omega - \omega). \tag{3.4.3}$$

I call this a "formal" solution because it is, of course, no solution at all, since the variables to be solved for, namely the full set of $h(\mathbf{p}, \Omega)$'s for all $\mathbf{p}$ and Ω, also appear on the right-hand side.

However, the structure of (3.4.3) does suggest a very natural perturbative solution. To zeroth order in the nonlinearity λ, we can just drop the λ term altogether. Doing this simply recovers the linear solution (3.1.3) of Section 3.1. To next order in perturbation theory, we can replace $h(\mathbf{p}, \omega)$ and $h(\mathbf{q} - \mathbf{p}, \Omega - \omega)$ with their linearized solutions. Then to next order, we can add these corrections to the solutions for $h(\mathbf{p}, \omega)$ and $h(\mathbf{q} - \mathbf{p}, \Omega - \omega)$, and so on.

For a discussion of this perturbative approach, see, e.g., [29]. My treatment here is stolen shamelessly from that paper.

This perturbative procedure is most easily done using Feynman diagrams. These start with a pictorial representation of our formal solution, as illustrated in Figure 3.4.1. In this picture, thick arrows represent the field h. The thin lines with a small arrow represent the propagator $G_0 = (-i\omega + \nu q^2)^{-1}$. The circle represents the noise. The three-pronged graph represents the λ nonlinearity in the equation of motion.

It is also useful to represent the original Fourier transformed equation of motion (3.4.1). Graphically, this amounts to "amputating" the leftmost leg of each of the Feynman diagrams in Figure 3.4.1, which gives Figure 3.4.2.

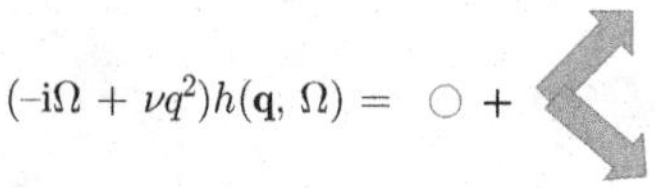

$$(-i\Omega + \nu q^2)h(\mathbf{q}, \Omega) = \bigcirc +$$

Fig. 3.4.2 Feynman diagram representation of the equation of motion itself, obtained by dividing the formal solution (3.4.3) of the equation of motion divided by $G_0(\tilde{\mathbf{q}})$. This amounts to "amputating" the leftmost leg of each of the Feynman diagrams in Figure 3.4.1.

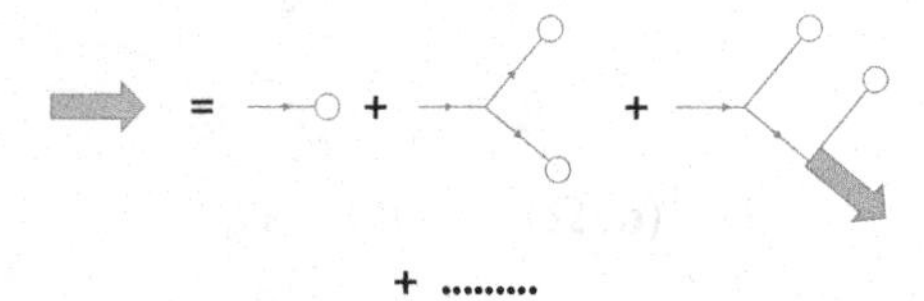

Fig. 3.4.3 Feynman diagram representation of the iterative perturbative solution of the equation of motion.

The iterative perturbative solution to the equation of motion can then be represented by the series of graphs shown in Figure 3.4.3.

Once this is done, one can see immediately that the graphs like the second term on the right-hand side of Figure 3.4.3, for which the incoming momentum and frequency on the left are respectively $\mathbf{q}$ and Ω, can be thought of as an extra contribution to the effective noise $f_{eff}(\mathbf{q}, \Omega)$, since this term is a random term on the right-hand side of the equation of motion. Their correlations, therefore, are renormalizations of the noise correlations D, represented by the Feynman diagrams shown in Figure 3.4.5.

In making this graph, I have used the fact that the averages of two noises are only nonzero if their wavevectors and frequencies are equal and opposite. Therefore, we represent this averaging by connecting lines that end in noises.

Hence, in this graph, the solid circles now represent two noises that have been correlated. Hence, each line that looks like Figure 3.4.4 now represents a correlation function $C(\mathbf{p}, \omega) = G(\mathbf{p}, \omega)\langle f_{eff}(\mathbf{q}, \Omega) f_{eff}(\mathbf{q}, \Omega)\rangle G(-\mathbf{p}, -\omega)$, where $\mathbf{p}$ and ω are the wavevectors and frequencies associated with the incoming G leg on the left. These can be determined for a graph like Figure 3.4.5 using the momentum sum rules at the vertices, as we'll see in a moment. Note also that, when I draw Feynman diagrams like this, the *shapes* of the lines are irrelevant; all that matters is the topology of the figure. In particular, this means that the lines with black circles on them in Figure 3.4.5, even though they're "bent" at the black dot, still represent a correlation function, identical to those in Figure 3.4.4. Sometimes I will draw these lines as bent, like those in Figure 3.4.5; at other times, I will draw them straight, as in Figure 3.4.4. On other occasions, I will draw them as circular arcs.

$$= \quad G(\mathbf{q},\Omega)\langle f(\mathbf{q},\Omega)f(-\mathbf{q},-\Omega)\rangle G(-\mathbf{q},-\Omega) = C(\mathbf{q},\Omega)$$

Fig. 3.4.4 Feynman diagram representation of $\langle |h(\mathbf{q},\Omega)|^2\rangle$ correlation function, obtained by connecting the noises at the end of two propagator legs that end in noises. The "connection" here represents the process of averaging over the noise. In this graph, the circle now represents two noises that have been correlated.

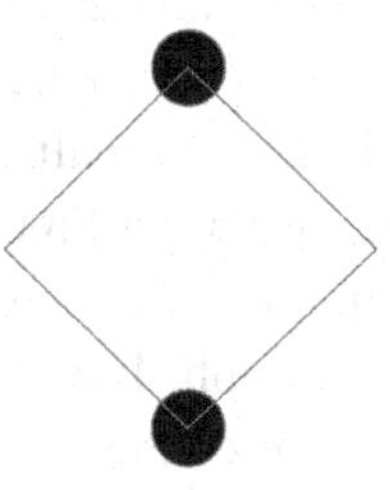

Fig. 3.4.5 Feynman diagram for the renormalization of the noise correlation D.

$$(-i\Omega + \nu q^2)h(\mathbf{q},\Omega) = \bigcirc + \quad + \quad \ldots\ldots\ldots$$

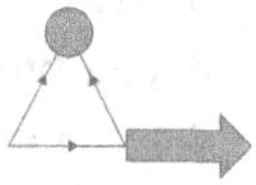

Fig. 3.4.6 Feynman diagram representation of the renormalized equation of motion. The second graph on the right-hand side (the triangle) will lead to a renormalization of the diffusion constant ν.

None of these differences matter: a line, bent, straight, or curved, with a black dot on it (or sometimes just an open circle) will *always* represent a correlation function $C(\mathbf{p},\omega)$, while a line with an arrow on it but no dot will *always* represent a propagator $G(\mathbf{p},\omega)$.

Note that the "legs" (i.e., the propagators) on both the left and the right have been "amputated," since we want to look directly at the noise correlation, rather than the change in the correlation function of $h(\mathbf{q},\Omega)$ itself.

Next, we average the full equation of motion, represented graphically by Figure 3.4.2, over the noises $f(\mathbf{p})$ for all wavevectors $\mathbf{p}$ other than the one ($\mathbf{q}$) in which we're interested. Since the averages of two noises are only nonzero if their wavevectors and frequencies are equal and opposite, we represent this averaging by connecting lines that end in noises, as I've already done in the graph for the noise renormalization.

Performing this connection for the graphs in Figure 3.4.2 yields graphs like Figure 3.4.6.

Note that these graphs are *linear* in $h(\mathbf{q})$, as can be seen by the fact that they have precisely one external line coming off to the right. The terms they represent can therefore be pulled over to the right-hand side of the equation schematically represented by Figure 3.4.6, and absorbed into renormalizations of the inverse propagator $G^{-1}(\mathbf{q}, \Omega) = -\mathrm{i}\Omega + \nu q^2$. From the form of $G^{-1}(\mathbf{q})$, we see that the part of such correction terms proportional to $q^2 h(\mathbf{q}, \Omega)$ can be absorbed into renormalizations of the diffusion constant ν.

So our perturbative solution can be entirely absorbed into "renormalizations" of the diffusion constant ν and the noise strength D. We'll now compute these, by evaluating the Feynman diagrams in Figures 3.4.5 and 3.4.6.

Let's begin with the noise D, whose renormalizations are represented, to leading order in perturbation theory, by the graph in Figure 3.4.5. The first thing we need to do is compute the so-called "combinatoric factor." That is, we need to figure out how many ways we can make such a graph. The answer, in this case, is two, because we can choose which pairs of legs we connect to make the "internal" correlation function lines in the loop.

Each of those internal lines represents a correlation function. Taking the momentum and frequency of one of them to be $\mathbf{p}$ and ω respectively, we see from the structure of the vertex – or, if you prefer, from the structure of the nonlinear term in the equation of motion (3.4.1) – that the momentum and wavevector of the other correlation function must be $\mathbf{q} - \mathbf{p}$ and $\omega - \Omega$, respectively. We also see that we must sum over $\mathbf{p}$ and ω. All of the other momenta and frequencies in the graphs are constrained by the Kronecker deltas in the correlation function, and by the "sum rule" that the ingoing momentum and frequency at each vertex must add up to the sum of the outgoing momenta and frequencies to take on the values illustrated in Figure 3.4.7. As a result, the correlation of this addition to the "effective noise" $f_{eff}(\mathbf{q}, \Omega)$ coming from this leading-order term in perturbation theory is

$$\Delta\langle f_{eff}(\mathbf{q}, \Omega) f_{eff}(\mathbf{q}', \Omega')\rangle = 2\left(\frac{1}{V}\right)\left(\frac{\lambda}{2}\right)^2 \sum_{\mathbf{p}} \sum_{\omega} C(\mathbf{p}, \omega) C(\mathbf{q}-\mathbf{p}, \Omega-\omega) \delta^K_{\mathbf{q}+\mathbf{q}'} \delta^K_{\Omega+\Omega'} . \tag{3.4.4}$$

Since this correction has exactly the same form as the original correlation (2.3.16) for our noises (that is, it multiplies the same Kronecker deltas, but otherwise has no dependence on $\mathbf{q}$ and Ω to leading (zeroeth) order in $\mathbf{q}$ and Ω), we can interpret this as a correction to $2D$ (note that tricky factor of 2; it's one of many such trivial but easy-to-miss factors in this calculation!). Turning the sums on $\mathbf{p}$ and ω into integrals the way we've done before yields

$$2\delta D = 2\left(\frac{\lambda}{2}\right)^2 \int \frac{d^d p}{(2\pi)^d} \int \frac{d\omega}{2\pi} \frac{(2D)^2 p^4}{(\omega^2 + \nu^2 p^4)^2} . \tag{3.4.5}$$

The integral over frequency in this expression is elementary, with the result

$$\int_{-\infty}^{\infty} \frac{d\omega}{2\pi} \frac{1}{(\omega^2 + \nu^2 p^4)^2} = \frac{1}{4\nu^3 p^6} . \tag{3.4.6}$$

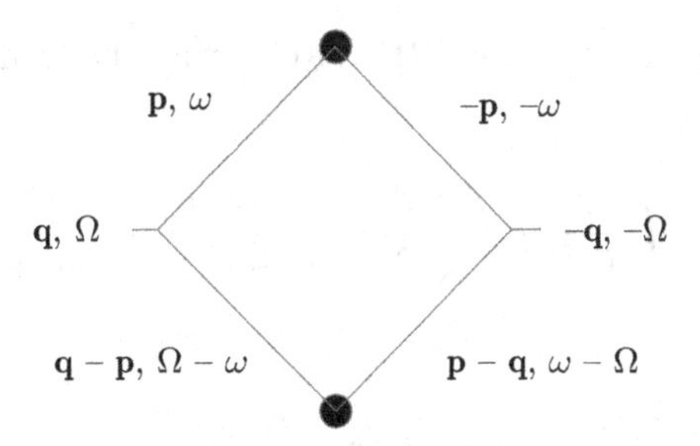

Fig. 3.4.7 Feynman diagram representation of the renormalization of the noise, with the momenta and frequencies labeled. Note that I have shown the "stumps" of the "amputated" external legs, just so that I can label them with the external momentum $\mathbf{q}$ and frequency Ω. We are free to choose the momentum $\mathbf{p}$ and the frequency ω running through the top left line. That momentum and frequency must therefore be summed over the entire Brillouin zone illustrated in Figure 2.3.1. All the other momenta and frequencies are then determined by the Kronecker deltas in the noise correlations, which force the momenta and frequencies on the upper-right and upper-left legs to be exactly opposite each other, and do the same for the lower-left and lower-right frequencies and momenta. In addition, the sum rule at the vertices imples that the lower-left momentum and frequency must be $\mathbf{q} - \mathbf{p}$ and $\Omega - \omega$ respectively, so that the outgoing momentum and frequency add up to the incoming $\mathbf{q}$ and Ω. This constraint is then automatically satisfied at the right-hand vertex, as you can see.

Using this in (3.4.5), we find

$$\delta D = \frac{g}{4}\left(\frac{I(\Lambda, L)}{K_d \Lambda^{d-2}}\right) D, \tag{3.4.7}$$

where

$$I(\Lambda, L) \equiv \int_{\frac{2\pi}{L}<p<\Lambda} \frac{\mathrm{d}^d p}{(2\pi)^d} \frac{1}{p^2} = \left(\frac{K_d}{d-2}\right)\left(\Lambda^{d-2} - \left(\frac{L}{2\pi}\right)^{2-d}\right) \tag{3.4.8}$$

is the integral we encountered earlier in our treatment of the XY model, and our calculations of the real-space correlations of h in the linear theory, and g is the dimensionless coupling defined earlier in (3.0.3).

Note that (3.4.8) goes to ∞ as $L \to \infty$ for $d \leq 2$. This says that perturbation theory fails in those dimensions, since perturbation theory assumes that the corrections it calculates are small.

The corrections to ν obtained from the graph in Figure 3.4.6 also diverge as $L \to \infty$ for $d \leq 2$.

That graph leads to a correction $\delta(\text{RHS})$ to the right-hand side of the equation of motion (3.4.1) given by

$$\delta(\text{RHS}) = \left\{4\left(\frac{\lambda}{2}\right)^2 \int \frac{d^d p}{(2\pi)^d} \int_{-\infty}^{\infty} \frac{d\omega}{2\pi} \frac{(2D)ip_i^+(-ip_i^-)ip_j^-}{(\omega^2 + \nu^2 p_-^4)(-i\omega + \nu p_+^2)}\right\} iq_j h(\mathbf{q}, \Omega), \tag{3.4.9}$$

where I've defined

$$\mathbf{p}_\pm \equiv \mathbf{p} \pm \frac{\mathbf{q}}{2}. \tag{3.4.10}$$

Gathering terms, and canceling off the "i"'s, I can rewrite this as

$$\delta(\text{RHS}) = -\{2D\lambda^2 I_j(\mathbf{q})\} q_j h(\mathbf{q}, \Omega), \tag{3.4.11}$$

where I've defined the (vector) integral

$$I_j(\mathbf{q}) = \int \frac{d^d p}{(2\pi)^d} \int_{-\infty}^{\infty} \frac{d\omega}{2\pi} \frac{\mathbf{p}_+ \cdot \mathbf{p}_- p_j^-}{(\omega^2 + \nu^2 p_-^4)(-i\omega + \nu p_+^2)}. \tag{3.4.12}$$

Multiplying the numerator and denominator of this expression by $i\omega + \nu p_+^2$ gives

$$I_j(\mathbf{q}) = \int \frac{d^d p}{(2\pi)^d} \int_{-\infty}^{\infty} \frac{d\omega}{2\pi} \frac{\mathbf{p}_+ \cdot \mathbf{p}_- p_j^- (i\omega + \nu p_+^2)}{(\omega^2 + \nu^2 p_-^4)(\omega^2 + \nu^2 p_+^4)}. \tag{3.4.13}$$

Now the piece of the integrand coming from the $i\omega$ term in the numerator is odd in ω, and so its integral over ω vanishes. Thus we are left with

$$I_j(\mathbf{q}) = \int \frac{d^d p}{(2\pi)^d} \int_{-\infty}^{\infty} \frac{d\omega}{2\pi} \frac{\mathbf{p}_+ \cdot \mathbf{p}_- p_j^- \nu p_+^2}{(\omega^2 + \nu^2 p_-^4)(\omega^2 + \nu^2 p_+^4)}. \tag{3.4.14}$$

As noted earlier, we are only interested in the value of this integral up to linear order in $\mathbf{q}$. Therefore we can drop terms of $O(q^2)$ in (3.4.14). For now doing this only in the numerator, I can rewrite (3.4.14) as

$$I_j(\mathbf{q}) = \nu \int \frac{d^d p}{(2\pi)^d} \int_{-\infty}^{\infty} \frac{d\omega}{2\pi} \frac{p^2(p^2 + \mathbf{p}\cdot\mathbf{q})(p_j - \frac{q_j}{2})}{(\omega^2 + \nu^2 p_-^4)(\omega^2 + \nu^2 p_+^4)}. \tag{3.4.15}$$

The integral over ω in this expression is elementary; performing it gives

$$I_j(\mathbf{q}) = \frac{1}{2\nu^2} \int \frac{d^d p}{(2\pi)^d} \frac{p^2(p^2 + \mathbf{p}\cdot\mathbf{q})(p_j - \frac{q_j}{2})}{(p_+^2 + p_-^2)p_+^2 p_-^2}. \tag{3.4.16}$$

It is straightforward to show that $p_+^2 p_-^2 = p^4 - (\mathbf{p}\cdot\mathbf{q})^2 + O(q^3) = p^4 + O(q^2)$ and $(p_+^2 + p_-^2 = 2p^2 + O(q^2)$. Dropping terms higher than linear order in q, as we've been doing all along, we thereby get

$$I_j(\mathbf{q}) = \frac{1}{4\nu^2} \int \frac{d^d p}{(2\pi)^d} \frac{1}{p^2}\left[\left(1 + \frac{\mathbf{p}\cdot\mathbf{q}}{p^2}\right)(p_j - \frac{q_j}{2})\right]. \tag{3.4.17}$$

Clearly, the cross-term between the 1 and p_j in the square brackets is odd in p_j, and will therefore integrate to 0. Furthermore, the cross-term between $\frac{\mathbf{p}\cdot\mathbf{q}}{p^2}$ and q_j is higher than linear order in q, and so can also be dropped. Thus we are left with

$$I_j(\mathbf{q}) = \frac{1}{4\nu^2}\int \frac{\mathrm{d}^d p}{(2\pi)^d}\frac{1}{p^2}\left[\frac{p_j p_k q_k}{p^2} - \frac{q_j}{2}\right], \tag{3.4.18}$$

where I have used the Einstein summation convention to rewrite the dot product $\mathbf{p}\cdot\mathbf{q}$ as $p_k q_k$.

We can deal with the dependence of the integrand in this expression on the *direction* of $\mathbf{p}$ by the following elegant trick, which I'll be using repeatedly throughout this book. Consider any integral J_{ij} of the form

$$J_{ij} = \int \frac{\mathrm{d}^d p}{(2\pi)^d} f(p) p_i p_j, \tag{3.4.19}$$

where the region of integration over $\mathbf{p}$ is *any* spherically symmetric volume centered on the origin. After a moment's reflection, it should be clear that this integral will vanish unless $i = j$, because it will be odd in the components p_i and p_j. A little more reflection will make it clear that, when i *does* $= j$, it doesn't matter *which* common value i and j have. That is, all d diagonal entries of J_{ij} are equal: $J_{xx} = J_{yy} = J_{zz} = \cdots = J_{ww}$. This is a simple and direct consequence of the spherical symmetry of both the integration region, and the factor $f(p)$ in (3.4.19).

We can summarize the above facts by writing

$$J_{ij} = J_0 \delta_{ij}, \tag{3.4.20}$$

where δ_{ij} is the usual Kronecker delta (which you can also think of as the tensor representation of the identity matrix), and J_0 is a constant that we have yet to determine.

We can make that determination by taking the trace J_{ii} (using the Einstein summation convention) on i and j of (3.4.20); this gives

$$J_{ii} = J_0 \delta_{ii} = J_0 d, \tag{3.4.21}$$

where in the second equality I've used the fact that the trace of a $d \times d$ identity matrix is just d, the number of diagonal entries.

On the other hand, if I take the same trace on i and j of equation (3.4.19), I get

$$J_{ii} = \int \frac{\mathrm{d}^d p}{(2\pi)^d} f(p) p_i p_i = \int \frac{\mathrm{d}^d p}{(2\pi)^d} f(p) p^2. \tag{3.4.22}$$

Equating (3.4.21) and (3.4.22) gives a simple equation for J_0:

$$J_0 d = \int \frac{\mathrm{d}^d p}{(2\pi)^d} f(p) p^2, \tag{3.4.23}$$

whose solution is trivially

$$J_0 = \frac{1}{d}\int \frac{d^d p}{(2\pi)^d} f(p)p^2 . \tag{3.4.24}$$

Using this in (3.4.20) gives my final general expression

$$\int \frac{d^d p}{(2\pi)^d} f(p)p_i p_j = \left(\int \frac{d^d p}{(2\pi)^d} f(p)p^2\right)\frac{\delta_{ij}}{d} . \tag{3.4.25}$$

Using this general expression for the integral of the first term in (3.4.18), and using the fact that $\delta_{jk}q_k = q_j$, I get

$$I_j = \left(\frac{2-d}{8d\nu^2}\right) q_j \int \frac{d^d p}{(2\pi)^d}\frac{1}{p^2} = \left(\frac{2-d}{8d\nu^2}\right) q_j I(\Lambda, L) . \tag{3.4.26}$$

Inserting this result into our expression (3.4.11) for the change δ(RHS) to the right-hand side of the equation of motion equation illustrated in Figure 3.4.6 gives

$$\delta(\text{RHS}) = -\delta\nu q^2 h(\mathbf{q}, \Omega) , \tag{3.4.27}$$

where I've defined

$$\delta\nu = \left(\frac{2-d}{4d}\right)\left(\frac{I(\Lambda, L)}{K_d \Lambda^{d-2}}\right) g\nu . \tag{3.4.28}$$

Pulling this term (3.4.27) over to the left-hand side of the equation of motion (3.4.1), we see that we can interpret this term as a renormalization of the inverse propagator $G(\mathbf{q}, \Omega)$, since it is linear in $h(\mathbf{q}, \Omega)$. Indeed, since it is also proportional to q^2, we see that we can be more specific, and interpret it as a renormalization of ν. That is, we can, at least to this order in perturbation theory, incorporate the effects of the nonlinearity on the equation of motion by replacing the bare value ν_0 of ν in the original theory by an effective or "renormalized" ν_R given by

$$\nu_R = \nu_0 + \delta\nu , \tag{3.4.29}$$

with $\delta\nu$ given by (3.4.28). Note that there is, implicitly, a ν dependence on the right-hand side of (3.4.28) (it's hidden in the g, with g given by equation (3.0.3)). To this order in perturbation theory, it is sufficient to replace ν with ν_0 in (3.0.3), and use the resulting g in (3.4.28).

Note that, if $d > 2$, so that the integral $I(\Lambda, L)$ given by equation (3.4.8) converges to a finite limit as $L \to \infty$, the corrections to both the noise strength D and the diffusion coefficient ν are small fractions of their bare values if the dimensionless nonlinear coupling g equation (3.0.3) is small (i.e., $g \ll 1$). To see this, note that for $d > 2$, we can drop the L^{2-d} term in my expression (3.4.8) for the integral $I(\Lambda, L)$ as $L \to \infty$, and write

$$I(\Lambda, L) \to I(\Lambda, L \to \infty) = \left(\frac{K_d}{d-2}\right)\Lambda^{d-2}. \tag{3.4.30}$$

Using this in my expressions (3.4.7) and (3.4.28) for the corrections δD and $\delta\nu$ to the noise strength D and the diffusion constant ν, I find

$$\frac{\delta D}{D} = \frac{g}{4(d-2)}, \quad \frac{\delta\nu}{\nu} = -\frac{g}{4d}. \tag{3.4.31}$$

Thus, the fractional change in the noise strength and diffusion constant are, indeed, small if $g \ll 1$, as I claimed earlier when I first introduced g.

However, for $d \leq 2$, *no matter how small g is*, the fractional changes to D and ν diverge as the system size $L \to \infty$. Since the original premise of the perturbation theory was that these corrections are *small*, this obviously means that the perturbation theory *fails* for $d \leq 2$.

So what do we do for $d \leq 2$? The answer is the renormalization group (RG), which I've already described. We can now use the perturbation theory I just developed to perform the step of the RG that I was vague about earlier: the averaging over short-wavelength degrees of freedom.

3.5 Perturbation Theory in the RG

The first step of the RG process – the averaging step described above – will lead to corrections to the diffusion constant ν and the noise strength D. These can be calculated exactly as we did when we were doing perturbation theory, except for one crucial difference: Because we are only integrating out those fields $h_>(\mathbf{p})$ with momenta $\mathbf{p}$ in the shell

$$b^{-1}\Lambda < |\mathbf{p}| < \Lambda, \tag{3.5.1}$$

all of the integrals over wavevector which in perturbation theory extended over all $\mathbf{p}$ will now only extend over the shell (3.5.1). This means that the problematical infrared (small $\mathbf{p}$) divergences that invalidated our perturbation theory will not be a problem here, since our wavevector integrals do not extend into the region of small $\mathbf{p}$ that leads to those divergences. Thus, perturbation theory will be valid on this step as long as the dimensionless coupling constant g is small. This is the great advantage of the renormalization group.

Applying this logic of simply using the perturbation theory results, but with the lower limit on all wavevector integrals replaced with $b^{-1}\Lambda$, I get for the correction

δD to D arising from the first step of the renormalization group process

$$\delta D = \frac{g}{4}\left(\frac{I_>(\Lambda,L)}{K_d\Lambda^{d-2}}\right)D, \tag{3.5.2}$$

where I've defined

$$I_>(\Lambda,L) \equiv \int_{b^{-1}\Lambda<p<\Lambda} \frac{d^dp}{(2\pi)^d}\frac{1}{p^2}. \tag{3.5.3}$$

For a number of computational reasons, it is convenient to choose the rescaling factor b in our renormalization group to be very close to 1. That is, I'll choose

$$b = 1 + d\ell \tag{3.5.4}$$

with $d\ell$ infinitesimal.

One of the conveniences of this choice is that, instead of labeling the renormalized parameters of our equation of motion by the number n of times we've iterated the two steps of the RG procedure, we can instead think of those parameters as continuous functions of the "renormalization group time" (which, I hasten to emphasize, has nothing to do with the real time in the equations of motion!)

$$\ell \equiv n d\ell. \tag{3.5.5}$$

Another convenience of this choice is that it makes many of the integrals we have to compute even easier. For example, it is straightforward to see that in this limit (3.5.4),

$$I_>(\Lambda,L) = K_d\Lambda^{d-2}d\ell. \tag{3.5.6}$$

Thus we have for the perturbative or, as it is often called in this business, the "graphical" correction to the noise strength D:

$$\delta D = \frac{g}{4}Dd\ell. \tag{3.5.7}$$

Similar reasoning leads to the graphical correction for ν:

$$\delta\nu = \left(\frac{2-d}{4d}\right)g\nu d\ell. \tag{3.5.8}$$

We can also make graphs like Figure 3.5.1. Since these have *two* external lines going off to the right, these represent corrections to the equation of motion that are quadratic in h. Unsurprisingly, these can be absorbed onto a renormalization of the one quadratic term we have in the original KPZ equation, namely λ.

In fact, these renormalizations of λ necessarily add up to zero. Furthermore, this is true *to all orders in perturbation theory*. That is, there is no correction to λ arising from the first step of the RG, on which we integrate out the short-wavelength degrees of freedom.

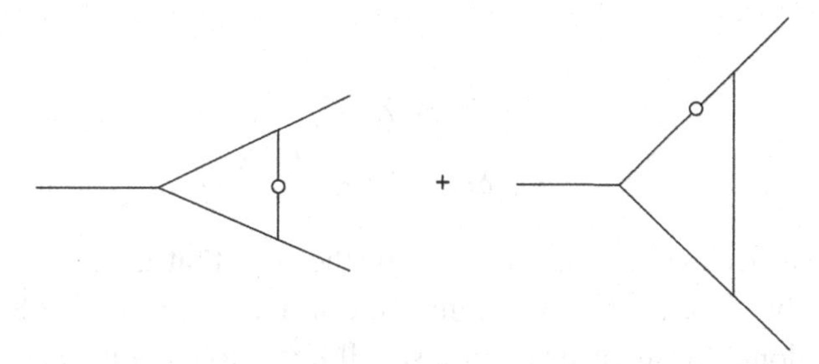

Fig. 3.5.1 Graphs that would appear to renormalize the nonlinear coupling λ. However, these graphs sum to zero, as do *all* graphs to *all* orders in perturbation theory. This cancellation is a consequence of the "pseudo-Galilean invariance" of the KPZ equation, as discussed in the text.

This seemingly miraculous cancellation is, in fact, no coincidence. In fact, it is an inevitable consequence of an "emergent symmetry" of the KPZ equation, and actually holds to all orders in perturbation theory. That is, there can be no renormalization of λ to *any* order in perturbation theory.

The symmetry in question is "pseudo-Galilean" invariance. What this means is that, if $h(\mathbf{r}, t)$ is a solution of the KPZ equation for a specific realization of the noise $f(\mathbf{r}, t)$, then

$$h'(\mathbf{r}, t) = h(\mathbf{r} + \lambda \mathbf{w}t, t) + \mathbf{w} \cdot \mathbf{r} + \frac{\lambda w^2 t}{2} \tag{3.5.9}$$

is also a solution, as can be verified by direct substitution. (This is called "pseudo-Galilean" invariance because, if we write (3.5.9) in terms of the *gradient* of h, and call that gradient $\mathbf{v}$, then (3.5.9) becomes

$$\mathbf{v}' = \nabla h' = \nabla h(\mathbf{r} + \lambda \mathbf{w}t, t) + \mathbf{w} = v(\mathbf{r} + \lambda \mathbf{w}t, t) + \mathbf{w}\,, \tag{3.5.10}$$

which looks like a Galilean "boost" for a fluid velocity field.)

This is a *symmetry* of the equation of motion. Therefore, it must be preserved by the renormalization group (except for trivial changes due to rescaling of the field h).

Hence, the renormalized equation of motion must, up to rescaling, have the same λ as the original equation, since the value of λ appears in the symmetry (3.5.9). Therefore, the graphical corrections to λ must be zero.

I can now summarize the results of the perturbative first step of the RG by giving the intermediate values ν_I, λ_I, and D_I obtained after that first step:

$$\nu_I = \nu + \delta\nu = \left(1 + \left(\frac{2-d}{4d}\right) g d\ell\right) \nu\,, \tag{3.5.11}$$

and

$$\lambda_I = \lambda\,, \tag{3.5.12}$$

$$D_I = D + \delta D = \left(1 + \frac{g}{4}\mathrm{d}\ell\right) D\,. \tag{3.5.13}$$

Now we must perform the second step of the RG; that is, rescaling the spatial coordinates $\mathbf{r}$, the time t, and the height field h according to (3.2.8) and (3.2.9). We've already done this analysis earlier, so I'll just quote the result:

$$\nu_R = b^{z-2}\nu_I = (1 + (z-2)\mathrm{d}\ell)(\nu + \delta\nu) = \left(1 + \left[z - 2 + \left(\frac{2-d}{4d}\right) g\right] \mathrm{d}\ell\right) \nu\,, \tag{3.5.14}$$

and

$$\lambda_R = b^{z+\chi-2}\lambda = (1 + [d - 2 + \chi]\mathrm{d}\ell)\lambda\,. \tag{3.5.15}$$

$$D_R = b^{z-2\chi-d} D_I = b^{z-2\chi-d}(D + \delta D) = \left(1 + \left[z - 2\chi - d + \frac{g}{4}\right]\mathrm{d}\ell\right) D\,. \tag{3.5.16}$$

In the second equality of all of these expressions, I have taken $b = 1 + \mathrm{d}\ell$, with $\mathrm{d}\ell$ infinitesimal, and therefore dropped terms of $O(\mathrm{d}\ell^2)$. (This is, of course, exactly the trick used in all of calculus.)

Now we wish repeatedly to iterate these two steps. Each time we do so, we will in general have different starting parameters ν, λ, and D. To keep track of these changes, we could label the parameters on each time step by n, the total number of times we have repeated this process. We can alternatively label the parameters with the RG "time" ℓ defined as

$$\ell \equiv n\mathrm{d}\ell\,. \tag{3.5.17}$$

Furthermore, since the increment $\mathrm{d}\ell$ of ℓ on each step is infinitesimal, we can think of the RG "time" ℓ (not to be confused with the real time t in the equation of motion) as a *continuous* variable. This means we can think of the parameters ν, λ, and D as *continuous* functions $\nu(\ell)$, $\lambda(\ell)$, and $D(\ell)$ of the RG time ℓ. We can also – and this is the most convenient feature of this approach – rewrite the recursion relations (3.5.14), (3.5.15), and (3.5.16) as *differential* equations.

To see this, consider the RG step that takes us from RG "time" ℓ to $\ell + \mathrm{d}\ell$. We can re-express the effect of the recursion relations (3.5.14), (3.5.15), and (3.5.16) just described as

$$\nu(\ell + \mathrm{d}\ell) = \left(1 + \left[z - 2 + \left(\frac{2-d}{4d}\right) g\right] \mathrm{d}\ell\right) \nu(\ell)\,, \tag{3.5.18}$$

$$\lambda(\ell + \mathrm{d}\ell) = (1 + [d - 2 + \chi]\mathrm{d}\ell)\lambda(\ell)\,, \tag{3.5.19}$$

and

$$D(\ell + \mathrm{d}\ell) = \left(1 + \left[z - 2\chi - d + \frac{g}{4}\right]\mathrm{d}\ell\right) D(\ell). \qquad (3.5.20)$$

Subtracting $\nu(\ell)$ from both sides of (3.5.18), and dividing both sides by $d\ell$, I get

$$\frac{\nu(\ell + \mathrm{d}\ell) - \nu(\ell)}{\mathrm{d}\ell} = \left[z - 2 + \left(\frac{2-d}{4d}\right) g\right] \nu(\ell). \qquad (3.5.21)$$

If you remember your introductory calculus course, you will immediately recognize the left-hand side of this expression as $\frac{\mathrm{d}\nu}{\mathrm{d}\ell}$. Hence, (3.5.21) can be rewritten as a differential equation:

$$\frac{\mathrm{d}\nu}{\mathrm{d}\ell} = \left[z - 2 + \left(\frac{2-d}{4d}\right) g(\ell)\right] \nu(\ell). \qquad (3.5.22)$$

Similar reasoning applied to the other two parameters leads to differential recursion relations for them as well:

$$\frac{\mathrm{d}\lambda}{\mathrm{d}\ell} = [z + \chi - 2]\lambda(\ell), \qquad (3.5.23)$$

and

$$\frac{\mathrm{d}D}{\mathrm{d}\ell} = \left[z - 2\chi - d + \frac{g(\ell)}{4}\right] D(\ell). \qquad (3.5.24)$$

The recursion relation (3.9.18) is *exact*, since, as discussed earlier, there are no graphical corrections to λ at any order in perturbation theory. The other two recursion relations (3.5.22) and (3.5.24) are accurate to linear order in the dimensionless coupling g. That is, there will be corrections of $O(g^2)$ to these coming from higher orders in perturbation theory. This can be summarized by the recursion relations:

$$\frac{\mathrm{d}\nu}{\mathrm{d}\ell} = \left[z - 2 + \left(\frac{2-d}{4d}\right) g(\ell) + O(g^2)\right] \nu(\ell), \qquad (3.5.25)$$

$$\frac{\mathrm{d}\lambda}{\mathrm{d}\ell} = [z + \chi - 2]\lambda(\ell), \qquad (3.5.26)$$

and

$$\frac{\mathrm{d}D}{\mathrm{d}\ell} = \left[z - 2\chi - d + \frac{g(\ell)}{4} + O(g^2)\right] D(\ell). \qquad (3.5.27)$$

We can derive a recursion relation for the coupling g itself by using its definition (3.0.3), which implies

$$\ln g = 2 \ln \lambda + \ln D - 3 \ln \nu + \text{constant}, \qquad (3.5.28)$$

where the "constant" involves genuine constants like S_d (the surface area of a unit d-dimensional sphere), π, and so forth, as well as the ultraviolet cutoff Λ. All of these "constants" are independent of RG time ℓ.

Therefore, the recursion relation for $\ln g$ follows simply by differentiating (3.5.28) with respect to ℓ:

$$\begin{aligned}\frac{d\ln g}{d\ell} &= 2\frac{d\ln\lambda}{d\ell} + \frac{d\ln D}{d\ell} - 3\frac{d\ln\nu}{d\ell}\\ &= \frac{2}{\lambda}\frac{d\lambda}{d\ell} + \frac{1}{D}\frac{dD}{d\ell} - \frac{3}{\nu}\frac{d\nu}{d\ell}\\ &= 2(z+\chi-2) + \left(z - 2\chi - d + \frac{g}{4}\right) - 3\left(z - 2 + \left(\frac{2-d}{4d}\right)g\right) + O(g^2)\\ &= 2 - d + \left(1 - \frac{3}{2d}\right)g + O(g^2),\end{aligned} \tag{3.5.29}$$

where in the second equality I have used the recursion relations (3.5.25), (3.5.26), and (3.5.27). This can obviously be rewritten as

$$\frac{dg}{d\ell} = \left[2 - d + \left(1 - \frac{3}{2d}\right)g\right]g + O(g^3), \tag{3.5.30}$$

where the $O(g^3)$ corrections, which I have not calculated, come from higher orders in perturbation theory.

Note that this recursion relation is *closed*; that is, it involves only the coupling constant g itself.

Note also that the behavior of g under renormalization depends crucially on the dimension of space. This is clear from the behavior of the recursion relation for small g. In that limit, the linear $(2-d)g$ term on the right-hand side of (3.5.30) dominates. Hence, if $d > 2$, initially small g's will get even smaller upon renormalization, and $g(\ell)$ will flow to zero as $\ell \to \infty$. Indeed, $g(\ell)$ will vanish very fast; exponentially, in fact:

$$g(\ell) = \text{constant} \times e^{(2-d)\ell}. \tag{3.5.31}$$

If the "bare" g – that is, $g(\ell = 0) \equiv g_0$ – is sufficiently small, then the "constant" in the above expression will very nearly equal g_0, since the $O(g^2)$ term in (3.5.30) will always be small.

Since, as I noted at the beginning of this discussion of the KPZ equation, g is a measure of the importance of the nonlinear term in the KPZ equation, one might reasonably expect, therefore, that, at least for initially small g_0, the long-wavelength, long-time behavior of the KPZ equation for $d > 2$ would be given correctly by the linear theory. One would be right, as I'll show in Section 3.6.

On the other hand, for $d < 2$, an initially small g will *grow*. One might now expect that, at sufficiently long distances, the correct scaling behavior of fluctuations in the KPZ equation would *never* be correctly given by the linear theory. And one would be right again.

Note that the critical dimension d_c that we've just identified by this argument – namely, $d_c = 2$ – is precisely the one we obtained by asking for the dimension below which the direct (non-RG) perturbation theory broke down. This is not a coincidence: the "relevance" of even an initially *small* g – that is, the fact that a small g grows upon renormalization – is a consequence of exactly the same physics: the effect of nonlinear couplings of fluctuations becoming important in sufficiently low dimensions due to the increasing size of those fluctuations as dimension is decreased.

So the behavior of the KPZ equation, like that of most condensed matter systems (indeed, of most physical systems of any type), is strongly dimension dependent. I'll therefore in the next few sections consider the cases $d > 2$, $d = 2$, and $d < 2$ separately. I will also focus specifically on the case $d = 2 + \epsilon$, with $\epsilon \ll 1$, which will give me a chance to illustrate the ϵ-expansion (and to discuss phase transitions), and the case $d = 1$, for which we can obtain exact exponents.

3.6 $d > 2$, and the $2 + \epsilon$ Expansion

As noted above, for $d > 2$, $g(\ell)$ flows to zero as $\ell \to \infty$, at least if the bare g_0 is sufficiently small. That is, $g = 0$ is a stable fixed point of the model. This implies that the long-wavelength scaling behavior of the system – e.g., the correlation function (3.2.21) – is given correctly by the linear theory of Section 3.1.

Note, however, that a system that starts with a sufficiently large g will *not* flow into this stable fixed point. If we drop the (uncalculated) $O(g^3)$ terms in the recursion relation (3.5.30) for g, it is easy to see (simply by setting the right-hand side of (3.5.30) to zero) that there is another fixed point at a value of g which I'll call g^* given by

$$g^* = \frac{2d(d-2)}{2d-3}. \tag{3.6.1}$$

This fixed point is *unstable*, and separates flows that go into the linear fixed point $g = 0$ from flows that run away to large g.

You might think that, since this fixed point is unstable, it's physically unimportant. After all, a generic system would have zero chance of actually winding up at this fixed point.

However, this is not the case, In fact, this fixed point controls a *phase transition* in the KPZ equation. To see this, imagine that an experimentalist working with a system that's described by the KPZ equation has a "knob" that she can "turn" to tune the parameter g.

There are a number of such knobs in realistic systems, the simplest of these being temperature. While the concept of temperature strictly speaking applies only to equilibrium systems, there are many nonequilibrium systems to which we routinely apply it. One of these is ourselves: We take our temperature to decide, e.g., if we may have contracted COVID-19. Yet human beings, like all living things, are nonequilibrium systems: hence the phrase "cold and dead."

So imagine, e.g., a sample cell containing a crystalline solid and vapor composed of the same substance, but out of equilibrium (either supersaturated, so the vapor condenses onto the crystal, causing the crystal to grow, or undersaturated, so that the crystal shrinks by evaporation). In either case, the fluctuations of the crystal surface shape are believed to be described by the KPZ equation. And much of the noise in this process is thermal noise, and hence can be increased (decreased) by raising or lowering the temperature of the system. So the experimentalist can, in this way, tune the noise strength D in the KPZ equation describing her system simply by raising or lowering temperature. The experimental "knob" in this case is literally the rheostat on the heater.

So now consider raising the temperature. This increases D, and, hence, all other parameters being equal, increases g, since $g \propto D$, as can be seen from its definition equation (3.0.3).

So let's consider a KPZ system a system in *more* than two dimensions. This admittedly is *not* a very practical experiment on a growing crystalline surface, since to have a surface of more than two dimensions, we would need to be living in a space of more than three dimensions. However, there could very well be three-dimensional systems described by the KPZ equation, so such experiments may be possible.

Now imagine tuning D in such a system by increasing temperature, and imagine that we start with a small enough D that $g < g^*$, where g^* is the fixed point value of g given by (3.6.1). Eventually, g will reach g^*. At that point, the long-distance scaling of the system will be controlled by the unstable fixed point.

Further increasing g puts it into the regime in which it grows upon renormalization. What happens then is unclear: The recursion relation (3.5.30), as it stands, predicts that g will grow without bound. However, we know that recursion relation is accurate only for small g; what happens for large g is unknown.

A reasonable speculation, however, is that higher-order terms in g in the recursion relation will eventually stop the growth, and that g will eventually reach a nonzero fixed-point value, as illustrated in Figure 3.6.1. This new, conjectured fixed point, called the "strong-coupling" fixed point, would, if this conjecture is correct, control the scaling properties of the KPZ equation in this regime. It would presumably have its own values of the scaling exponents z and χ, different from the values $z = 2$ and $\chi = (2 - d)/2$ that hold at the linear fixed point.

Thus, when g is experimentally tuned through g^*, the system undergoes an instantaneous change in its long-distance scaling properties, from those of the

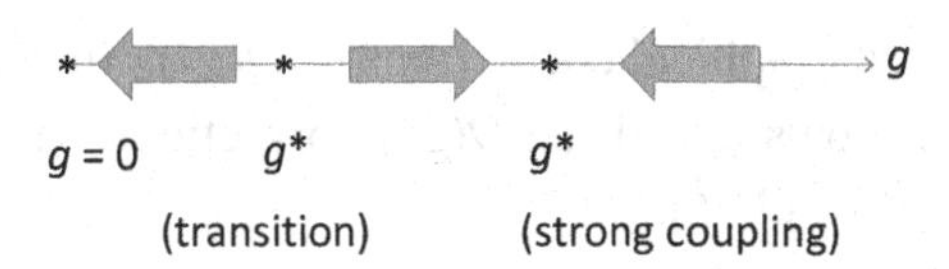

Fig. 3.6.1 Conjectured RG flows of the dimensionless coupling g for the KPZ equation in $d > 2$. There are three fixed points: a stable one at $g = 0$, which controls a phase in which the KPZ equation is described by the linear theory at long distances and times, an *un*stable one at $g = g^*_{\text{transition}}$, which controls the transition between the linear and strong-coupling phases, and a stable one at $g = g^*_{\text{strong coupling}}$, which controls the strong-coupling phase. Only the "linear" fixed point at $g = 0$ is actually accessible in perturbation theory in $d = 3$. The transition fixed point at at $g = g^*_{\text{transition}}$ *is* accessible in spatial dimensions $d = 2 + \epsilon$, with $\epsilon \ll 1$.

linear approximation, to the (unknown) scaling laws of the strong-coupling fixed point. That is, we've undergone a *phase transition* from a zero-coupling phase to a strong-coupling phase.

Now one could, with considerable justice, question whether we should believe in the existence of the critical fixed point at g^*, given that we only find it by neglecting the $O(g^2)$ terms in the recursion relation (3.5.30) for g. This is only a reliable approximation for $g \ll 1$, which is clearly not true of the g^* we've just found.

Unless, that is, we play a game which sounds extremely silly if you haven't heard of it before, but which has actually become a standard practice in the use of the renormalization group. This particular game is called the "ϵ-expansion" [30]. In this approach, we simply pretend that the dimension of space d, which is, of course, really an integer,[3] is a continuous variable, and ask: For what range of d is the fixed point g^* of equation (3.6.1) actually small enough that we trust the result? The answer is immediately obvious from inspecting equation (3.6.1): when d is close to $d = 2$.

So what we do is take

$$d = 2 + \epsilon \,, \tag{3.6.2}$$

where ϵ is assumed to be small (that is, $\epsilon \ll 1$). Then equation (3.6.1) reads

$$g^* = 4\epsilon + O(\epsilon^2) \,. \tag{3.6.3}$$

In this equation, we should now think of the $O(\epsilon^2)$ correction as incorporating *both* the corrections arising from the fact that the $2d$ factor in the numerator of equation (3.6.1) is $4 + O(\epsilon)$, and that the denominator factor of $2d - 3$ is $1 + O(\epsilon)$, *and* the

[3] I am not talking about fractals here, which have nothing whatsoever to do with the ϵ-expansion. Rather, this expansion should be thought of purely as a mathematical contrivance, to enable us to leverage our perturbation theory in g to obtain results in physical (i.e., integer) values of the spatial dimension d.

fact that we have ignored graphical corrections of $O(g^3)$. Since, as equation (3.6.3) shows, g^* is $O(\epsilon)$, it follows that these $O(g^3)$ corrections will, at our fixed point, lead to corrections to g^* of $O(\epsilon^2)$.

Thus, at least in this admittedly somewhat contrived continuation to spatial dimensions near $d = 2$, there definitely *is* a critical fixed point, and, hence, a phase transition between a small g, "weak-coupling" phase, whose large distance and time scaling properties are exactly those of the linear theory, and a large g, "strong-coupling" phase, whose large distance and time scaling properties are different, and, unfortunately, impossible to determine by the perturbation theory approach we've used here.

The speculation is that, for $g > g^*$, g flows into another, "strong-coupling" fixed point, which is stable, and controls a new, distinct "universality class" for the KPZ equation, with different scaling exponents. These conjectured flows are illustrated in Figure 3.6.1.

Unfortunately, we can say little about this strong-coupling fixed point analytically, since, being at large g, it is not accessible to our perturbation theory. Indeed, as can easily be seen by inspection of our one-loop recursion relation (3.5.30), there is no sign of such a fixed point in the one-loop calculation. On the contrary, the recursion relation (3.5.30) predicts that, if the bare g is greater than g^*, g will flow to infinity (and that it will do so in a finite renormalization group "time" ℓ).

The conventional wisdom about the KPZ equation is that this behavior, which strictly speaking we've only derived for spatial dimensions d near 2, persists all the way up to three dimensions. It is certainly true that, at small g, there is a weak-coupling phase exhibiting the linear behavior. It is somewhat more speculative to assert that there is a phase transition to a strong coupling phase at larger g, but that is what is generally believed, and there is considerable experimental and numerical evidence to support this [31].

So in three dimensions, there is, if this conjecture is right (that is, if three dimensions looks qualitatively like 2.01 dimensions), a range of parameters (namely, small g) for which we can analytically determine the scaling laws (which are just those of the linearized model) and a region (large g) in which we can say little other than that the scaling laws will be different, but, presumably, still universal. So we have a complete analytic understanding of part of the parameter space, but not the entire parameter space.

3.7 $d = 2$

In $d = 2$, however we have essentially nothing analytic: *All* initial g's, no matter how small, flow to large g, and out of the region that's accessible to our perturbative calculation. Thus, there is *no* regime of parameter space in which we can analytically predict the scaling laws.

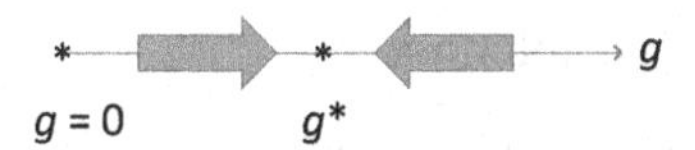

Fig. 3.7.1 Conjectured RG flows of the dimensionless coupling g for the KPZ equation in $d \leq 2$. There are now only *two* fixed points: an unstable one at $g = 0$, which controls nothing, and a stable one at $g = g^*_{\text{strong coupling}}$, which controls the behavior of all systems. In $d = 1$, the scaling exponents z and χ are known to be $z = 3/2$ and $\chi = 1/2$, *exactly*. In $d = 2$, the scaling exponents are known only from simulations.

However, the speculation, which is widely believed, is that the RG flow for $d = 2$ looks like Figure 3.7.1. The exponents of the only stable fixed point – namely, the one at $g = g^*_{\text{strongcoupling}}$, are not analytically known.

3.8 $d = 1$

In one spatial dimension $d = 1$, we can say more. Indeed, we can obtain the exact values of the dynamical exponent z and the roughness exponent χ. We'll do this (slavishly following KPZ) by mapping the KPZ equation onto an *equilibrium* problem. This mapping (as you'll see) only works in $d = 1$. But in that dimension, this mapping implies a fluctuation–dissipation theorem, which in turn implies an exact relation between the graphical corrections to the diffusion constant ν and the noise strength D. This, combined with the equally exact statement (which is valid in *all* spatial dimensions) that λ gets no graphical renormalization, proves to give us three very simple (and independent!) linear relations between the unknown exponents z and χ, and the corrections to ν and D at the fixed point. Solving these three simple linear equations then gives us z and χ exactly, and with amazingly little effort (what, after all, could be easier than solving three coupled linear equations in three unknowns?).

To prove all this, let's start by writing the KPZ equation in $d = 1$. This is simple because now the gradient vector has only one component – that is, it's a scalar. Calling the one spatial dimension x, we have

$$\partial_t h(x,t) = \nu\partial_x^2 h(x,t) + \frac{\lambda}{2}(\partial_x h(x,t))^2 + f(x,t)\,. \tag{3.8.1}$$

The big step is to now differentiate this equation with respect to x. This is easily seen to give

$$\partial_t u(x,t) = \nu\partial_x^2 u(x,t) + \lambda u\partial_x u(x,t) + f_u(x,t)\,, \tag{3.8.2}$$

where I've defined a new field u and new noise f_u via

$$u \equiv \partial_x h, \quad f_u \equiv \partial_x f\,. \tag{3.8.3}$$

Note that the new noise has correlations

$$\langle f_u(x,t) f_u(x',t')\rangle = 2D\partial_x\partial'_x\delta(x-x')\delta(t-t')\,. \qquad (3.8.4)$$

The significance of this peculiar correlation – in particular, the derivatives acting on the delta functions – is most easily seen in Fourier space, where it reads

$$\langle f_u(q,\omega) f_u(q',\omega')\rangle = 2Dq^2\delta^K_{q+q'}\delta^K_{\omega+\omega'}\,. \qquad (3.8.5)$$

Note that this is "blue noise": The noise amplitude is larger at large q (shorter wavelengths).

One final change of variables allows us to cast the equation of motion (3.8.2) in a more familiar form: Replace u with a new variable v given by

$$v \equiv -\lambda u\,. \qquad (3.8.6)$$

With this substitution, and pulling the nonlinear term over to the left-hand side of the equation, this gives us the equation of motion

$$\partial_t v(x,t) + v\partial_x v(x,t) = \nu\partial_x^2 v(x,t) + f_v(x,t)\,, \qquad (3.8.7)$$

with the noise f_v defined by

$$f_v \equiv -f_u/\lambda\,. \qquad (3.8.8)$$

This definition implies that the correlations of the noise f_v are also blue:

$$\langle f_v(x,t) f_v(x',t')\rangle = 2\left(\frac{D}{\lambda^2}\right)\partial_x\partial'_x\delta(x-x')\delta(t-t')\,. \qquad (3.8.9)$$

Equation (3.8.7) is known as the "Burgers equation." Those of you familiar with the theory of dislocations in solids may think that this equation is named after the Burgers who invented the famous "Burgers construction" for dislocations, but it is not. It's named after that Burgers' brother! (The dinner table discussions in that house during the brothers' childhoods must have been amazing!)

Equation (3.8.7) can also be thought of as the "infinitely compressible" limit of the Navier–Stokes equations of fluid mechanics in one dimension. To see this, recall that, in arbitrary dimension, the Navier–Stokes equation can be written [29] as

$$\partial_t\mathbf{v} + (\mathbf{v}\cdot\nabla)\mathbf{v} = -\nabla P + \nu_B\nabla(\nabla\cdot\mathbf{v}) + \nu_S\nabla^2\mathbf{v} + \mathbf{f}\,, \qquad (3.8.10)$$

where $\mathbf{v}(\mathbf{r},t)$ is the local velocity of the fluid, $\nu_{B,S}$ are the shear and bulk viscosities of the fluid, and P is the pressure. The combination $\partial_t\mathbf{v} + (\mathbf{v}\cdot\nabla)\mathbf{v}$ on the left-hand side of this equation is the so-called "convective acceleration": that is, it is the acceleration a small element of the fluid will experience as it moves along with the fluid velocity field $\mathbf{v}$.

If we consider an "infinitely compressible" fluid (i.e., one for which $P = 0$ always, which is a strange limit, but a well-behaved one), and go to one dimension, where the vectors velocity $\mathbf{v}$ and noise $\mathbf{v}$ become scalars, and the vector gradient operator ∇ becomes the scalar operator ∂_x, this equation clearly reduces to (3.8.7), with $\nu \equiv \nu_B + \nu_S$.

The most important point about (3.8.7) is that it is an *equilibrium* equation of motion. To see this, note that it can be written in the form

$$\partial_t v + v \partial_x v = \nu \partial_x^2 \frac{\delta H}{\delta v} + f_v \tag{3.8.11}$$

with the Hamiltonian H just being the kinetic energy

$$H = \frac{1}{2} \int dx \, v^2(x,t) \,. \tag{3.8.12}$$

Note that this Hamiltonian is completely trivial – i.e., quadratic – so there are no parameters in it that can get renormalized. That is, the coefficient of v^2 always remains equal to $1/2$, and no new terms can be generated in the Hamiltonian upon renormalization. Therefore, the only parameters in this problem that can get any renormalization are the diffusion constant (here the viscosity) ν and the noise strength D. Furthermore, the diffusion constant ν is strictly a kinetic coefficient; it gets no corrections from direct renormalization of the Hamiltonian.

All of these facts taken together imply that this equation of motion satisfies a fluctuation–dissipation theorem

$$\langle f_v(x,t) f_v(x',t') \rangle = 2 k_B T \Gamma \delta(x - x') \delta(t - t') \,, \tag{3.8.13}$$

where Γ is the operator

$$\Gamma \equiv \nu \partial_x^2 \tag{3.8.14}$$

acting on $\frac{\delta H}{\delta v}$, and I've defined the effective temperature of the model so that

$$k_B T = \frac{D}{\nu \lambda^2} \,. \tag{3.8.15}$$

Any model with the structure (3.8.13) – that is, with the damping (operator) and the noise correlations related in this way – obeys the fluctuation–dissipation theorem. This implies in particular that (3.8.15) always holds, *with same value of* $k_B T$, even after renormalization. In addition, the triviality of the Hamiltonian (3.8.12) guarantees that the Hamiltonian itself gets no graphical corrections. Hence, the graphical corrections $\delta\nu$, δD, and $\delta\lambda$ to ν, D, and λ must obey

$$\frac{D + \delta D}{(\nu + \delta\nu)(\lambda + \delta\lambda)^2} = \frac{D}{\nu \lambda^2} \,, \tag{3.8.16}$$

exactly, to all orders in g.

Since we already know that, in any spatial dimension, and therefore in $d = 1$, the graphical corrections $\delta\lambda$ to λ vanish exactly, we can rewrite this as

$$\frac{D + \delta D}{(\nu + \delta\nu)} = \frac{D}{\nu} \,. \tag{3.8.17}$$

Multiplying both sides of this expression by $\frac{\nu+\delta\nu}{D}$, and subtracting 1 from both sides, gives our final result

$$\frac{\delta D}{D} = \frac{\delta\nu}{\nu} \,. \tag{3.8.18}$$

This means that if we write the recursion relations *to all orders in* g in the form

$$\frac{d\nu}{d\ell} = [z - 2 + \eta_\nu(g)]\, \nu(\ell)\,, \tag{3.8.19}$$

$$\frac{d\lambda}{d\ell} = [z + \chi - 2]\lambda(\ell)\,, \tag{3.8.20}$$

and

$$\frac{dD}{d\ell} = [z - 2\chi - d + \eta_D(g)]\, D(\ell)\,, \tag{3.8.21}$$

which we can always do, as I proved earlier, then (3.8.18) implies that, in $d = 1$, we must have

$$\eta_\nu(g) = \eta_D(g)\,. \tag{3.8.22}$$

Note that my earlier lowest order in g – i.e., one-loop – calculation equations (3.5.22) and (3.5.24) are consistent with this result. To see this, compare those one-loop recursion relations with the general form (3.8.19), and (3.8.21); this comparison says that

$$\eta_\nu(g) = \left(\frac{2-d}{4d}\right) g(\ell) + O(g^2), \quad \eta_D(g) = \frac{g(\ell)}{4} + O(g^2)\,. \tag{3.8.23}$$

Obviously, these two expressions become identical in $d = 1$, as they must, since if two functions are equal, their Taylor series must be equal term by term.

Using (3.8.22) in (3.8.19), (3.8.20), and (3.8.21), and setting the spatial dimension $d = 1$ in those expressions, gives

$$\frac{d\nu}{d\ell} = [z - 2 + \eta_D(g)]\, \nu(\ell)\,, \tag{3.8.24}$$

$$\frac{d\lambda}{d\ell} = [z + \chi - 2]\lambda(\ell)\,, \tag{3.8.25}$$

and

$$\frac{dD}{d\ell} = [z - 2\chi - 1 + \eta_D(g)]\, D(\ell)\,. \tag{3.8.26}$$

From these relations, we can now determine the universal physical exponents z and χ that govern the scaling of the one-dimensional KPZ equation simply by looking for the values of z and χ that will keep ν, λ, and D fixed under the renormalization group at the fixed point $g = g_*$. Requiring this in (3.8.24), (3.8.25), and (3.8.26) clearly leads to three simple linear equations for the three unknowns z, χ, and $\eta_D(g_*)$, which read

$$z - 2 + \eta_D(g_*) = 0\,, \tag{3.8.27}$$

which follows from setting $\frac{d\nu}{d\ell} = 0$ in the recursion relation (3.8.24) for ν,

$$z + \chi - 2 = 0\,, \tag{3.8.28}$$

which follows from setting $\frac{d\lambda}{d\ell} = 0$ in the recursion relation (3.8.25) for λ, and

$$z - 2\chi - 1 + \eta_D(g_*) = 0\,, \tag{3.8.29}$$

which follows from setting $\frac{dD}{d\ell} = 0$ in the recursion relation (3.8.26) for D.

The solution of these three linear equations (3.8.27), (3.8.28), and (3.8.29) is obviously trivial, and gives

$$z = \frac{3}{2}, \quad \chi = \frac{1}{2}, \quad \eta_D(g_*) = \frac{1}{2}\,. \tag{3.8.30}$$

Using these results for z and χ in our general expression (3.2.21) for the two-point correlation function gives, in $d = 1$,

$$C_h(x,t) \equiv \langle (h(X,T) - h(X+x, T+t))^2 \rangle = |x| f(t/|x|^{3/2}) \propto \begin{cases} |x|, & |x|^{3/2} \gg t\,, \\ |t|^{2/3}, & |x|^{3/2} \ll t\,. \end{cases} \tag{3.8.31}$$

So, amazingly, we've been able to learn essentially everything we'd want to know about the one-dimensional KPZ equation simply by solving a trio of linear algebraic equations. Life is not always so good: You'll see that there are many problems for which we don't have nice arguments like the "pseudo-Galilean invariance," and the even nicer "equilibrium model" argument that we've used here to obtain additional relationships like equation (3.8.22) above, that enable us to determine the exponents so easily. But one *does* get lucky more often than one might expect (or deserve!), as we'll also see in the chapters that follow.

3.8.1 Alternative Interpretation: Diverging Effective Parameters and Trajectory Integral Matching Formalism

The scaling laws in one dimension that we've just found are different from those of the linear theory. We can interpret this difference as arising from infinite renormalizations of two of the parameters that appear in the linear theory: the diffusion constant ν, and the noise strength D. This is also true in two dimensions, and in three (or more) dimensions in the "strong-coupling" phase described earlier.

I can be much more precise than the phrase "infinite renormalizations." In fact, in any case in which the scaling departs from that predicted by the linear theory, we can express the spatio-temporally Fourier-transformed correlation function

$$C(\mathbf{q},\omega;\nu_0,D_0,\lambda_0) \equiv \langle h(\mathbf{q},\omega)h(-\mathbf{q},-\omega)\rangle \tag{3.8.32}$$

using the *same* expression (3.1.6) that we obtained in the linear theory, *except* that the bare ν_0 and D_0 must be replaced in those expressions by *strongly* wavevector-dependent functions $\nu(\mathbf{q})$ and $D(\mathbf{q})$, which diverge as *universal* powers of q:

$$\nu(\mathbf{q}) = \nu_s\left(\frac{\Lambda}{q}\right)^{\eta_\nu^*}, \quad D(\mathbf{q}) = D_s\left(\frac{\Lambda}{q}\right)^{\eta_D^*}, \tag{3.8.33}$$

where D_s and ν_s are nonuniversal constants, and the universal exponents η_D^* and η_ν^* are simply the fixed-point values $\eta_D(g^*)$ and $\eta_\nu(g^*)$ of the graphical corrections $\eta_D(g)$ and $\eta_\nu(g)$ appearing in the general recursion relations (3.8.19) and (3.8.21) for ν and D.

Of course, as discussed earlier, we don't actually *know* these values for the $d = 2$ problem, or for the strong-coupling fixed point of the $d > 2$ problem. However, we *do* know them for the $d = 1$ KPZ equation, for which we previously found

$$\eta_D^* = \eta_\nu^* = 1/2, \quad d = 1\,. \tag{3.8.34}$$

I'll show later in this section that using these values in the expressions (3.8.33) and using those in the manner just described – i.e., replacing ν and D in the expression for $C(\mathbf{q},\omega)$ of the linear theory with $\nu(\mathbf{q})$ and $D(\mathbf{q})$ as given by (3.8.33) – yields the values $z = 3/2$ and $\chi = 1/2$ for the dynamic and roughness exponents that we found earlier.

The easiest way to see this is to use the so-called "trajectory integral matching formalism" [32], which is really just a fancy way of saying that we keep track of the rescalings (3.2.8) and (3.2.9) that we performed while doing the RG , and thereby relate correlation functions in the original problem to those of the renormalized system. We've already done this in deriving the scaling law (3.2.21) for the real-space two-point correlation function. To extract the renormalized ν and D, it is easiest to use the spatio-temporally Fourier-transformed equal-time correlations

$$\langle h(\mathbf{q}, \Omega) h(\mathbf{q}', \Omega')\rangle \equiv C(\mathbf{q}, \Omega)\delta^K_{\mathbf{q}+\mathbf{q}'}\delta^K_{\Omega+\Omega'} \,. \tag{3.8.35}$$

To do this, we first need to relate the spatially Fourier transformed field $h(\mathbf{q}, \omega)$ to the rescaled field $h'(\mathbf{q}', \omega')$. This is straightforwardly done simply by recalling the rescalings of the spatial coordinates $\mathbf{r}$ and the field $h(\mathbf{r}, t)$. On *one step* of the RG, after RG "time" ℓ, this transformation is:

$$\mathbf{r} = (1 + \mathrm{d}\ell)br', \quad t = (1 + z(\ell)\mathrm{d}\ell), \quad h(\mathbf{r}, t) = (1 + \chi(\ell)\mathrm{d}\ell)h'(\mathbf{r}', t) \,. \tag{3.8.36}$$

Note that I have now generalized the rescaling to allow me to choose different field rescalings $\chi(\ell)$ and time rescalings $z(\ell)$ at different RG times ℓ. You'll see the reason I want this additional freedom in a moment. Repeating this process $n = \ell/\mathrm{d}\ell$ times, and using the fact that

$$\prod_{m=1}^{n}(1 + \chi(\ell_m)\mathrm{d}l) = \prod_{m=1}^{n} \mathrm{e}^{\chi(\ell_m)\mathrm{d}l + O(d\ell^2)} = \exp\left[\sum_{m=1}^{n} \chi(\ell_m)\mathrm{d}l + O(\mathrm{d}\ell)\right]$$
$$= \exp\left[\int_0^\ell \chi(\ell')\,\mathrm{d}\ell'\right], \tag{3.8.37}$$

along with the similar relations

$$\prod_{m=1}^{n}(1 + z(\ell_m)\mathrm{d}l) = \exp\left[\int_0^\ell z(\ell')\mathrm{d}\ell'\right], \quad \prod_{m=1}^{n}(1 + \mathrm{d}l) = \exp\left[\int_0^\ell \mathrm{d}\ell'\right] = \mathrm{e}^\ell \,, \tag{3.8.38}$$

I obtain, after RG "time" ℓ,

$$\mathbf{r} = \mathrm{e}^\ell \mathbf{r}', \quad t = \exp\left[\int_0^\ell z(\ell')\,\mathrm{d}\ell'\right]t', \quad h(\mathbf{r}, t) = \exp\left[\int_0^\ell \chi(\ell')\,\mathrm{d}\ell'\right]h'(\mathbf{r}', t') \,. \tag{3.8.39}$$

The rescaling of $\mathbf{r}$ implies a relation between the volume V of the original system and that V' of the renormalized system:

$$V = \mathrm{e}^{\ell d} V' \,. \tag{3.8.40}$$

Likewise, the rescaling of time implies a relation between the period T of the original system and that T' of the renormalized system:

$$T = \exp\left[\int_0^\ell z(\ell')\,\mathrm{d}\ell'\right]T' \,. \tag{3.8.41}$$

Putting all of these results together, and using them in the definition (3.1.2) of the spatio-temporally Fourier-transformed field $h(\mathbf{q}, \omega)$, I can relate $h(\mathbf{q}, \omega)$

to $h(\mathbf{q}', \omega')$, where $\mathbf{q}'$ and ω' are rescaled wavevectors and frequencies that I'll calculate precisely. I get:

$$
\begin{aligned}
h(\mathbf{q}, \omega) &= \frac{1}{\sqrt{VT}} \int_V \mathrm{d}^d r \int_0^T \mathrm{d}t\, h(\mathbf{r}, t) \mathrm{e}^{-\mathrm{i}(\mathbf{q}\cdot\mathbf{r} - \omega t)} \\
&= \exp\left[\int_0^\ell [\chi(\ell') + z(\ell')/2]\, d\ell' + \ell d/2\right] \\
&\quad \times \left(\frac{1}{\sqrt{V'T'}} \int_V \mathrm{d}^d r'\, h'(\mathbf{r}', t') \exp\left\{-\mathrm{i}\left(\mathrm{e}^\ell \mathbf{q}\cdot\mathbf{r}' - \exp\left[\int_0^\ell z(\ell')\,\mathrm{d}\ell'\right]\omega t'\right)\right\}\right) \\
&= \exp\left[\int_0^\ell [\chi(\ell') + z(\ell')/2]\,\mathrm{d}\ell' + \ell d/2\right] h'(\mathbf{q}', \omega'),
\end{aligned}
\tag{3.8.42}
$$

where I've defined

$$
\mathbf{q}' \equiv \mathrm{e}^\ell \mathbf{q}, \quad \omega' \equiv \exp\left[\int_0^\ell z(\ell')\,\mathrm{d}\ell'\right]\omega\,. \tag{3.8.43}
$$

With this result in hand, we can relate the spatio-temporally Fourier-transformed correlation function

$$
C(\mathbf{q}, \omega; \nu_0, D_0, \lambda_0) \equiv \langle h(\mathbf{q}, \omega) h(-\mathbf{q}, -\omega)\rangle \tag{3.8.44}
$$

to that of the renormalized system:

$$
\begin{aligned}
C(\mathbf{q}, \omega; \nu_0, D_0, \lambda_0) &= \langle h(\mathbf{q}, \omega) h(-\mathbf{q}, -\omega)\rangle \\
&= \exp\left[\int_0^\ell [2\chi(\ell') + z(\ell)]\,\mathrm{d}\ell' + \ell d\right] \langle h'(\mathbf{q}', \omega') h'(-\mathbf{q}', -\omega')\rangle \\
&= \exp\left[\int_0^\ell [2\chi(\ell') + z(\ell)]\,\mathrm{d}\ell' + \ell d\right] C(\mathbf{q}', \omega'; \nu(\ell), D(\ell), \lambda(\ell))\,,
\end{aligned}
\tag{3.8.45}
$$

where I've defined

$$
\mathbf{q}' \equiv \mathrm{e}^\ell \mathbf{q}, \quad \omega' \equiv \omega \mathrm{e}^\ell\,. \tag{3.8.46}
$$

These relations (3.8.45) and (3.8.46) hold for any value of the renormalization group "time" ℓ. We are completely free to choose that "RG time" to be anything we want. One convenient choice is

$$
\mathrm{e}^\ell q = \Lambda \Rightarrow \ell = \ln\left(\frac{\Lambda}{q}\right). \tag{3.8.47}
$$

With this choice, $|\mathbf{q}'| = \Lambda$, the ultraviolet cutoff. At such a large value of $\mathbf{q}'$, the correlation function $C(\mathbf{q}', \omega'; \nu(\ell), D(\ell), \lambda(\ell))$ should be given accurately by the linear theory, with the renormalized values $\nu(\ell), D(\ell), \lambda(\ell)$ of ν and D, since, as we saw in our discussion of perturbation theory, the perturbative corrections to the linear results only become important at *small* **q**.

Therefore, we have

$$C(\mathbf{q}', \omega'; \nu(\ell), D(\ell), \lambda(\ell)) = \frac{2D(\ell)}{\omega'^2 + \nu^2(\ell)\Lambda^4}, \tag{3.8.48}$$

where I've used the fact that $|\mathbf{q}'| = \Lambda$.

We can obtain $\nu(\ell)$ and $D(\ell)$ by solving their recursion relations (3.8.21) and (3.8.19):

$$D(\ell) = \exp\left[\int_0^\ell [z(\ell') - 2\chi(\ell') + \eta_D(g(\ell'))]\, d\ell' - \ell d\right] D_0, \tag{3.8.49}$$

$$\nu(\ell) = \exp\left[\int_0^\ell [z(\ell') + \eta_\nu(g(\ell'))]\, d\ell' - 2\ell\right] \nu_0. \tag{3.8.50}$$

Plugging these expressions for $D(\ell)$ and $\nu(\ell)$ into our expression (3.8.48) for the renormalized correlation function, and using the result and (3.8.46) in our expression (3.8.45) for the true correlation function of the original system gives

$$\begin{aligned} &C(\mathbf{q}, \omega; \nu(\ell), D(\ell), \lambda(\ell)) \\ &= \frac{2D_0 \exp\left[\int_0^\ell [2z(\ell') + \eta_D(g(\ell'))]\, d\ell'\right]}{\exp\left[\int_0^\ell [2z(\ell')]\, d\ell'\right]\left(\omega^2 + \nu_0^2 \exp\left[\int_0^\ell [\eta_\nu(g(\ell'))]\, d\ell'\right] q^4\right)} \\ &= \frac{2D_0 \exp\left[\int_0^\ell \eta_D(g(\ell'))\, d\ell'\right]}{\left(\omega^2 + \nu_0^2 \exp\left[\int_0^\ell 2\eta_\nu(g(\ell'))\, d\ell'\right] q^4\right)}. \end{aligned} \tag{3.8.51}$$

Making the **q**-dependent choice (3.8.47) of the RG time ℓ allows us to rewrite this result in a form that looks exactly like the result of the linear theory, *except* that the diffusion *constant* ν_0 and the noise correlation D_0 must be replaced by effective, *wavevector-dependent* quantities $\nu(\mathbf{q})$ and $D(\mathbf{q})$ given by

$$\nu(\mathbf{q}) \equiv \nu_0 \exp\left[\int_0^\ell \eta_\nu(g(\ell'))\, d\ell'\right], \quad D(\mathbf{q}) \equiv D_0 \exp\left[\int_0^\ell \eta_D(g(\ell'))\, d\ell'\right], \tag{3.8.52}$$

where the wavevector dependence comes in because $\ell = \ln\left(\frac{\Lambda}{q}\right)$.

These wavevector-dependent quantities $\nu(\mathbf{q})$ and $D(\mathbf{q})$ are often referred to as the "renormalized" parameters. Note that the term "renormalized" is here being used in a somewhat different sense than I used it earlier.

These wavevector-dependent quantities have a very simple wavevector dependence at small **q**: They are, asymptotically, simple power laws in $q = |\mathbf{q}|$.

I'll show this explicitly for $D(\mathbf{q})$; the argument for $\nu(\mathbf{q})$ is identical, with D replaced everywhere by ν.

I start by reorganizing the expression (3.8.52) for $D(\mathbf{q})$ as

$$D(\mathbf{q}) = D_0 \exp\left[\int_0^{\ell} (\eta_D(g(\ell')) - \eta_D^*) + \eta_D^* \, d\ell'\right]$$
$$= D_0 \exp\left[\int_0^{\ell} (\eta_D(g(\ell')) - \eta_D^*) \, d\ell'\right] e^{\eta_D^* \ell}$$
$$\equiv D_1(\mathbf{q}) \left(\frac{\Lambda}{q}\right)^{\eta_D^*}, \tag{3.8.53}$$

where I've defined

$$D_1(\mathbf{q}) \equiv D_0 \exp\left[\int_0^{\ell} (\eta_D(g(\ell')) - \eta_D^*) \, d\ell'\right], \tag{3.8.54}$$

with $\ell = \ln\left(\frac{\Lambda}{q}\right)$. I've also used this value of ℓ to write $e^{\eta_D^* \ell} = \left(\frac{\Lambda}{q}\right)^{\eta_D^*}$ in the last equality above.

The important point here is that $D_1(\mathbf{q})$ goes to a finite limit as $\mathbf{q} \to \mathbf{0}$, as I'll prove in a moment. Given this, it follows that the leading $\mathbf{q}$ dependence of $D(\mathbf{q})$ is the universal power law dependence on q embodied explicitly by the $\left(\frac{\Lambda}{q}\right)^{\eta_D^*}$ factor in equation (3.8.53), with $D_1(\mathbf{q})$ replaced by its finite, nonzero limiting value as $\mathbf{q} \to \mathbf{0}$.

An essentially identical argument applies to $\nu(\mathbf{q})$.

Thus, we can conclude, once I actually prove that $D_1(\mathbf{q} = \mathbf{0})$ and $\nu_1(\mathbf{q} = \mathbf{0})$ are finite, that

$$D(\mathbf{q}) = D_1(\mathbf{0}) \left(\frac{\Lambda}{q}\right)^{\eta_D^*}, \quad \nu(\mathbf{q}) = \nu_1(\mathbf{0}) \left(\frac{\Lambda}{q}\right)^{\eta_\nu^*}. \tag{3.8.55}$$

Now to the proof that $D_1(\mathbf{q} = \mathbf{0})$ and $\nu_1(\mathbf{q} = \mathbf{0})$ are finite. I'll present the detailed proof for $D_1(\mathbf{q} = \mathbf{0})$; the proof for $\nu_1(\mathbf{q} = \mathbf{0})$ is identical, except for the replacement of "D" by "ν" everywhere.

We start by noting that, when $\mathbf{q} = \mathbf{0}$, $\ell = \ln\left(\frac{\Lambda}{q}\right) = \infty$. Hence

$$D_1(\mathbf{q} = \mathbf{0}) \equiv D_0 \exp\left[\int_0^{\infty} (\eta_D(g(\ell)) - \eta_D^*) \, d\ell\right]. \tag{3.8.56}$$

Thus, to prove that $D_1(\mathbf{q} = \mathbf{0})$ is finite, we need only prove that

$$\int_0^{\infty} (\eta_D(g(\ell)) - \eta_D^*) \, d\ell \tag{3.8.57}$$

converges as $\ell \to \infty$. This can be done by noting that, at large ℓ, the dimensionless coupling $g(\ell)$ is close to g^*, its fixed-point value. Therefore, we can linearize the recursion relation for g near g^*. Using our general expressions (3.8.19), (3.8.20), and (3.8.21), we can easily show that the general recursion relation for g is

$$\frac{dg}{d\ell} = [2 - d + \eta_D(g) - 3\eta_\nu(g)]g\,. \tag{3.8.58}$$

At any fixed point with a nonzero g^*, this clearly implies that

$$2 - d + \eta_D(g^*) - 3\eta_\nu(g^*) = 0\,. \tag{3.8.59}$$

This in turn implies that if we linearize about g^*, that is, write

$$g(\ell) = g^* + \delta g(\ell)\,, \tag{3.8.60}$$

and rewrite the recursion relation (3.8.58) to linear order in $\delta g(\ell)$, we obtain

$$\begin{aligned}\frac{d\delta g}{d\ell} &= \left[\left\{\left(\frac{d\eta_D(g)}{dg}\right)\bigg|_{g=g^*} - 3\left(\frac{d\eta_\nu(g)}{dg}\right)\bigg|_{g=g^*}\right\}\delta g(\ell)\right]g^* \\ &= -\xi_g \delta g(\ell)\,,\end{aligned} \tag{3.8.61}$$

where I've defined the eigenvalue (actually minus the eigenvalue) of g

$$\xi_g \equiv \left(3\left(\frac{d\eta_\nu(g)}{dg}\right)\bigg|_{g=g^*} - \left(\frac{d\eta_D(g)}{dg}\right)\bigg|_{g=g^*}\right)g^*\,. \tag{3.8.62}$$

Note that ξ_g must be positive, since the eigenvalue $-\xi_g$ of the g recursion relation at the fixed point must be negative, because the fixed point we're considering is stable.

It is straightforward to solve (3.8.61) for $\delta g(\ell)$:

$$\delta g(\ell) = A e^{-\xi_g \ell}\,, \tag{3.8.63}$$

where A is a finite, $O(1)$ constant that depends on the initial conditions. This solution shows that δg vanishes exponentially at large ℓ. Therefore, so does the difference between $\eta_D(g(\ell))$ and η_{g^*} that appears in the integral (3.8.57):

$$\eta_D(g(\ell)) - \eta_D^* \approx \left(\frac{d\eta_D(g)}{dg}\right)\bigg|_{g=g^*} \delta g(\ell) \propto e^{-\xi_g \ell}\,. \tag{3.8.64}$$

Therefore, the integral in (3.8.57) will converge, since it is the integral of a decaying exponential. Hence, $D_1(\mathbf{q} = \mathbf{0})$ is finite.

As mentioned earlier, the same conclusion holds for $\nu_1(\mathbf{q} = \mathbf{0})$.

This completes the proof that (3.8.55) holds, with $D_1(\mathbf{q} = \mathbf{0})$ and $\nu_1(\mathbf{q} = \mathbf{0})$ finite, nonzero constants.

Note that those two constants are nonuniversal, both because the bare values D_0 and ν_0 of D and ν are, and because the bare value of g is. Changing the bare value of g will also change $D_1(\mathbf{q} = \mathbf{0})$and $\nu_1(\mathbf{q} = \mathbf{0})$, since it will change the values of $g(\ell)$ appearing in the integrals (3.8.52) giving D and ν. Thus, we change $D_1(\mathbf{q} = \mathbf{0})$ and $\nu_1(\mathbf{q} = \mathbf{0})$ by changing, e.g., λ or the ultraviolet cutoff Λ, even *without* changing the bare values D_0 and ν_0 of D and ν at all.

What we can *not* change is the scaling form (3.8.55) of the effective ν and D: The power law dependence on wavenumber q is universal, with universal exponents η_D^* and η_ν^*.

I will henceforth define $D_1(\mathbf{q} = \mathbf{0}) = D_s$ and $\nu_1(\mathbf{q} = \mathbf{0}) = \nu_s$ (“s” standing for “scaling”). I can then rewrite (3.8.55) as

$$D(\mathbf{q}) = D_s \left(\frac{\Lambda}{q}\right)^{\eta_D^*}, \quad \nu(\mathbf{q}) = \nu_s \left(\frac{\Lambda}{q}\right)^{\eta_\nu^*}. \tag{3.8.65}$$

This result means that we can now summarize the solution of the nonlinear problem by saying that it can be obtained from the solution of the linear problem, with the sole modification that the parameters D and ν of the linear theory be replaced everywhere by the wavevector-dependent functions (3.8.65).

Using this fact, I can derive universal scaling forms for the spatio-temporally Fourier-transformed correlation functions.

I can now recover the scaling form of the Fourier-transformed spatio-temporal correlations

$$\begin{aligned} C(\mathbf{q}, \omega; \nu(\ell), D(\ell), \lambda(\ell)) &= \frac{2D(\mathbf{q})}{\omega^2 + \nu^2(\mathbf{q})q^4} \\ &= q^{2\eta_\nu^* - \eta_D^* - 4}\left(\frac{2D_s}{\left(\frac{\omega}{q^{2-\eta_\nu}}\right)^2 + \nu_s^2}\right) \\ &\equiv q^{2\eta_\nu^* - \eta_D^* - 4} f_{FT}\left(\frac{\omega}{q^z}\right), \end{aligned} \tag{3.8.66}$$

where I’ve defined the scaling function

$$f_{FT}(x) \equiv \frac{2D_s}{x^2 + \nu_s^2}, \tag{3.8.67}$$

and the dynamical exponent

$$z = 2 - \eta_\nu. \tag{3.8.68}$$

Note that this value of z is exactly what I would have gotten directly from the sort of RG argument I presented earlier, in which we chose the rescaling exponents to keep parameters fixed. In particular, if I had chosen z to keep ν fixed, inspection of the recursion relation (3.8.19) reveals that I’d have to have chosen (3.8.68).

From the scaling form (3.8.66), I can derive the full scaling form (3.1.15) of the real-space correlations simply by Fourier transforming (3.8.66) back to real space:

$$C(\mathbf{r},t) = \int \frac{\mathrm{d}^d q\, \mathrm{d}\omega}{(2\pi)^{d+1}} C(\mathbf{q},\omega) \mathrm{e}^{\mathrm{i}(\mathbf{q}\cdot\mathbf{r}-\omega t)}$$
$$= \int \frac{\mathrm{d}^d q\, \mathrm{d}\omega}{(2\pi)^{d+1}} q^{\eta_\nu^* - \eta_D^* - 4} f_{FT}\left(\frac{\omega}{q^z}\right) \mathrm{e}^{\mathrm{i}(\mathbf{q}\cdot\mathbf{r}-\omega t)}. \tag{3.8.69}$$

Making the following changes of variables of integration from $\mathbf{q}$ and ω to $\mathbf{Q}$ and Ω:

$$\mathbf{q} = \mathbf{Q}/r, \quad \omega = \Omega/r^z, \tag{3.8.70}$$

we can rewrite this in a scaling form:

$$C(\mathbf{r},t) = r^{2\chi} f_{RS}\left(\frac{t}{r^z}\right), \tag{3.8.71}$$

where I've defined the roughness exponent

$$\chi = (z - d + \eta_D^*)/2 \tag{3.8.72}$$

and the scaling function

$$f_{RS}\left(\frac{t}{r^z}\right) = \int \frac{\mathrm{d}^d Q\, \mathrm{d}\Omega}{(2\pi)^{d+1}} Q^{\eta_\nu^* - \eta_D^* - 4} f_{FT}\left(\frac{\Omega}{Q^z}\right) \exp\left[\mathrm{i}\left(\mathbf{Q}\cdot\hat{\mathbf{r}} - \Omega\left(\frac{t}{r^z}\right)\right)\right]. \tag{3.8.73}$$

Note that this value of χ is again exactly what I would have gotten directly from the sort of RG argument I presented earlier. Had I chosen χ to keep D fixed, inspection of (3.8.21) reveals that I'd have to have chosen (3.8.72).

The above results (3.8.68) and (3.8.72) for z and χ are completely general; i.e., they hold in all spatial dimensions. Now note that if we specialize to $d = 1$, use the values

$$\eta_D^* = \eta_\nu^* = 1/2 \tag{3.8.74}$$

that we found earlier for η_D^* and η_ν^* in $d = 1$ in those general expressions, and also set $d = 1$ in the expression (3.8.72) for χ, we recover the results $z = 3/2$ and $\chi = 1/2$ that we obtained earlier.

In the discussions of flocking systems that follow, I will sometimes use this approach of expressing correlation functions in terms of wavevector-dependent parameters to obtain scaling laws, while on other occasions I will use the sort of RG scaling argument presented earlier. As I hope this example has made clear, these two approaches are completely equivalent.

3.9 "Irrelevant" Extra Terms, and Universality

So far, I've shown how the dynamical renormalization group can be used to determine the long-distance, long-time behavior of a particular dynamical equation. While tremendously useful, this is *not* the most important application of renormalization group ideas. Not only does the RG help us to *solve* a problem once we have an equation of motion for it, but it actually is what enables us to formulate the equation of motion in the first place. Indeed, as noted earlier, it is ultimately the answer to Einstein's question about why we can do physics at all. That question can be rephrased somewhat less elegantly as "Why do these simple hydrodynamic equations (the KPZ equation here, the Navier–Stokes equations for simple fluids, and the hydrodynamic equations for flocking that I'll discuss throughout the remainder of this book) provide an accurate description of *all* systems of the appropriate symmetry?"

The idea is extremely simple: When formulating an equation of motion to describe a system of a particular symmetry, we write down in its equation (or, more generally, equation*s*) of motion *only* the terms that are most relevant in the RG sense.

I'll illustrate how we can use this logic to construct the KPZ equation (3.0.1).

The symmetry we want is that the equation of motion be invariant after a uniform upward translation of the surface; i.e., the transformation

$$h(\mathbf{r}, t) \rightarrow h(\mathbf{r}, t) + h_0, \tag{3.9.1}$$

where h_0 is *any* constant. This simply means we expect a surface of a given shape to evolve in exactly the same way regardless of its mean vertical position – an obvious consequence of the translation invariance of space.

This means that $\partial_t h$ can only depend on spatial derivatives of h.

Obviously, all of the h-dependent terms on the right-hand side of our original KPZ equation (3.0.1) satisfy this requirement. But equally obviously, there are *infinitely* many other terms that *also* satisfy it. For example, one might well ask why we don't include a term $-\nu_4 \nabla^4 h(\mathbf{r}, t)$ on the right-hand side of (3.0.1).

Now that we know the RG, however, we can immediately answer this question: While we *could*, of course, include such a term, it is *irrelevant* in the RG sense. That is, it flows to zero upon renormalization. This can easily be seen from the recursion relation for ν_4, which reads:

$$\frac{d\nu_4}{d\ell} = [z - 4 + \text{graphs}]\nu_4(\ell). \tag{3.9.2}$$

If we're in $d > 2$, where the nonlinear coupling g flows to zero under renormalization, the graphical corrections will vanish, and we'll just have

$$\frac{\mathrm{d}\nu_4}{\mathrm{d}\ell} = [z - 4]\nu_4(\ell)\,. \tag{3.9.3}$$

Comparing this with the recursion relation for ν in the original KPZ equation:

$$\frac{\mathrm{d}\nu}{\mathrm{d}\ell} = [z - 2 + \text{graphs}]\nu(\ell)\,, \tag{3.9.4}$$

which, in $d > 2$ at the linear fixed point where g vanishes, becomes

$$\frac{\mathrm{d}\nu}{\mathrm{d}\ell} = [z - 2]\nu(\ell)\,. \tag{3.9.5}$$

If we now choose z to keep ν fixed, which clearly means we have to choose $z = 2$, then the recursion relation for our proposed new coupling ν_4 becomes

$$\frac{\mathrm{d}\nu_4}{\mathrm{d}\ell} = -2\nu_4(\ell)\,, \tag{3.9.6}$$

which clearly shows that $\nu_4(\ell \to \infty) \to 0$ exponentially fast.

The reason for this, of course, is obvious: The ν_4 term has more spatial derivatives than the ν term, and is, therefore, less relevant than that term, and therefore *irrelevant* if we choose our rescaling to keep ν fixed.

We can also understand this result in Fourier space, where it simply amounts to saying that we're looking at small q. To see this, let's for the moment keep the ν_4 term, and work out the linear theory including it. We have the equation of motion

$$\partial_t h(\mathbf{r}, t) = (\nu\nabla^2 - \nu_4\nabla^4)h(\mathbf{r}, t) + f(\mathbf{r}, t)\,. \tag{3.9.7}$$

Fourier transforming this gives

$$-\,\mathrm{i}\Omega h(\mathbf{q}, \Omega) = -(\nu q^2 + \nu_4 q^4)h(\mathbf{q}, \Omega) + f(\mathbf{q}, \Omega)\,, \tag{3.9.8}$$

which can readily be solved for $(\mathbf{q}, \Omega)$ in terms of $f(\mathbf{q}, \Omega)$:

$$h(\mathbf{q}, \Omega) = G(\mathbf{q}, \Omega)f(\mathbf{q}, \Omega)\,, \tag{3.9.9}$$

where the "propagator" is now given by

$$G(\mathbf{q}, \Omega) = \frac{1}{-\mathrm{i}\Omega + \nu q^2 + \nu_4 q^4}\,. \tag{3.9.10}$$

Obviously, in the limit $q \to 0$, this reduces to exactly the propagator we found ignoring the ν_4 term.

This statement follows through to the correlation function:

$$C(\mathbf{q}, \Omega) = 2D|G(\mathbf{q}, \omega)|^2 = \frac{2D}{(\Omega^2 + (\nu q^2 + \nu_4 q^4)^2)}\,, \tag{3.9.11}$$

which obviously also reduces to our earlier result in the limit $q \to 0$.

Following through to the equal-time correlation function

$$C_{\mathrm{ET}}(\mathbf{q}) = \int \frac{\mathrm{d}\Omega}{2\pi} \frac{2D}{(\Omega^2 + (\nu q^2 + \nu_4 q^4)^2)} = \frac{D}{\nu q^2 + \nu_4 q^4}, \tag{3.9.12}$$

we see that this also reduces to our earlier result as $q \to 0$.

Finally, if we calculate the mean squared real-space height fluctuations using this result, we get:

$$\langle (h(\mathbf{r},t))^2 \rangle = \int_{\Lambda > |\mathbf{q}| > 1/L} \frac{\mathrm{d}^d q}{(2\pi)^d} C_{ET}(\mathbf{q}) = \int \frac{\mathrm{d}^d q}{(2\pi)^d} \frac{D}{\nu q^2 + \nu_4 q^4}, \tag{3.9.13}$$

where L is the system size. Writing

$$\frac{1}{\nu q^2 + \nu_4 q^4} = \frac{1}{\nu q^2} + \left(\frac{1}{\nu q^2 + \nu_4 q^4} - \frac{1}{\nu q^2} \right) \tag{3.9.14}$$

and noting that the term in parentheses on the right-hand side goes to a finite constant $-\frac{\nu_4}{\nu^2}$ as $q \to 0$, we see that the leading dependence of the real-space fluctuations on system size L is dominated by the first term, which is, of course, exactly what we got when we ignored the ν_4 term altogether. Thus, our conclusion that

$$\langle (h(\mathbf{r},t))^2 \rangle \propto \begin{cases} L^{2-d}, & d < 2, \\ \ln\left(\frac{\Lambda L}{2\pi}\right), & d = 2, \\ \text{finite (independent of } L), & d > 2 \end{cases} \tag{3.9.15}$$

is completely robust against the inclusion of the ν_4 term, as the RG argument also told us.

We can eliminate an infinity of terms involving higher derivatives by this argument. Adding terms involving arbitrary integer powers of the Laplacian operator acting on h, we get the generalized KPZ equation

$$\partial_t h(\mathbf{r},t) = \left(\nu\nabla^2 + \sum_{n=1}^{\infty} (-1)^n \nu_{2n+2} \nabla^{2n+2} \right) h(\mathbf{r},t) + f(\mathbf{r},t). \tag{3.9.16}$$

Fourier transforming this gives

$$-\mathrm{i}\omega h(\mathbf{r},t) = -\left(\nu q^2 + \sum_{n=1}^{\infty} \nu_{2n+2} q^{2n+2} \right) h(\mathbf{r},t) + f(\mathbf{r},t), \tag{3.9.17}$$

which again clearly reduces to the original linearized KPZ equation as $q \to 0$. The reader can check for herself that the rest of the analysis done above for the ν_4 term goes through for this infinitely more general model as well, with the same conclusion: All of the large distance properties of the KPZ equation are completely unaffected by including these extra ν_{2n+2} terms.

All of those terms are simply terms that, like $\nu\nabla^2 h$, are linear in h, but involve more spatial derivatives. The above analysis, which can obviously easily be generalized to *any* problem, leads to the first RG-based rule for formulating an equation of motion for any system: When writing down the equation of motion, if there are many terms involving equal powers of the fields, we need only keep the term (or terms) that have the smallest number of spatial derivatives.

What about terms that have the same number of spatial derivatives, but different powers of the fields? We've actually already dealt with such a term in our treatment of the KPZ equation: the $\lambda|\nabla h(\mathbf{r},t)|^2$ term, which, like the ν term, has two spatial derivatives. It is, nonetheless, irrelevant for $d > 2$. Let's review the reason for this.

Recall that the RG recursion relation of *this* term is:

$$\frac{d\lambda}{d\ell} = [z - 2 + \chi]\lambda(\ell). \tag{3.9.18}$$

In $d > 2$, $\chi < 0$, and so, if we again choose $z = 2$ to keep ν fixed, we see that λ has eigenvalue χ, which is less than zero, which was our argument earlier that this term could be neglected in $d > 2$.

For the KPZ equation, this means we can also neglect terms like $\lambda_4|\nabla h(\mathbf{r},t)|^4$ whose recursion relation reads

$$\frac{d\lambda}{d\ell} = [z - 4 + 3\chi + \text{graphs}]\lambda(\ell). \tag{3.9.19}$$

This term has a double whammy: The -2 in the recursion relation for λ has become a -4, because of the two extra spatial derivatives, and the χ (which is negative for $d < 2$) has been replaced by 3χ, because there are two more h fields in this term. Hence, this term is even more irrelevant than the λ term for $d > 2$.

One might well ask what happens for $d < 2$, for which the λ term becomes relevant. The situation now becomes slightly less clear, because of the graphical corrections to the various terms, which no longer vanish. However, in *most* problems – although *not* for the KPZ equation, which is unusual in this respect – when one goes below the critical dimension at which one or more nonlinear terms become relevant, they typically flow to values that, at least for spatial dimensions close to the critical dimension, are small (typically, $O(\epsilon)$, where $\epsilon \equiv d_c - d$). (Recall our earlier discussion of the ϵ-expansion for the KPZ critical point.) If this is the case, then terms that originally had eigenvalues that were negative and $O(1)$ will, in $d = d_c - \epsilon$ dimensions, have eigenvalue $O(1) + O(\epsilon)$, which will still be negative, at least for small ϵ. The standard assumption, which can be justified by a sort of "Occam's razor" argument, is that if a new stable fixed point appears as we lower the spatial dimension d below some critical dimension d_c, and a particular coupling is irrelevant at that fixed point, then it will remain irrelevant all the way down to the dimension of physical interest, even if ϵ is not small there. Otherwise, one would have a more elaborate scenario in which a once-stable fixed point became unstable at least twice as dimension was lowered. Old William of Occam would reject such a scenario as too complex, and opt for the simpler scenario of a single instability of the fixed point as dimension was lowered.

The upshot of this argument is a second rule for deciding which terms to keep in writing a long-wavelength, long-time – or, to summarize in a single word, a *hydrodynamic* – theory of any *ordered* system. Since, if a model is *ordered*, then the fluctuations in its fields must necessarily be finite (as in, e.g., the XY model in $d > 2$). Hence, in any dimension in which a system *is* ordered, when comparing terms with equal numbers of derivatives, we need only keep the one(s) with the fewest fields.

To summarize these RG-motivated rules for formulating hydrodynamic equations: At any order in the fluctuating fields, we always keep only the terms with the smallest number of derivatives. And to any order in spatial derivatives, we keep only the terms with the smallest number of fluctuating fields.

There is, of course, one final rule: Our hydrodynamic equations must respect all of the symmetries of the system.

These rules can be applied to a physical system of *any* symmetry. Indeed, they can be thought of as the basis of all of physics. Take, for example, electrostatics. Why is the electrostatic potential ϕ in empty space governed by Laplace's equation

$$\nabla^2 \phi = 0 ? \tag{3.9.20}$$

The answer is that we're looking for an equation that has the following symmetries: rotation invariance, translation invariance, and $\phi(\mathbf{r}) \to \phi(\mathbf{r}) + \text{constant}$. We'd also like to derive this by minimizing some local energy. The $\phi(\mathbf{r}) \to \phi(\mathbf{r}) + $ constant symmetry means that the energy can only depend on gradients of ϕ.

These are exactly the same constraints that hold for the XY model written in terms of the angle field θ. As we saw in Chapter 2, these constraints force the energy to have the form

$$H_{\mathrm{XY}} = \frac{K}{2} \int \mathrm{d}^d r |\nabla \phi(\mathbf{r})|^2, \tag{3.9.21}$$

where I've simply replaced the angle field θ with the electrostatic potential ϕ. Minimizing the energy (3.9.21) immediately implies (3.9.20).

Now, of course, there are many other terms we could have included in the Hamiltonian (3.9.21) aside from the $|\nabla \phi(\mathbf{r})|^2$ term we did include. We could, for example, have added a $|\nabla^2 \phi(\mathbf{r})|^2$ term. Indeed, we could have included *all* terms $|\nabla^{2n} \phi(\mathbf{r})|^2$ for all integer n. However, we know, from the RG, that all such terms (including $n = 1$) will be irrelevant at sufficiently long length scales L, and so can be ignored. The same applies to all other terms one can think of that respect $\phi(\mathbf{r}) \to \phi(\mathbf{r}) =$ constant symmetry (and translation and rotation invariance).

In Electricity and Magnetism (E&M), "sufficiently long" length scales L are scales longer than the Compton wavelength of the electron: $L \gg \lambda = \frac{h}{mc} = 2.42631 \times 10^{-12}$ m, where m is the mass of the electron and here h is Planck's constant (not to be confused with the height field in the KPZ equation!). Since λ is much smaller than the length scales encountered in most table-top experiments (indeed, it's much smaller than a hydrogen atom!), Laplace's equation (3.9.20) is an *extremely* good approximation for such experiments.

This argument also explains why the gravitational potential in Newton's theory of gravity is also governed by (3.9.20).

This is why I agree with Paul Goldbart that the RG is the answer to Einstein's question "Why is the universe comprehensible?": Because almost all of the terms that might describe a real system at short distances are "irrelevant" at long distances. As a result, our physical models at long distances contain very few terms, and, so, are very simple (and hence comprehensible).

I'll now turn to one particular system that is comprehensible for precisely this reason: "dry active matter"; that is, flocks without momentum conservation.

4

Formulating the Hydrodynamic Model for Flocking

4.1 General Ideas of the Hydrodynamic Approach

We'll now apply the techniques and concepts that we've just learned to the flocking problem. As just discussed, we will use these ideas not only to analyze the equations of motion for that problem, but even to *formulate* them. This "hydrodynamic" reasoning, as should be clear from the discussion of Chapter 3, does *not* come from solving the many (*very* many!) body problem of computing the time-dependent positions $\mathbf{r}_i(t)$ of the 10^{23} constituent molecules of a fluid subject to intermolecular forces from all of the other 10^{23} molecules. Such an approach is analytically intractable even if one knew what the intermolecular forces were. Trying to compute analytically the behavior of, e.g., Vicsek's algorithm directly would be the corresponding, and equally impossible, approach to the flocking problem.

Instead, the way we understand fluid mechanics is by writing down a set of continuum equations – the Navier–Stokes equations – for continuous, smoothly varying number density $\rho(\vec{r}, t)$ and velocity $\mathbf{v}(\mathbf{r}, t)$ fields describing the fluid.

Although we know that fluids are made out of atoms and molecules, we can define "coarse-grained" number density $\rho(\vec{r}, t)$ and velocity $\mathbf{v}(\mathbf{r}, t)$ fields by averaging over "coarse-graining" volumes large compared with the intermolecular or, in the flocks, "interbird" spacing. On a large scale, even discrete systems *look* continuous, as we all know from close inspection of newspaper photographs and television images.

In writing down the Navier–Stokes equations, one "buries one's ignorance"[1] of the detailed microscopic dynamics of the fluid in a few phenomenological parameters, namely the mean density ρ_0, the bulk and shear viscosities η_B and η_S, the thermal conductivity κ, the specific heat c_v, and the compressibility χ. Once these

[1] This description of the fundamental logic of the hydrodynamic approach and the lovely phrase "bury your ignorance" that so nicely summarizes it, are taken from D. Forster, *Hydrodynamic Fluctuations, Broken Symmetry, and Correlation Functions* (Benjamin, Reading, 1975), a book that I heartily recommend to anyone seeking a deeper understanding of hydrodynamic reasoning.

have been deduced from experiment (or, occasionally, and at the cost of immense effort, calculated from a microscopic model), one can then predict the outcomes of all experiments that probe length scales much greater than a spatial coarse-graining scale ℓ_0 and time scales $\gg t_0$, a corresponding microscopic time, by solving these continuum equations, a far simpler task than solving the microscopic dynamics.

But how do we write down these continuum equations? The answer to this question is, in a way, extremely simple: We write down every *relevant* term that is not ruled out by the symmetries and conservation laws of the problem. In the case of the Navier–Stokes equations, the symmetries are rotational invariance, space and time translation invariance, and Galilean invariance (i.e., invariance under a boost to a reference frame moving at a constant velocity), while the conservation laws are conservation of particle number, momentum, and energy.

"Relevant," in this specification means terms that are important at large length scales and long time scales. The renormalization group approach presented in the previous chapter provides a precise criterion for deciding which terms are relevant. Among other things, as we saw, it implies a "gradient expansion": We keep in the equations of motion only terms with the smallest possible number of space and time derivatives. Recall that we have already used such a gradient expansion to formulate the XY Hamiltonian of Chapter 2. We also used it on our discussion of the formulation of the KPZ equation at the end of Chapter 3.

The same approach works in dynamics. For example, in the Navier–Stokes equations, we keep a viscous term $\eta_s \nabla^2 \mathbf{v}$, but not a term $\gamma \nabla^4 \mathbf{v}$, though the latter is also allowed by symmetry, because the $\gamma \nabla^4 \mathbf{v}$ term involves more spatial derivatives, and hence is smaller, for slow spatial variation, than the viscous term we've already got. As we saw in our discussion of the KPZ equation in Chapter 3, this heuristic argument can be completely justified by the renormalization group, which always finds that terms with more spatial derivatives are less relevant than those with fewer, *all other things being equal.*

By "all other things," I mean the number and type of fields. So, for example, in the KPZ equation, if symmetry had allowed a term proportional to h^2 in the equation of motion, then our $|\nabla h|^2$ would have been irrelevant compared to that h^2 term, since both terms have the same number of fields h (that is, both scale with h like h^2). However, since translation invariance symmetry forbids an h^2 term in the KPZ equation, we have no justification for neglecting the $|\nabla h|^2$ term. However, we *can* immediately throw out a $|\nabla^2 h|^2$ term, because it will be negligible relative to the $|\nabla h|^2$ term, since it has the same number of powers of h (two), but more spatial derivatives (four versus two).

Our current theoretical understanding of both dry and wet active matter is based largely on applying the hydrodynamic approach I've just outlined to those systems. The rest of the book will demonstrate how that is done for the specific case of dry polar active fluids, for which the only symmetry is rotation invariance ("dry" means no momentum conservation, while energy conservation doesn't apply to any active system, since the very term "active" implies the existence of an energy source for each particle or flocker).

4.2 Formulating the Hydrodynamic Model for Flocking

In this section, I'll review the derivation and analysis of the hydrodynamic model of polar ordered dry active fluids, which I'll also refer to as "ferromagnetic flocks." More details can be found in references [33, 34, 35, 36, 37].

While I share Mark Twain's belief that the only people who should refer to themselves as "we" are pregnant women and people with tapeworms, I will use "we" in most of the discussion in this chapter, since all of this work was done in collaboration with Yuhai Tu.

4.2.1 Symmetries and Conservation Laws

As discussed in the Introduction, the system we wish to model is any collection of a large number N of organisms (or other self-propelled things, all of which I will hereafter intermittently refer to as "birds," "boids," or "flockers") in a d-dimensional space, with each organism seeking to move in the same direction as its immediate neighbors.

We further assume that each organism has no "compass," in the sense defined in the Introduction, i.e., no intrinsically preferred direction in which it wishes to move. Rather, it is equally happy to move in any direction picked by its neighbors. However, the navigation of each organism is not perfect; it makes some errors in attempting to follow its neighbors. We consider the case in which these errors have zero mean; e.g., in two dimensions, a given bird is no more likely to err to the right than to the left of the direction picked by its neighbors. We also assume that these errors have no long temporal correlations; e.g., a bird that has erred to the right at time t is equally likely to err either left or right at a time t' much later than t.

The continuum model will describe the long distance behavior of *any* flock satisfying the symmetry conditions I'll specify in a moment. The automaton studied by Vicsek et al. [18, 19] described in the Introduction provides one concrete realization of such a model. Adding "bells and whistles" to this model by, e.g., including purely attractive or repulsive interactions between the birds, restricting their field of vision to those birds ahead of them, giving them some short-term memory, etc., will not change the hydrodynamic model, but can be incorporated simply into a change of the numerical values of a few phenomenological parameters in the model, in much the same way that all simple fluids are described by the Navier–Stokes equations, and changing fluids can be accounted for simply by changing, e.g., the viscosity that appears in those equations.

This model should also describe real flocks of real living organisms, provided that the flocks are large enough, and that they have the same symmetries and conservation laws that, e.g., Vicsek's algorithm does.

So, given this lengthy preamble, what *are* the symmetries and conservation laws of flocks?

The only symmetries of the system are invariance under rotations and translations.

Translation invariance simply means that displacing the positions of the whole flock rigidly by a constant amount has no physical effect, since the space the flock moves through is assumed to be on average homogeneous.[2] Since we are not considering translational ordering, this symmetry remains unbroken.

Rotation invariance simply says the "birds" lack a compass, so that all directions of space are equivalent. Thus, the "hydrodynamic" equation of motion we write down cannot have built into it any special direction picked a priori; all directions must be spontaneously picked out by the motion and spatial structure of the flock. As we shall see, this symmetry *severely* restricts the allowed terms in the equation of motion.

Note that the system does *not* have Galilean invariance: Changing the velocities of all the birds by some constant boost $\mathbf{v}_b$ does *not* leave the system invariant. Indeed, such a boost is *impossible* in a system that strictly obeys Vicsek's rules, since the *speeds* of all the birds will not remain equal to v_0 after the boost. One could imagine relaxing this constraint on the speed, and allowing birds occasionally to speed up or slow down, while tending on average to move at speed v_0. Then the boost just described would be possible, but clearly would change the subsequent evolution of the flock.

Another way to say this is that birds move through a resistive medium, which provides a special Galilean reference frame, in which the dynamics are particularly simple, and different from those in other reference frames. Since real organisms in flocks always move through such a medium (birds through the air, fish through the sea, wildebeest through the arid dust of the Serengeti), this is a very generic feature of real systems.[3]

[2] Strictly speaking, the medium through which the flock moves must be inhomogeneous (and dynamical) in order to exert a dissipative force and thus define a preferred frame. I assume here that the medium is *statistically* homogeneous, so that its only effect is to provide such a damping, and that a rigid, static displacement of the flock results in no restoring force. It is in this sense that I assume the medium to be translation-invariant.

[3] This lack of Galilean invariance is appropriate if only the background dissipative medium through which the flock moves is static. The fully Galilean-invariant treatment of the coupled dynamics of a flock and a dynamical fluid medium through which it moves is discussed in [3, 4, 5, 6, 36].

As we shall see shortly, this *lack* of Galilean invariance *allows* terms in the hydrodynamic equations of birds that are *not* present in, e.g., the Navier–Stokes equations for a simple fluid, which *must* be Galilean invariant, due to the absence of a luminiferous ether.

The sole conservation law for flocks is conservation of birds: We do not allow birds to be born or die "on the wing."

In contrast to the Navier–Stokes equation, we here consider systems without momentum, due to the presence of the resistive background medium which breaks Galilean invariance.

4.2.2 What Are the Hydrodynamic Variables?

Having established the symmetries and conservation laws constraining our model, we need now to identify the hydrodynamic variables.

What do I mean by "hydrodynamic"? I mean variables that evolve slowly at long wavelength. More precisely, I mean variables whose evolution rate goes to zero as the length scale on which they are probed goes to infinity.

When one first hears this concept, it is natural to wonder why there should be *any* such variables. For example, in a flock consisting of millions of organisms, wouldn't one expect all variables to relax on some "microscopic" time scale, such as the mean time scale of interaction between neighboring birds?

This reasoning is *almost* correct: *Almost* any variable one can think of in any system with an enormous number of degrees of freedom will relax back, on a microscopic time scale, to a value determined by the local values of the few "slow" or "hydrodynamic" variables. But again, why should *any* variable be slow?

There are two possible reasons a variable will be *generically*[4] slow:

(1) conservation laws, and
(2) spontaneously broken continuous symmetries.

The density ρ is an example of a variable which is hydrodynamic for the first reason. Variables that are slow for the second reason are called "Goldstone modes." In our problem, the Goldstone mode associated with the spontaneous breaking of rotation invariance is $\delta\mathbf{v}_\perp$, which, as illustrated in Figure 4.2.1, is the component of the local velocity $\mathbf{v}(\mathbf{r},t)$ perpendicular to the mean value $\langle\mathbf{v}(\mathbf{r},t)\rangle \equiv v_0\hat{\mathbf{x}}_\parallel$ of that velocity averaged over the whole flock. Mathematically,

$$\mathbf{v}(\mathbf{r},t) = v_0\hat{\mathbf{x}}_\parallel + \delta\mathbf{v}(\mathbf{r},t) = (v_0 + \delta v_\parallel(\mathbf{r},t))\hat{\mathbf{x}}_\parallel + \delta\mathbf{v}_\perp(\mathbf{r},t)\,. \tag{4.2.1}$$

[4] A third type of slow variable, namely "critical" variables, only become slow at continuous phase transitions, which are not "generic," in the sense that they require one to fine-tune some parameter (often temperature) to put the system right at some phase transition boundary. I will not discuss such critical slow variables in this book; readers interested in this topic should see, e.g., [30].

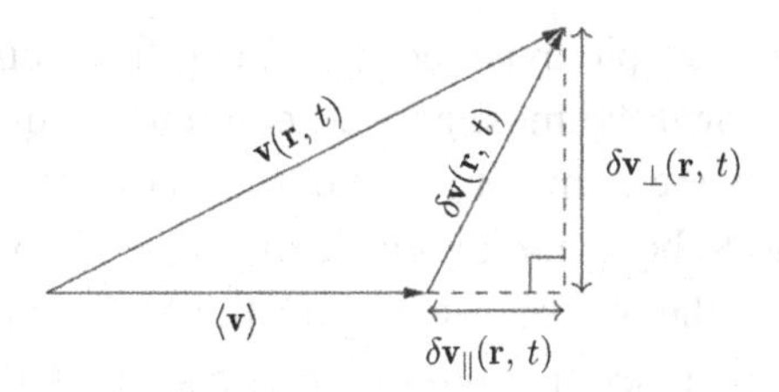

Fig. 4.2.1 Breaking the local velocity $\mathbf{v}$ into its mean value $\langle \mathbf{v}(\mathbf{r}, t)\rangle \equiv v_0 \hat{\mathbf{x}}_\parallel$, and the local fluctuation $\delta\mathbf{v}$ away from that average. The vector $\delta\mathbf{v}$ itself can then be further decomposed into its component $\delta v_\parallel$ parallel to the mean velocity, and its projection $\delta\mathbf{v}_\perp(\mathbf{r}, t)$. Reproduced from [7] by permission of Oxford University Press.

The vector $\delta\mathbf{v}_\perp$ is a Goldstone mode because a constant $\delta\mathbf{v}_\perp$ amounts just to a rotation, if $\delta v_\parallel$ relaxes back to the value $\delta v_\parallel = \sqrt{v_0^2 - |\delta\mathbf{v}_\perp|^2} - v_0$ required to keep $|\mathbf{v}| = v_0$. Since the system is rotation invariant, such a spatially uniform variation of $\delta\mathbf{v}_\perp$ can *never* relax; i.e., it has an infinite lifetime. Therefore, by continuity, if the field $\delta\mathbf{v}_\perp$ varies slowly in space, it must relax very slowly. More precisely, the relaxation time of such a distortion in $\delta\mathbf{v}_\perp$ must go to infinity as the length scale on which it varies does.

We've already seen an illustration of this for the pointer problem: As distance $r \to \infty$, the time t required for the field θ to equilibrate over that distance r diverges like r^2. This is because θ is the Goldstone mode, in the sense just described, for the pointer problem. The broken continuous symmetry with which θ is associated; that is, the symmetry that guarantees that θ will be "slow" at long wavelengths (which is precisely what I mean by "hydrodynamic") is just rotation invariance.

Note that although $\delta\mathbf{v}_\perp$ is a hydrodynamic variable, $\delta v_\parallel$ is not, since there is no symmetry that forbids the *speed* of the flockers from relaxing back to the preferred speed v_0 in a finite time, *even if* the fluctuation of the speed away from v_0 is spatially uniform. Nonetheless, because it is far simpler to see the consequences of rotation invariance for the full velocity field $\mathbf{v}$ than it is for the perpendicular component $\delta\mathbf{v}_\perp$ of $\mathbf{v}$ alone, I will initially formulate hydrodynamic equations of motion for the full velocity $\mathbf{v}$, even though this will include the nonhydrodynamic variable $\delta v_\parallel$. Once I have the equations of motion, it is then conceptually straightforward (although algebraically fairly monstrous, as we'll see) to eliminate $\delta v_\parallel$ and rewrite the equations of motion entirely in terms of the hydrodynamic variables $\delta\mathbf{v}_\perp$ and ρ.

4.2.3 Deriving the Hydrodynamic Equations

I will follow the historical precedent of the Navier–Stokes equation [29] by deriving our continuum, long-wavelength description of the flock *not* by explicitly coarse-graining the microscopic dynamics (a *very* difficult procedure in practice), but, rather, by writing down the most general continuum equations of motion for $\mathbf{v}$ and ρ consistent with the symmetries and conservation laws of the problem. This approach allows us to bury our ignorance in a few phenomenological parameters (e.g., the viscosity in the Navier–Stokes equation) whose numerical values will depend on the detailed microscopic rules of individual bird motion. What terms can be present in the equations of motion, however, should depend only on symmetries and conservation laws, and *not* on other aspects of the microscopic rules.

To reduce the complexity of our equations of motion still further, I will perform a spatio-temporal gradient expansion, and keep only the lowest-order terms in gradients and time derivatives of $\mathbf{v}$ and ρ. This is motivated and justified by our desire to consider *only* the long-distance, long-time properties of the flock. Higher-order terms in the gradient expansion are "irrelevant": They can lead to *finite* "renormalization" of the phenomenological parameters of the long-wavelength theory, but *cannot* change the type or scaling of the allowed terms.

So let's begin.

Rotation invariance implies that $\partial_t \mathbf{v}$, being a vector itself, must equal a sum of some other vectors. So, what vectors can we make out of $\mathbf{v}$, the scalar ρ, and the gradient operator?

Terms with No Gradients: the Mexican Hat

Well, the most obvious vector is $\mathbf{v}$ itself. More generally, we can multiply $\mathbf{v}$ by any scalar function of the speed $|\mathbf{v}|$ and the density ρ:

$$(\partial_t \mathbf{v})_1 = U(|\mathbf{v}|, \rho)\mathbf{v}. \tag{4.2.2}$$

This looks like a conventional frictional drag coefficient, except for the crucial difference that, while the drag coefficient of a passive system must always be negative (friction slows down a passive particle), for an active system, we'll allow $U > 0$, at least for small $|\mathbf{v}|$. This means the "flockers" will move; which is how we make our system active. We don't want U to be positive for all $|\mathbf{v}|$; if it was, the speed of the flock would grow without bound, which is clearly unphysical. So we will assume that U, plotted as a function of the speed $|\mathbf{v}|$, is positive for small speeds $|\mathbf{v}|$, and turns negative for large speeds $|\mathbf{v}|$. This leads to the acceleration in the direction of motion illustrated in Figure 4.2.2.

The effect of such a form for U is clearly the following: An initially slowly moving flock (or region thereof) will increase its speed until it reaches the speed v_0 at which $U(|\mathbf{v}|)$ vanishes. Likewise, a flock (or region thereof) that is moving faster than v_0 will slow down until its speed again reaches v_0. Thus the speed $|\mathbf{v}|$ is *not* a hydrodynamic variable; it relaxes back in a finite time $\tau = 1/(\partial U/\partial |\mathbf{v}|)_\rho$ to v_0.

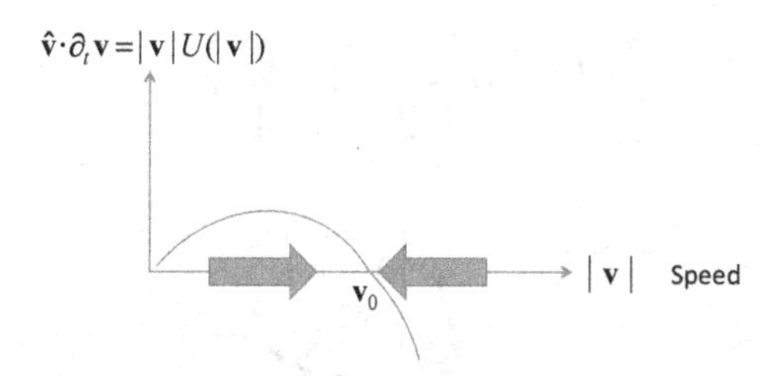

Fig. 4.2.2 Plot of the acceleration along $\mathbf{v}$ arising from the $\mathbf{v}U(|\mathbf{v}|)$ term in the equation of motion. This is the simplest qualitative form that can lead (for sufficiently small noise) to an ordered, moving flock. Reproduced from [7] by permission of Oxford University Press.

There are, obviously, infinitely many functions of the speed $|\mathbf{v}|$ and ρ that have the properties just described. Fortunately, since the speed always adjusts itself to be close to v_0, there prove to be only three parameters that we need to extract from U for our hydrodynamic theory: the steady-state speed v_0, and the derivatives $\partial U/\partial|\mathbf{v}|$ and $\partial U/\partial\rho$ evaluated at $|\mathbf{v}| = v_0$ and $\rho = \rho_0$, where ρ_0 is the mean density.

One popular choice for the function U (indeed, the choice Yuhai and I made in our early papers on this problem) is the "ψ^4" theory:

$$U = \alpha(\rho) - \beta(\rho)|\mathbf{v}|^2 . \tag{4.2.3}$$

The reason this is called "ψ^4" theory is that with this choice, we can write

$$U\mathbf{v} = -\partial V(\mathbf{v})/\partial \mathbf{v}, \tag{4.2.4}$$

where the "potential"

$$V(\mathbf{v}) = -\frac{1}{2}\alpha(\rho)|\mathbf{v}|^2 + \frac{1}{4}\beta(\rho)|\mathbf{v}|^4 \tag{4.2.5}$$

takes the form of the famous "Mexican hat," as shown in Figure 4.2.3. The dynamical effect of the $U(|\mathbf{v}|)$ term is then simply to make the velocity $\mathbf{v}$ evolve down towards the circular (or, in three dimensions, spherical) ring of identical minima at $|\mathbf{v}| = v_0 = \sqrt{\alpha/\beta}$.

This form is widely used in condensed matter physics and field theory. It is also the form most appropriate for studying the order–disorder transition. However, since here I'm just interested in the behavior of the flock deep inside its ordered phase, I will not restrict myself to this form. It is useful, however, to keep Figure 4.2.3 in mind, as we can construct a potential via (4.2.4) for any U, and, if that $|\mathbf{v}|U$ has the form plotted in Figure 4.2.2, the associated V will look qualitatively like Figure 4.2.3. And we can imagine the flock velocity $\mathbf{v}$ evolving by seeking the minimum of this potential, or, more precisely, the ring of minima of this potential.

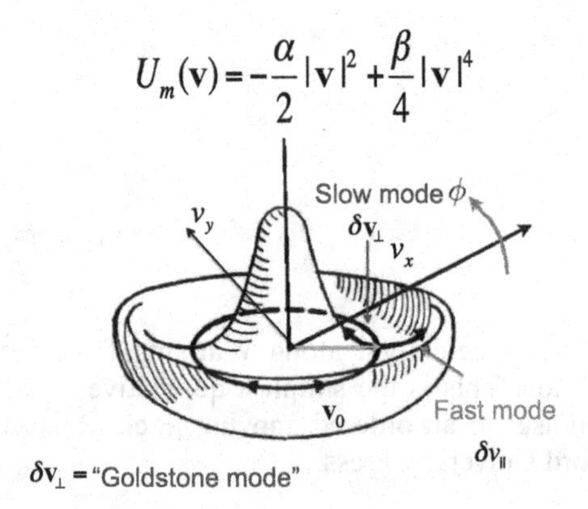

Fig. 4.2.3 The "Mexican hat" potential. Note the circular ring of minima. Fluctuations $\delta \mathbf{v}_\perp$ of the velocity that move it around this ring are "Goldstone modes," which experience no restoring force from this "potential." In contrast, fluctuations $\delta v_\parallel$ that carry one up the brim of the hat, or towards the crown (i.e., radially), are "fast"; i.e., they will relax quickly due to this potential, which simply reflects the existence of a preferred speed for the flockers' motion through the frictional medium. Reproduced from [7] by permission of Oxford University Press.

Indeed, the "spontaneous symmetry breaking" of a flock can be thought of as the system's settling into one of these degenerate minima.

Note that if we use the expansion (4.2.1) around the mean velocity in (4.2.2), we get no linear term in $\delta \mathbf{v}$ in the equation of motion for $\delta \mathbf{v}_\perp$. To see this, note that the speed

$$|\mathbf{v}| = \sqrt{(v_0 + \delta v_\parallel)^2 + |\delta \mathbf{v}_\perp(\mathbf{r}, t)|^2} = v_0 + \delta v_\parallel + \mathcal{O}(|\delta \mathbf{v}_\perp(\mathbf{r}, t)|^2) . \tag{4.2.6}$$

Using this in the projection of (4.2.2) perpendicular to the mean velocity, we obtain

$$\begin{aligned}(\partial_t \delta \mathbf{v}_\perp)_1 &= U(v_0 + \delta v_\parallel, \rho_0)\delta \mathbf{v}_\perp + \mathcal{O}(|\delta \mathbf{v}_\perp|^3)\\ &= U(v_0, \rho_0)\delta \mathbf{v}_\perp + (\partial U/\partial|\mathbf{v}|)\,|_{v_0,\rho_0} \delta v_\parallel \delta \mathbf{v}_\perp + (\partial U/\partial \rho)\,|_{v_0,\rho_0} \delta\rho \delta \mathbf{v}_\perp + \mathcal{O}(|\delta \mathbf{v}_\perp|^3)\\ &= U(v_0, \rho_0)\delta \mathbf{v}_\perp + \mathcal{O}(|\delta \mathbf{v}_\perp|^3, \delta v_\parallel \delta \mathbf{v}_\perp, \delta\rho \delta \mathbf{v}_\perp) = \mathcal{O}(|\delta \mathbf{v}_\perp|^3, \delta v_\parallel \delta \mathbf{v}_\perp, \delta\rho \delta \mathbf{v}_\perp) ,\end{aligned} \tag{4.2.7}$$

where in the last equality I have used the fact that $U(v_0, \rho_0) = 0$, which is just a consequence of the definition of v_0 as the steady-state speed.

This vanishing of all terms linear in $\delta \mathbf{v}_\perp$ in this term is no coincidence; rather, it is a consequence of the fact that $\delta \mathbf{v}_\perp$ is the Goldstone mode for this problem, so any term in its time derivative *must* have some spatial gradients, so that it vanishes when $\delta \mathbf{v}_\perp$ is spatially uniform.

This means if we want to include terms that depend on $\delta\mathbf{v}_\perp$ (which we certainly do!), then we need to look at terms involving the gradient operator ∇.

So let's look at the following.

Terms with One Gradient: the Convective Nonlinearity, Pressure Forces, and Their Weird Sisters

Consider combinations of velocities and one gradient operator. We need at least two velocities. Why? Well, can we make anything with a single gradient operator and a single velocity that transforms like a vector? The answer is no, as can most easily be seen using the Einstein summation convention: If we write

$$(\partial_t v_i)_{\text{trial}} = \text{constant} \times \partial_j v_k, \tag{4.2.8}$$

there's no choice of the indices j and k on the right-hand side that will make this equation make sense. If we take $j = i$, we have an extra index k running loose on the right-hand side. If we try to get rid of it by instead taking $j = k$, then we've made the right-hand side into a scalar (in fact, into $\nabla \cdot \mathbf{v}$), which we can't equate to a vector. So we need at least two velocities to combine with our gradient operator.

So let's try

$$(\partial_t v_i)_2 = \text{constant} \times v_\ell \partial_j v_k. \tag{4.2.9}$$

This can be made to work with a suitable choice of the three indices ℓ, j, and k. What we need to do is make two of them equal, so that they are summed over by the Einstein summation convention. This leaves one free index, which we must choose to be i, the free index on the right-hand side. Basically, we use the Einstein summation convention to "eat" the extra indices on the right-hand side. In fact, there are three ways to make this work:

(i) Take $k = i$ and $\ell = j$. The right-hand side has only one free index, so it's a vector, and it's the same free index as on the left-hand side, so the equation makes sense in the Einstein summation convention. We can write this term as

$$(\partial_t v_i)_{2.1} = -\lambda_1 \mathbf{v} \cdot \nabla v_i, \tag{4.2.10}$$

where I've arbitrarily defined the "constant" in (4.2.9) to be $-\lambda_1$. (The minus sign is chosen to make the resulting equation look as much like the Navier–Stokes equation as possible, as you'll see). Rewriting this in full glorious vector notation,

$$(\partial_t \mathbf{v})_{2.1} = -\lambda_1 \mathbf{v} \cdot \nabla \mathbf{v}. \tag{4.2.11}$$

(ii) Take $k = j$ and $\ell = i$. Once again, the right-hand side has only one free index, so it's a vector, and it's the same free index as on the left-hand side, so the equation makes sense in the Einstein summation convention. We can write this term as

$$(\partial_t v_i)_{2.2} = -\lambda_2 v_i \nabla \cdot \mathbf{v}, \tag{4.2.12}$$

where I've called the "constant" $-\lambda_2$. In vector notation,

$$(\partial_t \mathbf{v})_{2.2} = -\lambda_2 \mathbf{v}(\nabla \cdot \mathbf{v}). \tag{4.2.13}$$

(iii) Take $j = i$ and $\ell = k$. This also makes sense in the Einstein summation convention. We can write this term as

$$(\partial_t v_i)_{2.3} = -2\lambda_3 v_j \partial_i v_j = -\lambda_3 \partial_i(|\mathbf{v}|^2), \tag{4.2.14}$$

where I've introduced the factor of 2 in this definition of λ_3 for convenience in writing the second equality. In vector notation,

$$(\partial_t \mathbf{v})_{2.3} = -\lambda_3 \nabla(|\mathbf{v}|^2). \tag{4.2.15}$$

There are also combinations of one gradient, $\mathbf{v}$, and the density ρ that do not have the structure of (4.2.9), particularly if we allow more than one power of $\mathbf{v}$. Note that there is no reason we should not include such higher powers: Because of the spontaneous ordering, $|\mathbf{v}|$ itself is not small (in fact, it's close to v_0^2, which need not be small). We *do* intend to expand in powers of the fluctuation $\delta\mathbf{v}$ of $\mathbf{v}$ away from its mean value $v_0\hat{\mathbf{x}}$, but that is *not* the same as an expansion in powers of $\mathbf{v}$, because of this spontaneous order.

Fortunately, it turns out that we can incorporate all such one-gradient terms into five terms, namely the following.

(I) A pressure term

$$(\partial_t \mathbf{v})_{\text{pressure}} = -\nabla P(|\mathbf{v}|, \rho). \tag{4.2.16}$$

Those of you familiar with the Navier–Stokes equation will recognize this as exactly the form of the pressure term in that equation, except for the peculiarity here that the pressure can depend not only on the density ρ, but also on the speed $|\mathbf{v}|$. Such dependence is forbidden in the Navier–Stokes equation by Galilean invariance; since we don't have Galilean invariance in our dry active fluid, this dependence is allowed, and hence will, in general, be present.

(II–IV) density and speed dependences of the $\lambda_{1,2,3}$ terms.

(V) An anisotropic pressure term P_2 of the form

$$(\partial_t \mathbf{v})_{\text{aniso pressure}} = -\mathbf{v}(\mathbf{v} \cdot \nabla) P_2(|\mathbf{v}|, \rho). \tag{4.2.17}$$

To see that these five terms exhaust all possibilities, consider, for example, the term

$$(\partial_t v_i)_3 = v_\ell v_k \partial_j f(|\mathbf{v}|, \rho). \tag{4.2.18}$$

Again choosing the indices so that two of them are eaten by the Einstein summation convention, while the remaining one is i, we see that there are two ways to do this.

(i) $j = i$, $k = \ell$. This choice gives

$$(\partial_t v_i)_{3.1} = |\mathbf{v}|^2 \nabla f = \nabla(|\mathbf{v}|^2 f) - f \nabla v^2 \equiv -\nabla \delta P - \delta\lambda_2 \nabla v^2 , \tag{4.2.19}$$

where I've defined a contribution $\delta P(|\mathbf{v}|, \rho) \equiv f(|\mathbf{v}|, \rho)|\mathbf{v}|^2$ to the "pressure" defined previously, and a contribution $\delta\lambda_2(|\mathbf{v}|, \rho) \equiv f(|\mathbf{v}|, \rho)$ to $\lambda_2(|\mathbf{v}|, \rho)$.

(ii) $j = k$, $\ell = i$ (note that $j = \ell$, $k = i$ gives the same term):

$$(\partial_t v_i)_{3.2} = v_i v_j \partial_j f(|\mathbf{v}|, \rho), \tag{4.2.20}$$

which in vector form is precisely the P_2 term (4.2.17) with $P_2(|\mathbf{v}|, \rho) = -f(|\mathbf{v}|, \rho)$.

The $\lambda_{1,2,3}$, P, and P_2 terms can between them incorporate every "relevant" (i.e., nonnegligible) term that involves one gradient and arbitrary powers on $\mathbf{v}$ and ρ. To see this, consider, for example, the following term with four velocities and one gradient:

$$(\partial_t v_i)_{\text{trial2}} = \lambda_4 v_\ell \partial_j (v_n v_m v_k). \tag{4.2.21}$$

We need to "eat" four of the five indices on the right-hand side, and set the remaining one equal to i. Let's consider the term we get if we choose $m = n$, $\ell = j$, and $k = i$. This gives

$$(\partial_t v_i)_4 = \lambda_4 \mathbf{v} \cdot \nabla \left(v_i |\mathbf{v}|^2\right) = \lambda_4 \left[|\mathbf{v}|^2 \mathbf{v} \cdot \nabla(v_i) + v_i \mathbf{v} \cdot \nabla |\mathbf{v}|^2\right] . \tag{4.2.22}$$

The first term on the right-hand side is immediately recognizable as a contribution to λ_1 proportional to $|\mathbf{v}|^2$, while the second is a contribution to P_2.

It is straightforward to check that all terms that involve only one gradient can likewise be incorporated into speed $|\mathbf{v}|$ and density ρ-dependent corrections to one of the five aforementioned quantities isotropic pressure P, anisotropic pressure P_2, and $\lambda_{1,2,3}$.

4.2.4 Terms with Two Gradients: Viscosities Both Isotropic and Anisotropic

Let's now consider terms with two gradients. One might think that we need not keep such terms, since they have more gradients than the one-gradient terms we've just considered. However, it turns out, as we'll see shortly, that none of the one-gradient terms we've just considered damps out velocity fluctuations to linear order in the velocity fluctuations $\delta\mathbf{v}$, which prove to be small. Instead, they just lead to propagation without dissipation. Therefore, if we do not include any two-gradient terms, our theory would (erroneously) predict that there would be no damping of the fluctuations induced by the noise, which would therefore grow without bound over time. To prevent such an unphysical result, we need to go to higher-order gradient terms. Second order proves, again with hindsight, to be sufficient.

So what can we make with two gradients that transforms like a vector? As before, let's proceed by writing out possible terms in Einstein summation convention, and figure out how the indices can get eaten. So let's start with terms with one velocity and two gradients. Generically, this can be written:

$$(\partial_t v_i)_{\text{trial3}} = \text{constant} \times \partial_j \partial_k v_\ell. \tag{4.2.23}$$

By now, you should be familiar enough with how this goes to see that there are two menu options for "index eating":

(i) $j = k$ and $\ell = i$. This gives

$$(\partial_t v_i)_{4.1} = D_T \nabla^2 v_i \tag{4.2.24}$$

or, in vector notation,

$$(\partial_t \mathbf{v})_{4.1} = D_T \nabla^2 \mathbf{v}. \tag{4.2.25}$$

(ii) $j = i$ and $\ell = k$ (or, equivalently, $k = i$ and $\ell = j$), which gives

$$(\partial_t v_i)_{4.2} = D_B \partial_i \partial_j v_j \tag{4.2.26}$$

or, in vector notation,

$$(\partial_t \mathbf{v})_{4.2} = D_B \nabla(\nabla \cdot \mathbf{v}). \tag{4.2.27}$$

That's it for terms with two spatial gradients and one velocity. These two terms also occur in the Navier–Stokes equations, where the coefficients D_T and D_B are usually denoted as ν_T and ν_B, and are called the shear and bulk viscosities, respectively.

Can we make terms with more velocities and two derivatives? Absolutely; indeed, an overabundance of them. We can, however, tremendously reduce the number of possibilities by noting (as we did for the one-gradient terms above) that when we expand about the state of uniform motion via (4.2.1), any velocity that a gradient acts on can be replaced by $\delta\mathbf{v}$, since $v_0\hat{\mathbf{x}}$ is a constant. Since $\delta\mathbf{v}$ is small, the dominant terms will be those with only one $\delta\mathbf{v}$. We can therefore restrict ourselves to terms with only one full velocity $\mathbf{v}$ acted upon by the two derivatives. Therefore, all possible "relevant" terms involving two gradients and an *arbitrary* number of velocities can be written

$$(\partial_t v_i)_{4\,\text{general}} = \text{constant} \times [v_p v_n \cdots v_s v_u]\partial_j \partial_k v_\ell, \tag{4.2.28}$$

where $[v_p v_n \cdots v_s v_u]$ is a product of an even number $2m$ of components of $\mathbf{v}$. This number must be even, so that there are an odd number of indices altogether on the right-hand side. This is necessary to allow us to pair all but one of them off, thereby producing a vector. There are now five ways we can do this pairing off.

(i) Pair all of the v's to the left of the derivatives off with themselves, and set $j = k$ and $\ell = i$. This gives

$$(\partial_t v_i)_{4.1} = \text{constant} \times |\mathbf{v}|^{2m}\partial_j \partial_j v_i = \text{constant} \times |\mathbf{v}|^{2m}\nabla^2 v_i \tag{4.2.29}$$

or, in vector notation,

$$(\partial_t \mathbf{v})_{4.1} = \text{constant} \times |\mathbf{v}|^{2m}\nabla^2 \mathbf{v}. \tag{4.2.30}$$

We can absorb this into a contribution to the "shear viscosity" D_T proportional to $|\mathbf{v}|^{2m}$. We can therefore incorporate all possible such terms, up to arbitrary even powers of $|\mathbf{v}|$, by making D_T a suitably chosen function of $|\mathbf{v}|$. We can generalize this even further by making D_T depend on the density ρ as well.

(ii) Pair all of the v's to the left of the derivatives off with themselves, and set $j = i$ and $\ell = k$ (or, equivalently, $k = i$ and $\ell = j$). This gives

$$(\partial_t v_i)_{4.2} = \text{constant} \times |\mathbf{v}|^{2m}\partial_i \partial_j v_j \tag{4.2.31}$$

or, in vector notation,

$$(\partial_t \mathbf{v})_{4.2} = \text{constant} \times |\mathbf{v}|^{2m}\nabla(\nabla \cdot \mathbf{v}), \tag{4.2.32}$$

which we can absorb into a contribution to the "bulk viscosity" D_B proportional to $|\mathbf{v}|^{2m}$. We can therefore incorporate all possible such terms, up to arbitrary even powers of $|\mathbf{v}|$, by making D_B a suitably chosen function of $|\mathbf{v}|$. As for D_T, we can generalize this even further by making D_B depend on the density ρ as well.

(iii) Pair all but two of the v's to the left of the gradients off with themselves, and pair the remaining two with the gradients; this forces $\ell = i$ (since there are no other free indices left). This gives

$$(\partial_t v_i)_{4.3} = \text{constant} \times |\mathbf{v}|^{2m-2} v_j v_k \partial_j \partial_k v_i \tag{4.2.33}$$

or, in vector notation,

$$(\partial_t \mathbf{v})_{4.3} = \text{constant} \times |\mathbf{v}|^{2m-2}(\mathbf{v} \cdot \nabla)^2 \mathbf{v}. \tag{4.2.34}$$

This is the first genuinely new term. I'll sum up all such terms into a function that I'll call $D_2(|\mathbf{v}|, \rho)$ of the speed $|\mathbf{v}|$ and the density ρ times the combination $(\mathbf{v} \cdot \nabla)^2 \mathbf{v}$. This term makes anisotropic diffusion possible: We can now have a different diffusion constant along the direction of flock motion than perpendicular to it, as we would expect, since we've broken (or, rather, the flock has broken) the symmetry between the direction of flock motion and directions perpendicular to it.

(iv) Pair all but two of the v's to the left of the gradients off with themselves, and pair one of the other with one of the gradients, and the other with the velocity to the *right* of the gradients; this forces one of the gradient indices to be i (since there are no other free indices left). This gives

$$(\partial_t v_i)_{4.4} = \text{constant} \times |\mathbf{v}|^{2m-2} v_j v_k \partial_i \partial_k v_j. \tag{4.2.35}$$

This contribution proves to be negligible compared with those we've already kept. To see this, consider the implied sum on j in (4.2.35). One term in this sum is that with index $j = \parallel$; i.e., the Cartesian component along the mean direction of motion. We can replace $v_\parallel$ with $\delta v_\parallel$ to the right of the gradient in (4.2.35), since the mean velocity contribution to this term $v_0 \hat{\mathbf{x}}$ is a constant, and hence has zero gradient. But, as we noted earlier, $\delta v_\parallel \ll |\mathbf{v}_\perp|$ since $\delta v_\parallel$ is not a Goldstone mode, so this contribution from the sum on j to this term is negligible compared with the two-gradient, one-$\mathbf{v}_\perp$ terms we found above.

The other terms in the sum on j will be proportional to two $\mathbf{v}_\perp$'s (one to the left of the gradient, and one to the right), and so will be negligible compared to the "one-$\mathbf{v}_\perp$, two-gradient" terms found previously if $\mathbf{v}_\perp$ is small, as it will be in an ordered state. So those terms in the sum are negligible as well. So the entire term (4.2.35) is negligible.

(v) Finally, we can pair all but *four* of the v's to the left of the gradients among themselves, set one of the remaining indices on those v's equal to i, and pair off the remaining three velocities with the two gradients and the velocity to the right of the gradient. This gives

$$(\partial_t v_i)_{4.5} = \text{constant} \times |\mathbf{v}|^{2m-4} v_i v_j v_k v_l \partial_j \partial_k v_\ell. \tag{4.2.36}$$

The sum on ℓ in this term can be shown to be negligible by an argument almost identical to the one we just used for the previous contribution (4.2.35). So we'll drop this as well.

4.2.5 And it Stirs Itself: the Noise Term

The only other term we need to include is a random noise term:

$$(\partial_t \mathbf{v})_{noise} = \mathbf{f}(\mathbf{r}, t). \tag{4.2.37}$$

It is assumed to be Gaussian with white noise correlations:

$$\langle f_i(\mathbf{r}, t) f_j(\mathbf{r}', t')\rangle = 2D\delta_{ij}\delta^d(\mathbf{r} - \mathbf{r}')\delta(t - t'), \tag{4.2.38}$$

where the "noise strength" D is a constant parameter of the system, and i, j denote Cartesian components.

Using the dynamical RG, one can show that small departures of the noise statistics from purely Gaussian have no effect on the long-distance physics.

4.2.6 The Full Equations of Motion

That's all, folks! Any other terms you construct will have more gradients, and so will be negligible at long distances compared with the terms we've already found.

Putting all of these terms together gives the equation of motion for $\mathbf{v}$:

$$\begin{aligned} &\partial_t \mathbf{v} + \lambda_1(\mathbf{v} \cdot \nabla)\mathbf{v} + \lambda_2(\nabla \cdot \mathbf{v})\mathbf{v} + \lambda_3 \nabla(|\mathbf{v}|^2) = \\ &\quad U(|\mathbf{v}|, \rho)\mathbf{v} - \nabla P - \mathbf{v}(\mathbf{v} \cdot \nabla P_2) + D_B \nabla(\nabla \cdot \mathbf{v}) + D_T \nabla^2 \mathbf{v} + D_2(\mathbf{v} \cdot \nabla)^2 \mathbf{v} + \mathbf{f}. \end{aligned} \tag{4.2.39}$$

Keep in mind that this equation is (even!) more complicated than it looks, because all of the parameters $\lambda_i (i = 1 \rightarrow 3)$, U, the "damping coefficients" $D_{B,T,2}$, the "isotropic pressure" $P(\rho, v)$, and the "anisotropic pressure" $P_2(\rho, v)$ are functions of the density ρ and the magnitude $v \equiv |\mathbf{v}|$ of the local velocity.

To close these equations of motion, we also need one for the density. The final equation (4.2.40) is just conservation of bird number (we don't allow our birds to reproduce or die on the wing). That equation is just the familiar "continuity equation":

$$\frac{\partial \rho}{\partial t} + \nabla \cdot (\mathbf{v}\rho) = 0. \tag{4.2.40}$$

The pair of equations (4.2.39) and (4.2.40) are, I'm embarrassed to confess, known in the field as "the Toner–Tu equations" (although Yuhai and I can obviously make no claim to priority for (4.2.40)!).

With these equations of motion (4.2.39) and (4.2.40) in hand, we can now use them to figure out how flocks actually behave, and in particular why they can order even in $d = 2$.

4.3 Expanding the Equations of Motion to "Relevant" Nonlinear Order

The hydrodynamic model embodied in equations (4.2.39) and (4.2.40) is equally valid in both the "disordered " (i.e., nonmoving) state, in which $U(|\mathbf{v}|, \rho)$ is negative for all $|\mathbf{v}|$, and in the moving or "ferromagnetically ordered" state, in which $|\mathbf{v}|U(|\mathbf{v}|)$ looks like Figure 4.2.2, with a positive region at small $|\mathbf{v}|$, which allows for the possibility of a moving state. In this section I'll focus on the "ferromagnetically ordered," broken-symmetry phase, and specifically on the question of whether fluctuations around the symmetry broken-ground state destroy the ordered phase (as in the analogous phase of the two-dimensional XY model). When $|\mathbf{v}|U(|\mathbf{v}|)$ looks like Figure 4.2.2, we can expand the velocity field as in (4.2.1), which I rewrite here for convenience:

$$\mathbf{v}(\mathbf{r}, t) = v_0\hat{x}_\parallel + \delta\mathbf{v}(\mathbf{r}, t) = (v_0 + \delta v_\parallel)\hat{x}_\parallel + \delta\mathbf{v}_\perp(\mathbf{r}, t), \tag{4.3.1}$$

where I remind you that $v_0\hat{x}_\parallel$ is the spontaneous average value of $\mathbf{v}$ in the ordered phase in the absence of fluctuations, whose magnitude v_0 is just that at which $U(|\mathbf{v}|, \rho) = 0$.

As I've discussed above, the fluctuation $\delta v_\parallel$ of the component of $\mathbf{v}$ along the mean direction of flock motion $\hat{x}_\parallel$ away from its preferred value v_0 is *not* a hydrodynamic variable of the system; rather, it relaxes back quickly to a value determined by the true hydrodynamic variables ρ and $\delta\mathbf{v}_\perp$. It therefore behooves us to eliminate it by solving for it in terms of those variables. Doing so is rather tricky – indeed, Yuhai and I got this slightly wrong in our earlier work on this problem [33, 34, 35, 36] – so I will go through the argument rather carefully and in some detail here. For further details, see [37].

Since we know fluctuations in the speed (i.e., the magnitude $|\mathbf{v}|$ of $\mathbf{v}$) will be fast, it is useful to turn our equation of motion (4.2.39) for the velocity into an equation of motion for that speed. This can be done by taking the dot product of both sides of equation (4.2.39) with $\mathbf{v}$ itself, which gives:

$$\begin{aligned}\frac{1}{2}&\left(\partial_t|\mathbf{v}|^2 + (\lambda_1 + 2\lambda_3)(\mathbf{v}\cdot\nabla)|\mathbf{v}|^2\right) + \lambda_2(\nabla\cdot\mathbf{v})|\mathbf{v}|^2\\ &= U(|\mathbf{v}|, \rho)|\mathbf{v}|^2 - \mathbf{v}\cdot\nabla P - |\mathbf{v}|^2\mathbf{v}\cdot\nabla P_2 + D_B\mathbf{v}\cdot\nabla(\nabla\cdot\mathbf{v}) + D_T\mathbf{v}\cdot\nabla^2\mathbf{v}\\ &\quad + D_2\mathbf{v}\cdot\left((\mathbf{v}\cdot\nabla)^2\mathbf{v}\right) + \mathbf{v}\cdot\mathbf{f}.\end{aligned} \tag{4.3.2}$$

In this hydrodynamic approach, we are interested only in fluctuations $\delta\mathbf{v}(\mathbf{r}, t)$ and $\delta\rho(\mathbf{r}, t)$ that vary slowly in space and time. (Indeed, the hydrodynamic equations (4.2.39) and (4.2.40) are valid only in this limit.) Hence, terms involving space and time derivatives of $\delta\mathbf{v}(\mathbf{r}, t)$ and $\delta\rho(\mathbf{r}, t)$ are always negligible, in the hydrodynamic limit, compared with terms involving the same number of powers of fields without any time or space derivatives.

Furthermore, the fluctuations $\delta\mathbf{v}(\mathbf{r},t)$ and $\delta\rho(\mathbf{r},t)$ can themselves be shown to be small in the long-wavelength limit. Hence, we need only keep terms in equation (4.3.2) up to linear order in $\delta\mathbf{v}(\mathbf{r},t)$ and $\delta\rho(\mathbf{r},t)$. The $\mathbf{v}\cdot\mathbf{f}$ term can likewise be dropped, since it only leads to a term of order $\mathbf{v}_\perp f_\parallel$ in the $\mathbf{v}_\perp$ equation of motion, which is negligible (since $\mathbf{v}_\perp$ is small) relative to the $\mathbf{f}_\perp$ term already there.

These observations can be used to eliminate many of the terms in equation (4.3.2), and solve for U; the solution is:

$$U = \lambda_2\nabla\cdot\mathbf{v} + \mathbf{v}\cdot\nabla P_2 + \frac{\sigma_1}{v_0}\partial_\parallel\delta\rho + \frac{1}{2v_0}\left(\partial_t + \gamma_2\partial_\parallel\right)\delta v_\parallel, \tag{4.3.3}$$

where I've defined

$$\gamma_2 \equiv (\lambda_1 + 2\lambda_3)v_0 \tag{4.3.4}$$

and

$$\sigma_1 \equiv \left(\frac{\partial P}{\partial\rho}\right)_0 . \tag{4.3.5}$$

Here and hereafter, superscripts or subscripts 0 denote functions of ρ and $|\mathbf{v}|$ evaluated at the steady-state values $\rho = \rho_0$ and $|\mathbf{v}| = v_0$.

Inserting the expression (4.3.3) for U back into equation (4.2.39), I find that P_2 and λ_2 cancel out of the $\mathbf{v}$ equation of motion, leaving

$$\begin{aligned}
&\partial_t\mathbf{v} + \lambda_1(\mathbf{v}\cdot\nabla)\mathbf{v} + \lambda_3\nabla(|\mathbf{v}|^2)\\
&= \frac{\sigma_1}{v_0}\mathbf{v}(\partial_\parallel\delta\rho) - \nabla P + D_1\nabla(\nabla\cdot\mathbf{v}) + D_T\nabla^2\mathbf{v} + D_2(\mathbf{v}\cdot\nabla)^2\mathbf{v}\\
&\quad + \left[\frac{1}{2v_0}\left(\partial_t + \gamma_2\partial_\parallel\right)\delta v_\parallel\right]\mathbf{v} + \mathbf{f}.
\end{aligned} \tag{4.3.6}$$

This can be made into an equation of motion for $\mathbf{v}_\perp$ involving only $\mathbf{v}_\perp(\mathbf{r},t)$ and $\delta\rho(\mathbf{r},t)$ by projecting perpendicular to the direction of mean flock motion $\hat{\mathbf{x}}_\parallel$, and eliminating $\delta v_\parallel$ using equation (4.3.3) and the expansion

$$U \approx -\Gamma_1\left(\delta v_\parallel + \frac{|\mathbf{v}_\perp|^2}{2v_0}\right) - \Gamma_2\delta\rho, \tag{4.3.7}$$

where I've defined

$$\Gamma_1 \equiv -\left(\frac{\partial U}{\partial|\mathbf{v}|}\right)_\rho^0, \quad \Gamma_2 \equiv -\left(\frac{\partial U}{\partial\rho}\right)_{|\mathbf{v}|}^0 . \tag{4.3.8}$$

I've also used the expansion (4.3.1) for the velocity in terms of the fluctuations $\delta v_\parallel$ and $\delta\mathbf{v}_\perp$ to write

$$|\mathbf{v}| = v_0 + \delta v_\parallel + \frac{|\mathbf{v}_\perp|^2}{2v_0} + O(\delta v_\parallel^2, |\mathbf{v}_\perp|^4), \tag{4.3.9}$$

and kept only terms that an RG analysis shows to be relevant in the long-wavelength limit. Inserting (4.3.9) into (4.3.7) gives:

$$-\Gamma_1\left(\delta v_\parallel + \frac{|\mathbf{v}_\perp|^2}{2v_0}\right) - \Gamma_2\delta\rho$$
$$= \lambda_2\nabla_\perp\cdot\mathbf{v}_\perp + \lambda_2\partial_\parallel\delta v_\parallel + \frac{(\mu_1 v_0^2+\sigma_1)}{v_0}\partial_\parallel\delta\rho + \frac{1}{2v_0}\left(\partial_t + \gamma_2\partial_\parallel\right)\delta v_\parallel, \tag{4.3.10}$$

where I've kept only linear terms on the right-hand side of this equation, since the nonlinear terms are at least of order derivatives of $|\mathbf{v}_\perp|^2$, and hence negligible, in the hydrodynamic limit, relative to the $|\mathbf{v}_\perp|^2$ term explicitly displayed on the left-hand side. In (4.3.10), I have defined

$$\mu_1 \equiv \left(\frac{\partial P_2}{\partial\rho}\right)_0 . \tag{4.3.11}$$

Equation (4.3.10) can be solved iteratively for $\delta v_\parallel$ in terms of $\mathbf{v}_\perp$, $\delta\rho$, and its derivatives. To lowest (zeroth) order in derivatives, $\delta v_\parallel \approx -\frac{\Gamma_2}{\Gamma_1}\delta\rho$. Inserting this approximate expression for $\delta v_\parallel$ into equation (4.3.10) everywhere $\delta v_\parallel$ appears on the right-hand side of that equation gives $\delta v_\parallel$ to first order in derivatives:

$$\delta v_\parallel \approx -\frac{\Gamma_2}{\Gamma_1}\left(\delta\rho - \frac{1}{v_0\Gamma_1}\partial_t\delta\rho + \frac{\lambda_4\partial_\parallel\delta\rho}{\Gamma_2}\right) - \frac{\lambda_2}{\Gamma_1}\nabla_\perp\cdot\mathbf{v}_\perp - \frac{|\mathbf{v}_\perp|^2}{2v_0}, \tag{4.3.12}$$

where I've defined

$$\lambda_4 \equiv \frac{(\mu_1 v_0^2+\sigma_1)}{v_0} - \frac{\Gamma_2}{\Gamma_1}\left(\lambda_2 + \frac{\gamma_2}{v_0}\right) = \frac{(\mu_1 v_0^2+\sigma_1)}{v_0} - \frac{\Gamma_2}{\Gamma_1}\left(\lambda_1+\lambda_2+2\lambda_3\right). \tag{4.3.13}$$

In deriving the second equality in (4.3.13), I've used the definition (4.3.4) of γ_2.

Inserting (4.3.1), (4.3.9), and (4.3.12) into the equation of motion (4.3.6) for $\mathbf{v}$, and projecting that equation perpendicular to the mean direction of flock motion $\hat{x}_\parallel$ gives, neglecting "irrelevant" terms:

$$\partial_t\mathbf{v}_\perp + \gamma\partial_\parallel\mathbf{v}_\perp + \lambda_1^0\left(\mathbf{v}_\perp\cdot\nabla_\perp\right)\mathbf{v}_\perp$$
$$= -g_1\delta\rho\partial_\parallel\mathbf{v}_\perp - g_2\mathbf{v}_\perp\partial_\parallel\delta\rho - \frac{c_0^2}{\rho_0}\nabla_\perp\delta\rho - g_3\nabla_\perp(\delta\rho^2)$$
$$+ D_B\nabla_\perp\left(\nabla_\perp\cdot\mathbf{v}_\perp\right) + D_T\nabla_\perp^2\mathbf{v}_\perp + D_\parallel\partial_\parallel^2\mathbf{v}_\perp + \nu_t\partial_t\nabla_\perp\delta\rho + \nu_\parallel\partial_\parallel\nabla_\perp\delta\rho + \mathbf{f}_\perp, \tag{4.3.14}$$

where I've defined

$$D_B \equiv D_1 + \frac{2v_0\lambda_3^0\lambda_2^0}{\Gamma_1}, \tag{4.3.15}$$

$$D_{\parallel} \equiv D_T + D_2 v_0^2, \tag{4.3.16}$$

$$\gamma \equiv \lambda_1^0 v_0, \tag{4.3.17}$$

$$g_1 \equiv v_0\left(\frac{\partial\lambda_1}{\partial\rho}\right)_0 - \frac{\Gamma_2\lambda_1^0}{\Gamma_1}, \tag{4.3.18}$$

$$g_2 \equiv \frac{\Gamma_2\gamma_2^0}{\Gamma_1 v_0} - \frac{\sigma_1}{v_0}, \tag{4.3.19}$$

$$g_3 \equiv \sigma_2 + \left(\frac{\Gamma_2}{\Gamma_1}\right)^2\lambda_3^0 - \left(\frac{\partial\lambda_3}{\partial\rho}\right)_0\frac{\Gamma_2 v_0}{\Gamma_1}, \tag{4.3.20}$$

$$c_0^2 \equiv \rho_0\sigma_1 - \frac{2\rho_0 v_0\lambda_3^0\Gamma_2}{\Gamma_1}, \tag{4.3.21}$$

$$\nu_t \equiv -\frac{2\Gamma_2\lambda_3^0}{\Gamma_1^2}, \tag{4.3.22}$$

and

$$\nu_{\parallel} \equiv \frac{2v_0\lambda_3^0\lambda_4^0}{\Gamma_1} + \frac{\Gamma_2 D_1}{\Gamma_1}. \tag{4.3.23}$$

Using (4.3.1) and (4.3.9) in the equation of motion (4.2.40) for ρ gives, again neglecting irrelevant terms:

$$\partial_t\delta\rho + \rho_o\nabla_\perp\cdot\mathbf{v}_\perp + w_1\nabla_\perp\cdot(\mathbf{v}_\perp\delta\rho) + v_2\partial_{\parallel}\delta\rho$$
$$= D_{\rho_\parallel}\partial_{\parallel}^2\delta\rho + D_{\rho\perp}\nabla_\perp^2\delta\rho + D_{\rho v}\partial_{\parallel}\left(\nabla_\perp\cdot\mathbf{v}_\perp\right) + \phi\partial_t\partial_{\parallel}\delta\rho + w_2\partial_{\parallel}(\delta\rho^2) + w_3\partial_{\parallel}(|\mathbf{v}_\perp|^2), \tag{4.3.24}$$

where I've defined:

$$v_2 \equiv v_0 - \frac{\rho_0\Gamma_2}{\Gamma_1}, \tag{4.3.25}$$

$$\phi \equiv \frac{\Gamma_2\rho_0}{v_0\Gamma_1^2}, \tag{4.3.26}$$

$$w_2 \equiv \frac{\Gamma_2}{\Gamma_1}, \tag{4.3.27}$$

$$w_3 \equiv \frac{\rho_0}{2v_0}, \tag{4.3.28}$$

$$D_{\rho_\parallel} \equiv \frac{\rho_0\lambda_4^0}{\Gamma_1}$$
$$= \frac{\rho_0}{\Gamma_1}\left(\frac{(\mu_1 v_0^2 + \sigma_1)}{v_0} - \frac{\Gamma_2}{\Gamma_1}\left(\lambda_1^0 + \lambda_2^0 + 2\lambda_3^0\right)\right), \tag{4.3.29}$$

and, last but by no means least,

$$D_{\rho v} \equiv \frac{\lambda_2^0 \rho_o}{\Gamma_1}. \tag{4.3.30}$$

The parameter $D_{\rho\perp}$ is actually zero at this point in the calculation, but I've included it in equation (4.3.24) anyway, because it is generated by the nonlinear terms under the renormalization group, as I'll discuss in Chapter 5. Likewise, the parameter $w_1 = 1$, but I've also included it for convenience in discussing the RG in that chapter.

I will henceforth focus my attention on the fluid, orientationally ordered state, in which all of the diffusion constants $D_{\rho\parallel}$, $D_{\rho\perp}$, $D_{\rho v}$, D_B, $D_{\parallel}$, and D_T are positive. I'll take them all to have their steady-state values D_T^0, etc., at $|\mathbf{v}| = v_0$ and $\rho = \rho_0$, since fluctuations away from that can be shown to be irrelevant.

4.4 Solving the Linearized Model

4.4.1 Linearizing the Equations of Motion

Expanding (4.3.14) and (4.3.24) to linear order in the small fluctuations $\mathbf{v}_\perp$ and $\delta\rho$ gives:

$$\begin{aligned}\partial_t \mathbf{v}_\perp + \gamma \partial_\parallel \mathbf{v}_\perp = &- \frac{c_0^2}{\rho_0} \nabla_\perp \delta\rho + D_B \nabla_\perp \left(\nabla_\perp \cdot \mathbf{v}_\perp\right) + D_T \nabla_\perp^2 \mathbf{v}_\perp + D_\parallel \partial_\parallel^2 \mathbf{v}_\perp + \nu_t \partial_t \nabla_\perp \delta\rho \\ &+ \nu_\parallel \partial_\parallel \nabla_\perp \delta\rho + \mathbf{f}_\perp ,\end{aligned} \tag{4.4.1}$$

and

$$\partial_t \delta\rho + \rho_o \nabla_\perp \cdot \mathbf{v}_\perp + v_2 \partial_\parallel \delta\rho = D_{\rho\parallel} \partial_\parallel^2 \delta\rho + D_{\rho\perp} \nabla_\perp^2 \delta\rho + D_{\rho v} \partial_\parallel \left(\nabla_\perp \cdot \mathbf{v}_\perp\right) + \phi \partial_t \partial_\parallel \delta\rho. \tag{4.4.2}$$

4.4.2 Mode Structure

These equations can now readily be solved for the mode structure and correlations by Fourier transforming in space and time; this gives

$$\left[-\mathrm{i}(\omega - \gamma q_\parallel) + \Gamma_L(\mathbf{q})\right] v_L + \left[\frac{\mathrm{i} c_0^2}{\rho_0} q_\perp - \nu_t q_\perp \omega - \nu_\parallel q_\perp q_\parallel\right] \delta\rho = f_L, \tag{4.4.3}$$

$$\left[-\mathrm{i}(\omega - \gamma q_\parallel) + \Gamma_T(\mathbf{q})\right] \mathbf{v}_T = \mathbf{f}_T, \tag{4.4.4}$$

$$\left[\mathrm{i}\rho_0 q_\perp + D_{\rho v} q_\perp q_\parallel\right] v_L + \left[-\mathrm{i}(\omega - v_2 q_\parallel) + \Gamma_\rho(\mathbf{q}) - \phi q_\parallel \omega\right] \delta\rho = 0, \tag{4.4.5}$$

and where I've defined the wavevector-dependent longitudinal, transverse, and ρ dampings $\Gamma_{L,T,\rho}$:

$$\Gamma_L(\mathbf{q}) = D_L q_\perp^2 + D_\parallel q_\parallel^2 , \tag{4.4.6}$$

$$\Gamma_T(\mathbf{q}) = D_T q_\perp^2 + D_\parallel q_\parallel^2 , \tag{4.4.7}$$

$$\Gamma_\rho(\mathbf{q}) = D_{\rho\parallel} q_\parallel^2 + D_{\rho\perp} q_\perp^2 , \tag{4.4.8}$$

with $D_L \equiv D_B + D_T$. I've also separated the velocity $\mathbf{v}_\perp$ and the noise $\mathbf{f}_\perp$ into components along and perpendicular to the projection $\mathbf{q}_\perp$ of $\mathbf{q}$ perpendicular to $\langle \mathbf{v} \rangle$ via

$$v_L \equiv \mathbf{v}_\perp \cdot \mathbf{q}_\perp / q_\perp , \; \mathbf{v}_T \equiv \mathbf{v}_\perp - v_L \frac{\mathbf{q}_\perp}{q_\perp} , \tag{4.4.9}$$

with f_L and $\mathbf{f}_T$ obtained from $\mathbf{f}$ in the same way.

These equations differ from the corresponding equations considered in [33, 34, 35, 36] only in the $v_{t,\parallel}$ terms in (4.4.3), and the $D_{\rho\parallel}$ and $D_{\rho v}$ terms in (4.4.5). These prove to lead only to minor changes in the propagation direction dependence, but not the scaling with wavelength, of the damping of the sound modes found in [33, 34, 35, 36], as I will now demonstrate.

I begin by determining the eigenfrequencies of the system, defined in the usual way as the complex, wavevector-dependent frequencies $\omega(\mathbf{q})$ at which the Fourier transformed hydrodynamic equations (4.4.3), (4.4.4), and (4.4.5) admit nonzero solutions for $\mathbf{v}_T$, $\delta\rho$, and v_L when the noise $\mathbf{f}$ is set to zero. Note that $\mathbf{v}_T$ is decoupled from v_L and ρ; this implies a pair of "longitudinal" eigenmodes involving just the longitudinal velocity v_L and ρ, and an additional $d-2$ "transverse" modes associated with the transverse velocity $\mathbf{v}_T$. The longitudinal modes are closely analogous to ordinary sound waves in a simple fluid, while the transverse modes are the analog of the diffusive shear modes in such a fluid.

In the hydrodynamic limit (i.e., when wavenumber $q \to 0$), the longitudinal eigenfrequencies become a pair of underdamped, propagating modes with complex eigenfrequencies

$$\omega_\pm(\mathbf{q}) = c_\pm(\theta_\mathbf{q}) q - \mathrm{i}\epsilon_\pm(\mathbf{q}) , \tag{4.4.10}$$

where the direction-dependent sound speeds $c_\pm(\theta_\mathbf{q})$ are given by exactly the same expression as found in previous work [33, 34, 35, 36]:

$$c_\pm(\theta_\mathbf{q}) = \left(\frac{\gamma + v_2}{2} \right) \cos(\theta_\mathbf{q}) \pm c_2(\theta_\mathbf{q}) , \tag{4.4.11}$$

where I've defined

$$c_2(\theta_\mathbf{q}) \equiv \sqrt{\frac{(\gamma - v_2)^2 \cos^2(\theta_\mathbf{q})}{4} + c_0^2 \sin^2(\theta_\mathbf{q})} , \tag{4.4.12}$$

where $\theta_\mathbf{q}$ is the angle between $\mathbf{q}$ and the direction of flock motion (i.e., the $x_\parallel$ axis).

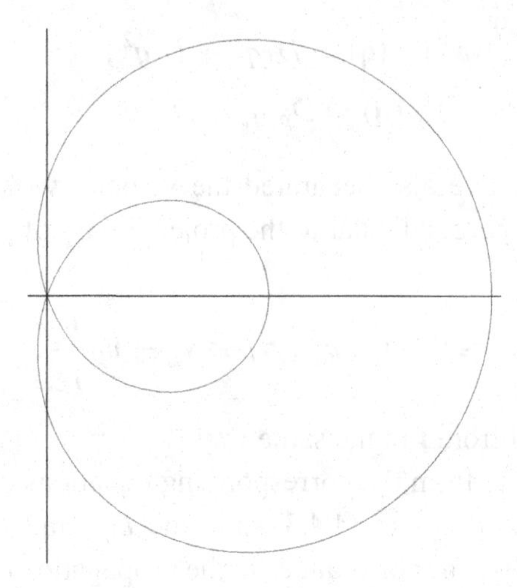

Fig. 4.4.1 Polar plot of the direction-dependent sound speeds $c_\pm\left(\theta_{\mathbf{q}}\right)$, with the horizontal axis along the direction of mean flock motion. Reproduced from [7] by permission of Oxford University Press.

A polar plot of this highly anisotropic sound speed is given in Figure 4.4.1. This plot can be used to determine the sound speed in some direction of propagation (relative to the direction of mean flock motion) by drawing a line from the origin that makes the same angle with the rightward horizontal axis in that figure as the direction of propagation. The distance from the origin to the crossing of the double-looped curve gives the speed of sound propagation in that direction. There are double crossings since there are two propagating sound modes for each direction of propagation (namely, c_+ and c_-, as given by equation (4.4.11)).

The plot in Figure 4.4.1 assumes γ and v_2 have the same sign, which is not required by symmetry. If they have opposite signs, the polar plot of the sound speeds now assumes the shape shown in Figure 4.4.2 (which looks remarkably like the Kuiper belt object 486958 Arrokoth!).

The existence of *two* sound modes for each direction of propagation may seem strange, but it is really no different (in this sense) from a simple isotropic fluid, for which there are *also* two sound speeds: $\pm c_0$, where here c_0 is the speed of sound in the fluid. That is, for every direction, there's one mode propagating forward, and one propagating backwards.

A polar plot like (4.4.11) for an isotropic fluid would simply be a circle of radius c_0, and the construction just described would yield both the forward and backward propagating mode.

While the *number* of sound speeds for a given direction of motion is the same for flocks, they are very different in most other respects. Not only is the sound speed *anisotropic*, but, as can be seen from (4.4.11), there are some directions (e.g., horizontally in Figure 4.4.1 – that is, along the direction of flock motion)

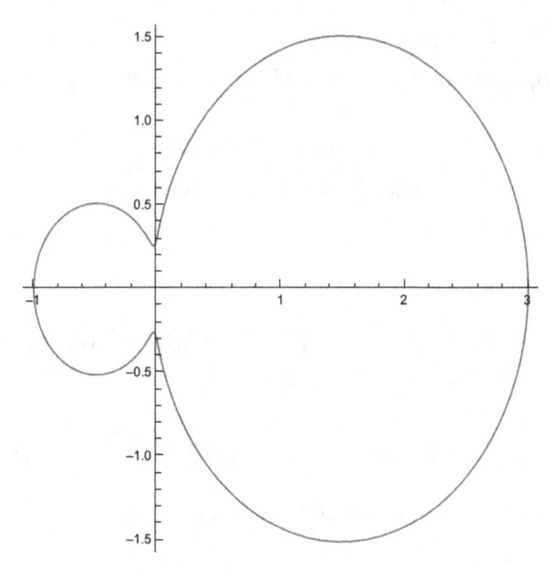

Fig. 4.4.2 Polar plot of the sound speeds when γ and v_2 have *opposite* signs. I know of no experiments or simulations in which γ and v_2 actually *do* have opposite signs, but the possibility is not forbidden by any symmetry (or other argument).

both sound modes propagate in the same direction,[5] while for others (e.g., vertically – that is, *perpendicular* to the direction of flock motion) they propagate in opposite directions. There are also some directions in which one of the sound speeds vanishes exactly; specifically when the propagation angle $\theta_{\mathbf{q}}$ satisfies

$$\tan(\theta_{\mathbf{q}}) = \frac{\sqrt{\gamma v_2}}{c_0} . \tag{4.4.13}$$

These directions of propagation prove to dominate the fluctuations when flocks move through a disordered medium (i.e., when there is "quenched disorder" present). This is discussed in [38, 39, 40, 41, 42]; I will not discuss it further here.

This prediction for the anisotropy of the sound speeds in flocks has recently been confirmed quantitatively in experiments on synthetic flockers (specifically, Quinke rotators) [13].

As mentioned earlier, the wavevector dependent dampings $\epsilon_{\pm}(\mathbf{q})$ of these propagating sound modes *are* altered slightly from the form found in [33, 34, 35, 36].

[5] That is, if the parameters v_2 and γ are both positive, as they are in the systems in which it has been measured (see ([34]). There is no symmetry or stability argument fixing the sign of those two parameters, so other possibilities also exist.

They remain of $O(q^2)$, as found in previous work, but with a slightly modified dependence on propagation direction $\hat{q}$. More precisely, they are given by:

$$\epsilon_\pm \equiv \frac{\text{Hideous Numerator}}{(2c_\pm(\theta_{\mathbf{q}}) - (v_2 + \gamma)\cos(\theta_{\mathbf{q}}))} \tag{4.4.14}$$

with

$$\begin{aligned}\text{Hideous Numerator} \equiv\ & (\Gamma_L(\mathbf{q}) + \Gamma_\rho(\mathbf{q}) - \phi c_\pm(\theta_{\mathbf{q}})\cos(\theta_{\mathbf{q}})q^2)c_\pm(\theta_{\mathbf{q}}) \\ & - v_2\Gamma_L(\mathbf{q})\cos(\theta_{\mathbf{q}}) \\ & - \gamma(\Gamma_\rho(\mathbf{q}) - \phi c_\pm(\theta_{\mathbf{q}})\cos(\theta_{\mathbf{q}})q^2)\cos(\theta_{\mathbf{q}}) + \frac{c_0^2}{\rho_0}D_{\rho v}\frac{q_\parallel q_\perp^2}{q} \\ & - \rho_0 q_\perp^2(\nu_t c_\pm(\theta_{\mathbf{q}}) + \nu_\parallel \cos(\theta_{\mathbf{q}})),\end{aligned} \tag{4.4.15}$$

where I remind the reader that the wavevector-dependent dampings $\Gamma_{L,\rho}$ are $O(q^2)$, and defined earlier in equations (4.4.6) and (4.4.8). Thus, the "Hideous Numerator," while indeed hideous in its angular dependence, is nonetheless simple in its scaling with the *magnitude* q of $\mathbf{q}$: It scales like q^2. This implies that the dampings $\epsilon_\pm \propto q^2$ as well.

The transverse modes have the far simpler character of simply convected anisotropic diffusion:

$$\omega_T(\mathbf{q}) = \gamma q_\parallel - i\Gamma_T(\mathbf{q}) \tag{4.4.16}$$

with the wavevector-dependent damping Γ_T also $O(q^2)$, and defined earlier in equation (4.4.7). This corresponds to simple anisotropic diffusion in a "pseudo-comoving" frame, by which I mean a frame that moves in the direction of mean flock motion, but with a speed γ that differs from the speed v_0 of the flock itself.

4.4.3 Correlation Functions

I now turn to the correlation functions in this linearized approximation. These are easily obtained by first solving the linear algebraic equations (4.4.3), (4.4.4), and (4.4.5) for the fields $v_L(\mathbf{q},\omega)$, $\mathbf{v}_T(\mathbf{q},\omega)$, and $\rho(\mathbf{q},\omega)$ in terms of the noises $f_L(\mathbf{q},\omega)$ and $\mathbf{f}_T(\mathbf{q},\omega)$. These solutions are, of course, linear in those noises. Hence, by correlating these solutions pairwise, one can obtain any two-field correlation function in terms of the correlations (4.2.38) of $\mathbf{f}$. The resulting correlation function for the velocity is:

$$
\begin{aligned}
C_{ij}(\mathbf{q},\omega) &\equiv \left\langle v_{\perp i}(\mathbf{q},\omega)\, v_{\perp j}(-\mathbf{q},-\omega)\right\rangle \\
&= \frac{\Delta\left(\omega - v_2 q_{\parallel}\right)^2 L_{ij}^{\perp}}{\left[(\omega - c_+(\theta_{\mathbf{q}})q)^2 + \epsilon_+^2(\mathbf{q})\right]\left[(\omega - c_-(\theta_{\mathbf{q}})q)^2 + \epsilon_-^2(\mathbf{q})\right]} \\
&\quad + \frac{\Delta P_{ij}^{\perp}}{\left[(\omega - \gamma q_{\parallel})^2 + \Gamma_T^2(\mathbf{q})\right]},
\end{aligned}
\tag{4.4.17}
$$

where I've defined the longitudinal (L) and transverse (T) projection operators in the $\perp$ plane

$$
L_{ij}^{\perp}(\hat{q}) \equiv \frac{q_{\perp i} q_{\perp j}}{q_{\perp}^2}, \quad P_{ij}^{\perp}(\hat{q}) \equiv \delta_{ij}^{\perp} - L_{ij}^{\perp}(\hat{q}),
\tag{4.4.18}
$$

where $\delta_{ij}^{\perp}$ is a Kronecker delta in the $\perp$ plane (i.e., it is equal to the usual Kronecker delta if $i \neq \parallel \neq j$, and zero otherwise). These operators project any vector first into the $\perp$ plane, and then either along (L) or orthogonal to (P) $\mathbf{q}_{\perp}$ within the $\perp$ plane.

The first term in equation (4.4.17) comes from the "longitudinal" component v_L while the second comes from the $d-2$ "transverse" components of $\mathbf{v}_{\perp}$. Clearly, in $d=2$, only the longitudinal component is present; the second (transverse) term in (4.4.17) vanishes in $d=2$.

The density autocorrelations obtained by the procedure described above are given, to leading order in wavevector and frequency, by:

$$
\begin{aligned}
C_{\rho\rho}(\mathbf{q},\omega) &\equiv \langle \rho(\mathbf{q},\omega)\, \rho(-\mathbf{q},-\omega)\rangle \\
&= \frac{\rho_0 q_{\perp}^2 \Delta}{\left[(\omega - c_+(\theta_{\mathbf{q}})q)^2 + \epsilon_+^2(\mathbf{q})\right]\left[(\omega - c_-(\theta_{\mathbf{q}})q)^2 + \epsilon_-^2(\mathbf{q})\right]}.
\end{aligned}
\tag{4.4.19}
$$

Both the velocity correlations (4.4.17) and the density correlations (4.4.19) have the same form, and the same scaling with frequency and wavevector, as those reported in earlier work [33, 34, 35, 36]. The only change from those earlier results is the slightly modified form in (4.4.14) and (4.4.15) of the sound dampings which appear in (4.4.17) and (4.4.19).

The same statement is true of the equal-time correlations of $\mathbf{v}$ and ρ, which can be obtained in the usual way by integrating the spatio-temporally Fourier transformed correlations (4.4.17) and (4.4.19) over all frequency ω. These equal-time correlations are important, because they determine the size of the velocity and density fluctuations. The size of the velocity fluctuations determines whether or not long-ranged order can exist in these systems, while the size of the density fluctuations determines the presence or absence of giant number fluctuations [36, 43, 44, 45, 46, 47].

Integrating (4.4.17) over all ω and tracing over the Cartesian components i,j gives the equal-time correlation of $\mathbf{v}$:

$$\left\langle |\mathbf{v}_\perp(\mathbf{q},t)|^2 \right\rangle = \frac{1}{2}\left(\frac{\Delta\left(c_+(\theta_\mathbf{q}) - v_2\cos(\theta_\mathbf{q})\right)^2}{\epsilon_+(\mathbf{q})\,[c_+(\theta_\mathbf{q}) - c_-(\theta_\mathbf{q})]^2} + \frac{\Delta\left(c_-(\theta_\mathbf{q}) - v_2\cos(\theta_\mathbf{q})\right)^2}{\epsilon_-(\mathbf{q})\,[c_+(\theta_\mathbf{q}) - c_-(\theta_\mathbf{q})]^2} + \frac{(d-2)\Delta}{\Gamma_T(\mathbf{q})} \right). \quad (4.4.20)$$

Note that this scales like $1/q^2$ for all directions of wavevector $\mathbf{q}$. That is, it can be rewritten as

$$\left\langle |\mathbf{v}_\perp(\mathbf{q},t)|^2 \right\rangle = \frac{g(\theta_\mathbf{q})}{q^2}, \quad (4.4.21)$$

where $g(\theta_\mathbf{q})$ is a horribly complicated function of the angle $\theta_\mathbf{q}$ which can be obtained from the almost as horrible angular dependences of $c_\pm(\theta_\mathbf{q})$, $\epsilon(\mathbf{q})$, and $\Gamma_T(\mathbf{q})$ found previously. All we'll need to know about $g(\theta_\mathbf{q})$ is that it is finite for all $g(\theta_\mathbf{q})$, so integrals of it over all directions of $\mathbf{q}$ are finite.

This scaling is precisely the same as that found in the linearized theory of [33, 34, 35, 36]; only the precise form of the dependence on the direction of $\mathbf{q}$ is slightly changed by the presence of the new linear terms v_t, $v_\parallel$, and ϕ that I've found here, which were missed in the treatment of [33, 34, 35, 36].

4.4.4 The Return of the Mermin-Wagner-Hohenberg Theorem, Damn it!

This $1/q^2$ scaling of $\mathbf{v}_\perp$ fluctuations with q in Fourier space implies that the real-space fluctuations

$$\left\langle |\mathbf{v}_\perp(\mathbf{r},t)|^2 \right\rangle = \int \frac{d^d q}{(2\pi)^d} \left\langle |\mathbf{v}_\perp(\mathbf{q},t)|^2 \right\rangle \quad (4.4.22)$$

diverge in the infrared ($q \to 0$ or system size $L \to \infty$) limit in all spatial dimensions $d \leq 2$. This in turn implies that long-ranged order (i.e., the existence of a nonzero $\langle \mathbf{v}_\perp(\mathbf{r},t) \rangle$) is not possible in $d = 2$, *according to the linearized theory*.

To see this divergence, note that we can rewrite (4.4.22) in hyperspherical coordinates as

$$\left\langle |\mathbf{v}_\perp(\mathbf{r},t)|^2 \right\rangle = \int \frac{d^d q}{(2\pi)^d} \frac{g(\theta_\mathbf{q})}{q^2} = \left(\int \frac{d\Omega}{(2\pi)^d} g(\theta_\mathbf{q}) \right) \int q^{d-3} dq, \quad (4.4.23)$$

where $\int d\Omega$ denotes an integral over the *directions* of $\mathbf{q}$, while $\int dq$ denotes an integral over the *magnitude* q of $\mathbf{q}$. That is, in two dimensions, we would have simply

$$\int d\Omega = \int_0^{2\pi} d\theta_{\mathbf{q}}\,, \tag{4.4.24}$$

while in three dimensions,

$$\int d\Omega = \int_0^{\pi} d\theta_{\mathbf{q}}\ \sin(\theta_{\mathbf{q}}) \int_0^{2\pi} d\phi_{\mathbf{q}}\,, \tag{4.4.25}$$

where $\theta_{\mathbf{q}}$ and $\phi_{\mathbf{q}}$ are the polar and azimuthal angles of $\mathbf{q}$, respectively. The generalization to higher dimensions is straightforward. Indeed, using the fact that $g(\theta_{\mathbf{q}})$ depends only on the polar angle $\theta_{\mathbf{q}}$, we can show that

$$\int d\Omega g(\theta_{\mathbf{q}}) = S_{d-2} \int_0^{\pi} d\theta_{\mathbf{q}}\ \sin^{d-2}(\theta_{\mathbf{q}}) g(\theta_{\mathbf{q}})\,, \tag{4.4.26}$$

where S_d is the surface hyperarea of a unit d-dimensional hypersphere. Note that by a "d-dimensional unit sphere," I mean the hypersurface $|\mathbf{r}| = 1$ of a d-component vector $\mathbf{r}$. I offer this clarification because mathematicians refer to such a surface, when $\mathbf{r}$ has three components, as a "two-sphere," because the *surface* of the sphere is two-dimensional.

In light of our earlier observation that $g(\theta_{\mathbf{q}})$ is finite for all $\theta_{\mathbf{q}}$, it is clear that the $\int d\Omega g(\theta_{\mathbf{q}})$ in (4.4.23) is finite. Thus, the only possible divergence must come from the integral over the magnitude q of $\mathbf{q}$ in that expression.

That integral is precisely the same as the integral we did to calculate the real-space fluctuations of the equilibrium XY model in Chapter 2. Therefore we have

$$\left\langle |\mathbf{v}_{\perp}(\mathbf{r},t)|^2 \right\rangle \propto \begin{cases} L^{2-d}, & d < 2, \\ \ln\left(\frac{\Lambda L}{2\pi}\right), & d = 2, \\ \text{finite (independent of } L), & d > 2. \end{cases} \tag{4.4.27}$$

This result, which is simply the Mermin–Wagner–Hohenberg theorem [25, 26, 27], is actually overturned by nonlinear effects, which stabilize the long-ranged order in $d = 2$ (i.e., make the existence of a nonzero $\langle \mathbf{v}(\mathbf{r},t) \rangle$ possible), as first noted by [33, 34, 35, 36]. I'll show in Section 5.1 that nonlinear effects still stabilize long-ranged order in this way even when the additional nonlinearities I've found here, which were missed in [33, 34, 35, 36], are included.

The equal-time density autocorrelations can likewise be obtained by integrating equation (4.4.19) over frequency ω; this gives

$$\left\langle |\delta\rho(\mathbf{q},t)|^2 \right\rangle = \frac{1}{2}\left(\frac{\Delta\rho_0 q_{\perp}^2}{[c_+(\theta_{\mathbf{q}}) - c_-(\theta_{\mathbf{q}})]^2 q^2}\right)\left(\frac{1}{\epsilon_+(\mathbf{q})} + \frac{1}{\epsilon_-(\mathbf{q})}\right). \tag{4.4.28}$$

This also scales like $1/q^2$ for all directions of $\mathbf{q}$. This divergence implies "Giant Number Fluctuations"[36, 43, 44, 45, 46, 47]: The root mean square (RMS) fluctuations $\sqrt{\langle \delta N^2 \rangle}$ of the number of particles within a large region of the system scale like the mean number of particles $\langle N \rangle$ faster than $\sqrt{\langle N \rangle}$; specifically, $\sqrt{\langle \delta N^2 \rangle} \propto \langle N \rangle^{\phi(d)}$, with $\phi(d) = 1/2 + 1/d$ in spatial dimension d. Note that this means in particular that $\sqrt{\langle \delta N^2 \rangle} \propto \langle N \rangle$ in $d = 2$.

Again, I emphasize that this is the prediction of the linearized theory. It once again coincides with the results of the linearized treatment of [33, 34, 35, 36].

Both the prediction that long-ranged orientational order is destroyed in $d = 2$, and the value $\phi(d) = 1/2 + 1/d$ of the exponent $\phi(d)$ for $d < 4$ prove, when nonlinear effects are taken into account, to be incorrect, as first noted by [33]. I now turn to the treatment of those nonlinear effects.

5

The Dynamical Renormalization Group Applied to the Flocking Problem

We have seen that the linearized theory does not explain the mystery that motivated my original interest in this problem: the persistence of long-ranged order in flocks even in $d = 2$. Fortunately, it turns out that the nonlinearities that we ignored in the previous section in fact completely change the scaling behavior of these systems at long distances, as first noted by [33, 34, 35, 36]. In this chapter, I'll deal with those nonlinearities. While a few of the precise quantitative conclusions of [33, 34, 35, 36] prove to be less certain than Yuhai and I originally thought, the essential conclusions that

(1) nonlinearities radically change the scaling behavior of these systems for all $d \leq 4$, and
(2) these changes in scaling stabilize long-ranged order in $d = 2$, remain valid.

Equally noteworthy are the nonlinear terms that are missing from (4.3.14) and (4.3.24): All nonlinearities arising from the anisotropic pressure P_2 and the λ_2 nonlinearity drop out of (4.3.14) and (4.3.24). This in particular has the very important consequence of saving the Mermin–Wagner–Hohenberg theorem. This is because the λ_2 term is allowed even in equilibrium systems [48]. The incorrect treatment in [33, 34, 35, 36] suggested that this term *by itself* could stabilize long-ranged order in $d = 2$. Given that this term is allowed in equilibrium, this would imply that the Mermin–Wagner–Hohenberg theorem would fail for such an equilibrium system. The correct treatment I've done here shows that this is not the case: The λ_2 term by itself cannot stabilize long-ranged order in $d = 2$, since the nonlinearities associated with it drop out of the long-wavelength description of the ordered phase.

Returning now to the nonlinearities in (4.3.14) and (4.3.24) that were missed by [33, 34, 35, 36], I will show in Section 5.1 that *all* of them become relevant, in the renormalization group (RG) sense of changing the long-distance, long-time behavior of flocks [30], for spatial dimensions $d \leq 4$.

5.1 The Linear Fixed Point, and Its Instability for $d < 4$

To assess the effect of the new nonlinear terms I've found here, I'll analyze equations (4.3.14) and (4.3.24) using the dynamical RG [29].

The dynamical RG starts by averaging the equations of motion over the short-wavelength fluctuations: i.e., those with support in the "shell" of Fourier space $b^{-1}\Lambda \leq |\mathbf{q}| \leq \Lambda$, where Λ is an "ultraviolet cutoff," and b is an arbitrary rescaling factor. Then, one rescales lengths, time, $\delta\rho$, and $\mathbf{v}_\perp$ in equations (4.3.14) and (4.3.24) according to $\mathbf{v}_\perp = b^\chi \mathbf{v}'_\perp$, $\delta\rho = b^\chi \delta\rho'$, $\mathbf{r}_\perp = b\mathbf{r}'_\perp$, $r'_\parallel = b^\zeta (r'_\parallel)'$, and $t = b^z t'$ to restore the ultraviolet cutoff to Λ.[1] This leads to a new pair of equations of motion of the same form as (4.3.14) and (4.3.24), but with "renormalized" values (denoted here by primes) of the parameters given by:

$$D'_{B,T} = b^{z-2}(D_{B,T} + \text{graphs}), \tag{5.1.1}$$

$$D'_\parallel = b^{z-2\zeta}(D_\parallel + \text{graphs}), \tag{5.1.2}$$

$$\Delta' = b^{z-\zeta-2\chi+1-d}(\Delta + \text{graphs}), \tag{5.1.3}$$

$$(\lambda_1^0)' = b^{z+\chi-1}(\lambda_1^0 + \text{graphs}), \tag{5.1.4}$$

$$g'_{1,2,3} = b^{z+\chi-1}(g_{1,2,3} + \text{graphs}), \tag{5.1.5}$$

$$w'_{1,2,3} = b^{z+\chi-1}(w_{1,2,3} + \text{graphs}), \tag{5.1.6}$$

where "graphs" denotes contributions from integrating out the short-wavelength degrees of freedom, which we could in principle evaluate by doing Feynman diagrams. (More on the difference between "in principle" and "in practice" in this context later.)

As we did for the KPZ equation in Chapter 3, I'll assess the importance of the various nonlinearities in the equations of motion (4.3.14) and (4.3.24) by considering the limit in which all of the nonlinearities are very small. In this limit, the graphical corrections (denoted "graphs" in equations (5.1.1)–(5.1.6)) vanish, since, without nonlinearities, Fourier modes at different wavevectors and frequencies do not interact. It is then straightforward to determine from equations (5.1.1)–(5.1.3) the values of the rescaling exponents z, ζ, and χ that will keep $D_{B,T,\parallel,\rho\parallel}$ and Δ (and, hence, the size of the fluctuations) fixed: Simply choose those which make the exponents of the b's in (5.1.1)–(5.1.3) vanish. That is, we must choose

$$z - 2 = 0 \quad \text{(linear fixed point)} \tag{5.1.7}$$

to keep D_B and D_T fixed,

$$z - 2\zeta = 0 \quad \text{(linear fixed point)}, \tag{5.1.8}$$

[1] One could more generally rescale $\delta\rho$ with a different rescaling exponent χ_ρ from the exponent χ used for $\mathbf{v}_\perp$. However, since fluctuations of $\delta\rho$ and $\mathbf{v}_\perp$ have the same scaling with distance as time, it is most convenient to rescale them the same way. None of my conclusions would be changed if I did not make this arbitrary choice.

to keep $D_{\parallel}$ and $D_{\rho\parallel}$ fixed, and

$$z - \zeta - 2\chi + 1 - d = 0 \text{ (linear fixed point)}, \tag{5.1.9}$$

to keep Δ fixed under the RG. The solutions to these three conditions (5.1.7)–(5.1.9) are trivially found to be:

$$z = 2 \text{ (linear fixed point)}, \tag{5.1.10}$$
$$\zeta = 1 \text{ (linear fixed point)}, \tag{5.1.11}$$

and

$$\chi = (2 - d)/2 \text{ (linear fixed point)}. \tag{5.1.12}$$

Let's now consider the stability of this linear fixed point against the effect of the nonlinear terms λ_1^0, $g_{1,2,3}$, and $w_{1,2,3}$. Because, as mentioned earlier, I have chosen the rescaling exponents so as to keep the magnitude of the fluctuations the same on all length scales, a given nonlinearity has important effects at long distances if it grows upon renormalization with this choice (5.1.10)–(5.1.12) of the rescaling exponents z, ζ, and χ. On the other hand, if it gets smaller upon renormalization with this choice of the rescaling exponents, it is unimportant at long distances.[2] Using the exponents (5.1.10)–(5.1.12) in the recursion relations (5.1.4) and (5.1.6), and ignoring the graphical corrections, which are higher than linear order in λ_1^0, $g_{1,2,3}$, and $w_{1,2,3}$, I find that all seven of these nonlinearities have identical renormalization group eigenvalues of $(4 - d)/2$ at the linearized fixed point.

Thus, for $d > 4$, all of the nonlinearities flow to zero, and so become unimportant, at long length and time scales. Hence, the linearized theory is correct at long length and time scales, for $d > 4$. For $d < 4$, however, all of these nonlinearities grow, and the linear theory breaks down at sufficiently long length and time scales. This forces us to try to find a *nonlinear* fixed point.

This is a formidable problem. To appreciate just *how* formidable it is, consider the calculation of the renormalization of these seven nonlinear vertices. These can be represented by graphs that look like Figure 3.5.1 in Chapter 3. Now, however, each three-legged vertex in the graph can be any one of seven possible vertices. So naively, there are $7^3 = 343$ Feynman diagrams to evaluate for this problem –

[2] As we saw in our discussion in Chapter 3 of the KPZ equation, other choices of rescaling force one to find suitable "dimensionless couplings"; that is, ratios of the nonlinear parameters (e.g., λ_1^0) and suitable powers of the linear parameters (e.g., Δ, D_B) that actually give the ratios of the renormalizations of the linear parameters by fluctuations to their linear values. These necessarily have the same RG eigenvalues as those we found here for λ_1^0, etc., alone, independent of the arbitrary choice of z, χ, and ζ.

and that's just to renormalize the nonlinear terms! Then we have to renormalize the linear terms as well!

As you'll see in Chapter 9, doing the dynamical renormalization group on a problem with only one nonlinear vertex can take as much as 26 pages of typeset calculations. The full problem we're considering here would therefore, as a rough estimate, take at least $26 \times 343 = 8918$ pages of typeset calculation. That's half the length of the *Oxford English Dictionary*! And of course, behind each page of typeset calculation lies at least 10 pages of handwritten calculations, with their various mistakes and crossings out, etc.

This is far too difficult a calculation for me to be willing to undertake (indeed, I'd be exceedingly unlikely to live long enough to finish it!). So instead, I will, in the following chapters, discuss some limits in which the problem becomes tractable, or, alternatively, some simpler variants of this original problem. I'll begin by considering the *incompressible* limit.

6

Incompressible Polar Active Fluids in the Moving Phase in Dimensions $d > 2$: the "Canonical" Exponents

In this chapter, I'll review the work of Chen, Lee, and Toner [49, 50] (collectively referred to as "we" for the remainder of this chapter), which shows how eliminating all of the nonlinearities associated with the density by making a flock incompressible makes it possible to obtain *exact* scaling laws for such systems [49, 50].

Incompressibility is *not* merely a theoretical contrivance; not only can it be readily simulated [51], it can arise in a variety of real experimental situations, such as the following.

(1) Systems with strong repulsive short-ranged interactions between the active particles. Incompressibility has, in fact, been assumed in, e.g., recent experimental studies on cell motility [52].
(2) Systems with long-ranged repulsive interactions; here, true incompressibility is possible. Long-ranged interactions are quite reasonable in certain contexts: birds, for example, can often see all the way across a flock [53]. One could also imagine charged active agents moving in a neutralizing background.

Alternatively, incompressibility could be experimentally realized by programming it into the rules for the motion of a collection of self-propelled robots (drones).

6.1 Hydrodynamic Model

We start with the hydrodynamic model for *compressible* polar active fluids without momentum conservation derived in Chapter 4 [33, 35, 36, 37]:

$$\partial_t \mathbf{v} + \lambda_1(\mathbf{v}\cdot\nabla)\mathbf{v} + \lambda_2(\nabla\cdot\mathbf{v})\mathbf{v} + \lambda_3\nabla(|\mathbf{v}|^2) = U\mathbf{v} - \nabla P - \mathbf{v}\,(\mathbf{v}\cdot\nabla P_2)$$
$$+\mu_B\nabla(\nabla\cdot\mathbf{v}) + \mu_T\nabla^2\mathbf{v} + \mu_2(\mathbf{v}\cdot\nabla)^2\mathbf{v} + \mathbf{f}, \qquad (6.1.1)$$

$$\partial_t\rho + \nabla\cdot(\mathbf{v}\rho) = 0, \qquad (6.1.2)$$

where I remind you that $\mathbf{v}(\mathbf{r}, t)$ and $\rho(\mathbf{r}, t)$ are, respectively, the coarse-grained continuous velocity and density fields. I also remind you that all of the parameters $\lambda_i(i = 1 \to 3)$, U, the "damping coefficients" $\mu_{B,T,2}$, the "isotropic pressure" $P(\rho, v)$, and the "anisotropic pressure" $P_2(\rho, v)$ are functions of the density ρ and the magnitude $v \equiv |\mathbf{v}|$ of the local velocity.

As discussed in Chapter 4's treatment of compressible flocks, the U term makes the local $\mathbf{v}$ have a nonzero magnitude v_0 in the ordered phase, by having $U > 0$ for $v < v_0$, $U = 0$ for $v = v_0$, and $U < 0$ for $v > v_0$.

Finally, recall that the $\mathbf{f}$ term is a random driving force representing the noise. As before, it is assumed to be Gaussian with white-noise correlations:

$$\langle f_i(\mathbf{r}, t) f_j(\mathbf{r}', t')\rangle = 2D\delta_{ij}\delta^d(\mathbf{r} - \mathbf{r}')\delta(t - t'), \tag{6.1.3}$$

where the "noise strength" D is a constant parameter of the system, and i, j denote Cartesian components.

We now take the incompressible limit by taking *the isotropic pressure P only* to be extremely sensitive to departures from the mean density ρ_0.Making $U(\rho, v)$ and $P_2(\rho, v)$ extremely sensitive to changes in ρ as well proves to destabilize the system by generating a "banding instability", similar to the instability found in compressible active fluids around the onset of collective motion [54, 55, 56, 57, 58, 59, 60]. Since we wished to focus on stable flocks, we did not consider this possibility further.

Focusing here on the case in which *only* the isotropic pressure P becomes extremely sensitive to changes in the density, we see that, in this limit, in which the isotropic pressure will suppress density fluctuations extremely effectively, changes in the density will be too small to affect $U(\rho, v)$, $\lambda_{1,2,3}(\rho, v)$, $\mu_{B,T,2}(\rho, v)$, and $P_2(\rho, v)$. As a result, all of them become functions only of the speed v; their ρ-dependence will drop out since ρ will be essentially constant.

The suppression of density fluctuations by the isotropic pressure P reduces the continuity equation (6.1.2) to the familiar condition for incompressible flow,

$$\nabla \cdot \mathbf{v} = 0\,, \tag{6.1.4}$$

which can, as in simple fluid mechanics, be used to determine the isotropic pressure P.

The result of the above observations is the equation of motion:

$$\partial_t \mathbf{v} + \lambda(\mathbf{v} \cdot \nabla)\mathbf{v} = U(v)\mathbf{v} - \nabla P - \mathbf{v}\,(\mathbf{v} \cdot \nabla P_2) + \mu_\perp \nabla_\perp^2 \mathbf{v} + \mu_x \partial_x^2 \mathbf{v} + \mathbf{f}\,, \tag{6.1.5}$$

where the statistics of the noise term are given by (6.1.3), and the pressure P, with the λ_3 term absorbed into it, is determined by the incompressibility condition (6.1.4). We've also defined $\lambda \equiv \lambda_1(|\mathbf{v}| = v_0)$, $\mu_\perp \equiv \mu_T(|\mathbf{v}| = v_0)$,

$\mu_x \equiv \mu_T(|\mathbf{v}| = v_0) + \mu_2(|\mathbf{v}| = v_0)v_0^2$, and dropped "irrelevant" terms arising from expanding $\lambda_1(|\mathbf{v}|)$ and $\mu_{T,2}(|\mathbf{v}|)$ to higher powers of $(|\mathbf{v}| - v_0)$ [29].

As in the compressible case, we are interested in the behavior of the state of collective motion, in which the velocity $\mathbf{v}$ acquires a nonzero average value. We choose our coordinates so that the collective motion is along the x-direction; with this choice, we can write the velocity field as

$$\mathbf{v}(\mathbf{r}, t) = (v_0 + u_x(\mathbf{r}, t))\hat{\mathbf{x}} + \mathbf{u}_\perp(\mathbf{r}, t), \tag{6.1.6}$$

where v_0 is the value of $|\mathbf{v}|$ at which $U(|\mathbf{v}|)$ vanishes.

6.2 Linear Theory

We first study the equations of motion to linear order in $\mathbf{u}$. The procedure is exactly the same as that used above for the compressible problem: First, we insert equation (6.1.6) into equation (6.1.5). Then we keep only terms linear in $\mathbf{u}$. Next, we make the extremely useful step of rewriting the resultant equation in Fourier space. Finally, we act on the equation of motion with the transverse projection operator (in tensor notation) $P_{ml} = \delta_{ml} - q_m q_l / q^2$ to eliminate the isotropic pressure term. The net result of all of these steps is

$$\begin{aligned} -\mathrm{i}(\omega - v_1 q_x) u_m(\mathbf{q}, \omega) = & -(2\alpha + \mathrm{i}\lambda_4 v_0^3 q_x) P_{mx} u_x(\mathbf{q}, \omega) \\ & - \Gamma(\mathbf{q}) u_m(\mathbf{q}, \omega) + P_{ml} f_l(\mathbf{q}, \omega), \end{aligned} \tag{6.2.1}$$

where $v_1 \equiv \lambda v_0$ and

$$\Gamma(\mathbf{q}) \equiv \mu_\perp q_\perp^2 + \mu_x q_x^2 \,. \tag{6.2.2}$$

The coefficient λ_4 and the longitudinal mass α are defined respectively by

$$\lambda_4 \equiv \frac{1}{v} \left[\frac{\mathrm{d}P_2(v)}{\mathrm{d}v} \right]_{v=v_0}, \quad \alpha \equiv \frac{v_0}{2} \left[\frac{\mathrm{d}U(v)}{\mathrm{d}v} \right]_{v=v_0}. \tag{6.2.3}$$

To proceed, we first eliminate u_x in terms of the other fields using the Fourier transform of the incompressibility condition ($\nabla \cdot \mathbf{u} = \nabla \cdot \mathbf{v} = 0$):

$$u_x(\mathbf{q}, t) = -\frac{\mathbf{q}_\perp \cdot \mathbf{u}_\perp(\mathbf{q}, t)}{q_x}. \tag{6.2.4}$$

Since we have chosen the x-direction to be the "stiff" direction, we expect that fluctuations of u_x are small. In addition, equation (6.2.4) shows that the component of $\mathbf{u}_\perp$ along the direction of $\mathbf{q}_\perp$ and u_x are locked together, which suggests the fluctuations of this component of $\mathbf{u}_\perp$ are small as well. To verify this, we further decompose $\mathbf{u}_\perp$ into components parallel and perpendicular to $\mathbf{q}_\perp$ (see Figure 6.2.1):

$$\mathbf{u}_\perp = u_P \hat{\mathbf{q}}_\perp + \mathbf{u}_T, \tag{6.2.5}$$

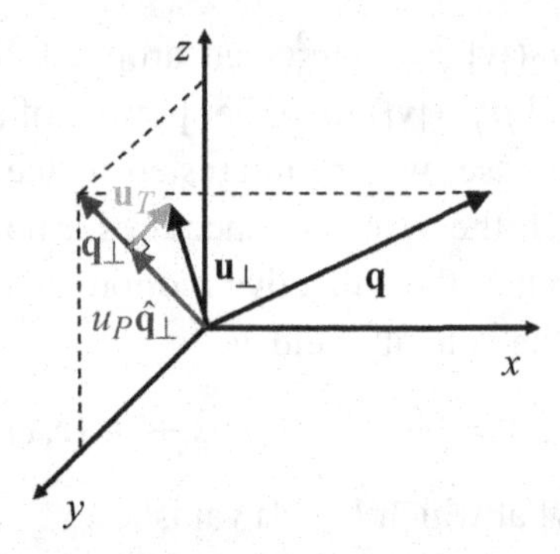

Fig. 6.2.1 Schematics of the vectorial decomposition discussed in this chapter. Note that all vectors shown are orthogonal to $\hat{\mathbf{x}}$ except $\mathbf{q}$. Because soft fluctuations must be orthogonal to both of the direction of collective motion (parallel to $\hat{\mathbf{x}}$) and $\mathbf{q}$ (the latter by the incompressibility condition), fluctuations of $\mathbf{u}_T$ dominate over those of other components of $\mathbf{u}$. Reproduced from [49].

where we use the subscript P (T) to denote the component parallel (transverse) to $\hat{\mathbf{q}}_\perp \equiv \mathbf{q}_\perp/|\mathbf{q}_\perp|$.

Next, we apply the projection operator

$$P^\perp_{mn}(\mathbf{q}_\perp) = \delta^\perp_{mn} - \frac{q^\perp_m q^\perp_n}{q^2_\perp} \tag{6.2.6}$$

to equation (6.2.1) and solve the resultant equation for $\mathbf{u}_T$ to obtain

$$u^T_m(\mathbf{q}, \omega) = \frac{P^\perp_{mn}(\mathbf{q}_\perp) f_n(\mathbf{q}, \omega)}{-\mathrm{i}(\omega - v_1 q_x) + \Gamma(\mathbf{q})} . \tag{6.2.7}$$

Having found $\mathbf{u}_T$, we now turn to u_x and u_P. Taking the x component of (6.2.1) and solving for u_x, we find

$$u_x(\mathbf{q}, \omega) = \frac{P_{xm}(\mathbf{q}) f_m(\mathbf{q}, \omega)}{-\mathrm{i}\left[\omega - c(\hat{\mathbf{q}})q\right] + \Gamma(\mathbf{q}) + 2\alpha\frac{q^2_\perp}{q^2}} , \tag{6.2.8}$$

where $c(\hat{\mathbf{q}})$ is defined as

$$c(\hat{\mathbf{q}}) \equiv v_1 \frac{q_x}{q} + \lambda_4 v_0^3 \frac{q^2_\perp q_x}{q^3} . \tag{6.2.9}$$

This, combined with the Fourier transform of the incompressibility condition, which reads $q_x u_x + q_\perp u_P = 0$, gives

$$u_P(\mathbf{q}, \omega) = -\left(\frac{q_x}{q_\perp}\right) \frac{P_{xm}(\mathbf{q}) f_m(\mathbf{q}, \omega)}{\left[-\mathrm{i}\left[\omega - c(\hat{\mathbf{q}})q\right] + \Gamma(\mathbf{q}) + 2\alpha\left(\frac{q^2_\perp}{q^2}\right)\right]} . \tag{6.2.10}$$

We can now autocorrelate these expressions (6.2.7), (6.2.8), and (6.2.10), using the force correlations (6.1.3), to obtain the correlations of the various components of $\mathbf{u}$. For example, for $\mathbf{u}_T$, we find

$$\begin{aligned}\langle u_m^T(\mathbf{q},\omega)u_i^T(-\mathbf{q},-\omega)\rangle &= \frac{P_{mn}^{\perp}(\mathbf{q}_\perp)P_{ij}^{\perp}(\mathbf{q}_\perp)\langle f_n(\mathbf{q},\omega)f_j(-\mathbf{q},-\omega)\rangle}{|-\mathrm{i}(\omega-v_1q_x)+\Gamma(\mathbf{q})|^2} \\ &= \frac{2DP_{mj}^{\perp}(\mathbf{q}_\perp)P_{ij}^{\perp}(\mathbf{q}_\perp)}{\left[(\omega-v_1q_x)^2+\Gamma^2(\mathbf{q})\right]} \\ &= \frac{2DP_{im}^{\perp}(\mathbf{q}_\perp)}{\left[(\omega-v_1q_x)^2+\Gamma^2(\mathbf{q})\right]}, \end{aligned} \tag{6.2.11}$$

where in the last equality we used the identity $P_{mj}^{\perp}(\mathbf{q}_\perp)P_{ij}^{\perp}(\mathbf{q}_\perp) = P_{im}^{\perp}(\mathbf{q}_\perp)$, which is a fairly obvious property of a projection operator, since once a vector has been projected perpendicular to some direction, projecting the result perpendicular to that direction a second time clearly changes nothing. The less geometrically minded can also prove this result algebraically directly from the definition (6.2.6), an exercise I will leave to the reader.

We can calculate the mean squared amplitude $\langle|\mathbf{u}_T(\mathbf{q},\omega)|^2\rangle$ of the full vector $\mathbf{u}_T(\mathbf{q},\omega)$ simply by taking the trace of (6.2.11). This gives

$$\langle|\mathbf{u}_T(\mathbf{q},\omega)|^2\rangle = \frac{2D(d-2)}{\left[(\omega-v_1q_x)^2+\Gamma^2(\mathbf{q})\right]}, \tag{6.2.12}$$

where we used the fact that the trace $P_{ii}^{\perp}(\mathbf{q}_\perp) = d-2$. The physical origin of the factor $d-2$ is very simple: It's just the number of components of the vector $\mathbf{u}_T(\mathbf{q},\omega)$. Since each component has the same autocorrelation, as (6.2.12) shows, the total vector has a mean squared amplitude proportional to that number of components, and, hence, to $d-2$.

An equivalent way to see the origin of this factor of $d-2$ is illustrated in Figure 6.2.1, which represents the fluctuations at some particular $\mathbf{q}$ in Fourier space. In order to be a Goldstone mode (which is what $\mathbf{u}_T$ is, which is why it has such large fluctuations), the Fourier components $\mathbf{u}(\mathbf{q},t)$ of the velocity fluctuation $\mathbf{u}(\mathbf{r},t) \equiv \mathbf{v}(\mathbf{r},t) - \langle\mathbf{v}\rangle$ must be orthogonal to $\langle\mathbf{v}\rangle$. At the same time, incompressibility requires that $\mathbf{u}(\mathbf{q},t)$ be orthogonal to $\mathbf{q}$. This is no problem in spatial dimensions $d > 2$, because then there are some directions ($d-2$ of them, to be precise) that *are* orthogonal to *both* $\langle\mathbf{v}\rangle$ and $\mathbf{q}$ (specifically, the direction(s) denoted by $\mathbf{u}_T$ in Figure 6.2.1); these are the degrees of freedom embodied in $\mathbf{u}_T$. But in $d = 2$, obviously, there are no such degrees of freedom, since $d-2 = 0$ there. That is, in $d = 2$, you're not usually allowed to point orthogonally to both $\langle\mathbf{v}\rangle$ and $\mathbf{q}$.

As a result, we find, as I'll show in more detail in Chapter 8, that, paradoxically, and contrary to all of our previous experience in this book (and almost everywhere else), fluctuations in incompressible flocks are actually *smaller* in $d = 2$ than in $d > 2$. Because of this, the case $d = 2$ requires separate, and special, treatment, which is why it gets a whole chapter to itself.

Deferring the $d = 2$ calculation to the next chapter, we'll continue here to calculate the autocorrelation of u_P. The calculation is very similar to that for $\mathbf{u}_T$, although the result is not: It proves to be much smaller. We get

$$\begin{aligned}\left\langle|u_P(\mathbf{q},\omega)|^2\right\rangle &= \frac{\left(\frac{q_x^2}{q_\perp^2}\right)P_{xm}(\mathbf{q})P_{xn}(\mathbf{q})\langle f_m(\mathbf{q},\omega)f_n(-\mathbf{q},-\omega)\rangle}{\left|-\mathrm{i}\left[\omega-c(\hat{\mathbf{q}})q\right]+\Gamma(\mathbf{q})+2\alpha\left(\frac{q_y}{q}\right)^2\right|^2}\\ &= \frac{2D\left(\frac{q_x^2}{q_\perp^2}\right)P_{xm}(\mathbf{q})P_{xm}(\mathbf{q})}{\left[\omega-c(\hat{\mathbf{q}})q\right]^2+\left[\Gamma(\mathbf{q})+2\alpha\left(\frac{q_y}{q}\right)^2\right]^2}\\ &= \frac{2D\left(\frac{q_x^2}{q_\perp^2}\right)P_{xx}(\mathbf{q})}{\left[\omega-c(\hat{\mathbf{q}})q\right]^2+\left[\Gamma(\mathbf{q})+2\alpha\left(\frac{q_y}{q}\right)^2\right]^2}\\ &= \frac{2D\left(\frac{q_x^2}{q^2}\right)}{\left[\omega-c(\hat{\mathbf{q}})q\right]^2+\left[\Gamma(\mathbf{q})+2\alpha\left(\frac{q_y}{q}\right)^2\right]^2}.\end{aligned} \tag{6.2.13}$$

Finally, for u_x we find

$$\left\langle|u_x(\mathbf{q},\omega)|^2\right\rangle = \left(\frac{q_\perp^2}{q_x^2}\right)\left\langle|u_P(\mathbf{q},\omega)|^2\right\rangle = \frac{2D\left(\frac{q_\perp^2}{q^2}\right)}{\left[\omega-c(\hat{\mathbf{q}})q\right]^2+\left[\Gamma(\mathbf{q})+2\alpha\left(\frac{q_y}{q}\right)^2\right]^2}. \tag{6.2.14}$$

The interested reader can easily verify that all cross-correlations between $\mathbf{u}_T$, u_P, and u_x vanish exactly.

Integrating these over over all frequency ω, and dividing by 2π, gives the equal time correlation functions:

$$\left\langle|\mathbf{u}_T(\mathbf{q},t)|^2\right\rangle = \frac{(d-2)D}{\Gamma(\mathbf{q})}, \tag{6.2.15}$$

$$\left\langle|u_x(\mathbf{q},t)|^2\right\rangle = \frac{Dq_\perp^2}{\Gamma(\mathbf{q})q^2+2\alpha q_\perp^2}, \tag{6.2.16}$$

$$\left\langle u_P(\mathbf{q},t)u_P(\mathbf{q}',t)\right\rangle = \frac{Dq_x^2}{\Gamma(\mathbf{q})q^2+2\alpha q_\perp^2}. \tag{6.2.17}$$

The expressions above show that, as expected, the fluctuations of $\mathbf{u}$ are dominated by those of $\mathbf{u}_T$ (except in $d = 2$, where they vanish). Specifically, as $\mathbf{q} \to \mathbf{0}$, $\langle |u_x(\mathbf{q},t)|^2 \rangle$ is always much smaller than $\langle |\mathbf{u}_T(\mathbf{q},t)|^2 \rangle$, and $\langle |u_P(\mathbf{q},t)|^2 \rangle$ is much smaller than $\langle |\mathbf{u}_T(\mathbf{q},t)|^2 \rangle$ unless $q_\perp \lesssim q_x^2$, which represents a tiny fraction in $\mathbf{q}$-space. In addition, the dominant field $\mathbf{u}_T$ has spatially isotropic fluctuations in this linear theory.

Now the real-space fluctuations can be readily calculated:

$$\begin{aligned}\langle |\mathbf{u}(\mathbf{r},t)|^2 \rangle &= \frac{1}{(2\pi)^d} \int \mathrm{d}^d q \langle |\mathbf{u}(\mathbf{q},t)|^2 \rangle \\ &\approx \frac{1}{(2\pi)^d} \langle |\mathbf{u}_T(\mathbf{q},t)|^2 \rangle \\ &= \frac{(d-2)}{(2\pi)^d} \int_{q \gtrsim \frac{1}{L}} \mathrm{d}^d q \, \frac{D}{\Gamma(\mathbf{q})} ,\end{aligned} \tag{6.2.18}$$

where in the "$\approx$" we have kept only the dominant fluctuations. This integral converges as $L \to \infty$ for $d > 2$, which implies long-ranged orientational order in the ordered phase in $d = 3$.

Of course, everything we've done up to here is based on the linear theory, which you'd be very well justified in doubting. We found that this conclusion remains valid even beyond the linear theory. I will present our demonstration of this, and our calculation of the true scaling laws, which *are*, indeed, modified by the nonlinearities, in Section 6.3.

6.3 Nonlinear Theory

Since the fluctuations of $\mathbf{u}_\perp$ dominate over those of u_x, we insert equation (6.1.6) into equation (6.1.5) and focus on the $\perp$ components of the resultant equations of motion. By power counting, we can show that all nonlinearities arising from $U(v)$ are irrelevant, as well as those arising from the P_2 term; details are given in Section 6.4. Dropping those nonlinearities, and boosting to a new Galilean frame via the change of variables $x = x' - v_1 t$, where $v_1 = \lambda v_0$, gives

$$\partial_t \mathbf{u}_\perp + \lambda(\mathbf{u}_\perp \cdot \nabla_\perp)\mathbf{u}_\perp = -\nabla_\perp P + \mu_\perp \nabla_\perp^2 \mathbf{u}_\perp + \mu_x \partial_x^2 \mathbf{u}_\perp + \mathbf{f}_\perp , \tag{6.3.1}$$

where all x-derivatives are now implicitly derivatives with respect to the "pseudo-comoving" coordinate x' defined previously; we have simply suppressed the primes for convenience.

I'll now present our calculation of the exact scaling exponents from equation (6.3.1), using the DRG approach described in Chapter 3, and [29].

Because of the anisotropy of the flocking problem, we slightly modified the approach used in Chapter 3, by taking the Brillouin zone to be anisotropic. Since short-wavelength physics will not affect the long-wavelength physics, we were free to choose any Brillouin zone that we liked, provided that it prevented any ultraviolet divergences. We therefore made the choice that proves most computationally convenient, which is a cylindrical Brillouin zone, rather than a spherical one. That is, we took our Brillouin zone to be defined by:

$$|\vec{q}_\perp| \leq \Lambda, \quad -\infty < q_x < \infty, \tag{6.3.2}$$

with Λ, as usual, an ultraviolet cutoff given by the inverse of some microscopic length.

The DRG, as always, now proceeds by averaging the equation of motion over the short-wavelength fluctuations. With our choice of a cylindrical Brillouin zone, this means integrating out degrees of freedom in the hypercylindrical shell of Fourier space $b^{-1}\Lambda \leq |\vec{q}_\perp| \leq \Lambda$, where Λ is an "ultraviolet cutoff" and b is an arbitrary rescaling factor. Then, in an additional new wrinkle, I'll *anisotropically* rescale lengths, time, and $\mathbf{u}_\perp$ in equation (6.3.1) according to $\mathbf{r}_\perp = b\mathbf{r}'_\perp$, $x = b^\zeta x'$, $t = b^z t'$, and $\mathbf{u}_\perp = b^\chi \mathbf{u}'_\perp$. Note that, by construction, the rescaling of $\mathbf{r}_\perp$ automatically restores the ultraviolet cutoff to Λ.

When evaluating the "graphical corrections" – that is, the renormalizations that arise due to averaging over the short-wavelength fluctuations – it is extremely useful to note that the $\lambda(\mathbf{u}_\perp \cdot \nabla_\perp)\mathbf{u}_\perp$ term can now, for this incompressible problem, be written as a total $\perp$ derivative:

$$\begin{aligned}(\mathbf{u}_\perp \cdot \nabla_\perp)u_m^\perp &= \nabla_n^\perp\left(u_n^\perp u_m^\perp\right) - u_m^\perp \nabla_\perp \cdot \mathbf{u}_\perp \\ &= \nabla_n^\perp\left(u_n^\perp u_m^\perp\right) + u_m^\perp \partial_x u_x \\ &\approx \nabla_n^\perp\left(u_n^\perp u_m^\perp\right),\end{aligned} \tag{6.3.3}$$

where in the second "=" we have replaced $\nabla_\perp \cdot \mathbf{u}_\perp$ with $-\partial_x u_x$ using the incompressibility condition, and in "$\approx$" we have ignored $u_m^\perp \partial_x u_x$ since it is much smaller by power counting than $\nabla_n^\perp\left(u_n^\perp u_m^\perp\right)$, since fluctuations of $u_\perp$ dominate over those of u_x.

This implies that, when averaging over short-wavelength fluctuations, the λ term can only renormalize terms that contain at least one $\perp$ spatial derivative. Since this term is the *only* relevant nonlinear term in the model, this means that only terms that contain at least one $\perp$ spatial derivative can get any graphical renormalization *at all*. In particular neither the "x viscosity" μ_x, nor the noise strength D, can get any graphical renormalization.

Furthermore, there is no graphical correction to λ, due to the pseudo-Galileo invariance of the equation of motion (6.3.1). That is, if we let $\mathbf{u}_\perp(\mathbf{x}, t) = \mathbf{u}_\perp(\mathbf{x}, t) + \mathbf{u}_0$ and simultaneously boost the coordinate $\mathbf{r} = \mathbf{r} - \lambda \mathbf{u}_0 t$, where $\mathbf{u}_0$ is an arbitrary position-$\mathbf{r}$ independent vector in the $\perp$ plane, the equation of motion (equation

(6.3.1)) remains invariant. Since this symmetry of the equation of motion involves λ, this implies that λ cannot be graphically renormalized.

Based on these arguments, the DRG flow equations can be written *exactly* as

$$\frac{d \ln \mu_x}{d\ell} = z - 2\zeta , \tag{6.3.4}$$

$$\frac{d \ln \mu_\perp}{d\ell} = z - 2 + \eta_\mu , \tag{6.3.5}$$

$$\frac{d \ln \lambda}{d\ell} = z - 1 + \chi , \tag{6.3.6}$$

$$\frac{d \ln D}{d\ell} = z - 2\chi - \zeta - (d-1) , \tag{6.3.7}$$

where η_μ represents graphical corrections to $\mu_\perp$, which is now the *only* parameter that actually gets any graphical corrections.

At a fixed point, the right-hand sides of equations (6.3.4), (6.3.6), and (6.3.7) must vanish. Since these don't involve any graphical corrections, we'll be able to determine the scaling exponents z, χ, and ζ *exactly*, and with extremely little work. The situation is comparable to (albeit, obviously, different in detail from) that in the one-dimensional KPZ equation, for which we could also obtain exact exponents, because of certain exact statements that could be made about the graphical corrections.

Setting the right-hand sides of equations (6.3.4), (6.3.6), and (6.3.7) all equal to zero leads to the three conditions:

$$z - 2\zeta = 0 , \tag{6.3.8}$$

$$z - 1 + \chi = 0, \tag{6.3.9}$$

$$z - 2\chi - \zeta - (d-1) = 0 , \tag{6.3.10}$$

which are readily solved to give the scaling exponents

$$\zeta = \frac{d+1}{5}, \quad z = \frac{2(d+1)}{5}, \quad \chi = \frac{3-2d}{5} \tag{6.3.11}$$

for spatial dimensions $2 < d \leq 4$. For $d > 4$, the anisotropy and dynamical exponents ζ and z lock on to their values in $d = 4$: $\zeta = 1, z = 2$, while the "roughness" exponent $\chi = \frac{2-d}{2}$ for $d > 4$. Note that all three exponents are continuous functions of d near $d = 4$, but have discontinuous slopes at $d = 4$. We emphasize that these exponents are exact in the sense that their derivation is not perturbative in nature, in contrast to, e.g., the ϵ-expansion method.

Note, however, that these exponents are not continuous at $d = 2$; that is, the exponents do not take on the values given by equations (6.3.11) in $d = 2$ [50]. This is because, as discussed above, and as I'll show in Chapter 8, and as also discussed in [50], the number of soft modes in incompressible flocks is $d-2$. Because this vanishes in $d=2$, the two-dimensional case is special, and is not characterized by the exponents (6.3.11). It *is*, however, possible to obtain exact results for the $d=2$ case. In fact, these results can be derived by a mapping to our old friend the KPZ equation, but in *one* spatial dimension.

Returning now to the case $d>2$, we noted that, with these exponents (6.3.11) in hand, we can make predictions for the scaling behavior of the velocity correlation functions

$$C(x,\mathbf{r}_\perp,t)\equiv\langle\mathbf{u}_\perp(x',\mathbf{r}'_\perp,t')\cdot\mathbf{u}_\perp(x'',\mathbf{r}''_\perp,t'')\rangle, \tag{6.3.12}$$

where $x=x''-x'$, $\mathbf{r}_\perp=\mathbf{r}''_\perp-\mathbf{r}'_\perp$, and $t=t''-t'$. The DRG analysis implies

$$C(x,\mathbf{r}_\perp,t)=b^{2\chi}C(|x|b^{-\zeta},r_\perp b^{-1},|t|b^{-z}). \tag{6.3.13}$$

Letting $b=r_\perp$ in the above equation, we obtain

$$C(x,\mathbf{r}_\perp,t)=r_\perp^{2\chi}g\left(\frac{|x|}{r_\perp^{\zeta}},\frac{|t|}{r_\perp^{z}}\right), \tag{6.3.14}$$

where

$$g\left(\frac{|x|}{r_\perp^{\zeta}},\frac{|t|}{r_\perp^{z}}\right)\equiv C\left(\frac{|x|}{r_\perp^{\zeta}},1,\frac{|t|}{r_\perp^{z}}\right) \tag{6.3.15}$$

is a scaling function. The scaling behavior of $g(X,T)$ can be deduced from three limiting cases. For $r_\perp\to\infty$, $x=0$, and $|t|=0$, $C(|x|,r_\perp,|t|)$ should only depend on $r_\perp$, which implies $g(X,T)\sim 1$. Likewise, $g(X,T)\sim X^{\frac{2\chi}{\zeta}}$ for $r_\perp=0$, $|x|\to\infty$, and $t=0$; $g(X,T)\sim T^{\frac{2\chi}{z}}$ for $r_\perp=0$, $x=0$, and $|t|\to\infty$. The crossover between these limiting cases can be worked out by connecting the three results of $g(X,T)$ in the parameter space of X and T. Finally we have

$$g(X,T)\sim\begin{cases}1, & X\ll 1, T\ll 1,\\ X^{\frac{2\chi}{\zeta}}, & X\gg 1, X\gg T^{\frac{1}{z}},\\ T^{\frac{2\chi}{z}}, & T\gg 1, T\gg X^{2}.\end{cases} \tag{6.3.16}$$

Plugging (6.3.16) into (6.3.14), we find the scaling behavior of the velocity correlation function:

$$C(|x|,r_\perp,|t|)\sim\begin{cases}r_\perp^{2\chi}, & |x|\ll r_\perp^{\zeta},\ |t|\ll r_\perp^{z},\\ |x|^{\frac{2\chi}{\zeta}}, & |x|\gg r_\perp^{\zeta}, |x|\gg |t|^{\frac{1}{z}},\\ |t|^{\frac{2\chi}{z}}, & |t|\gg r_\perp^{z}, |t|\gg |x|^{2}.\end{cases} \tag{6.3.17}$$

Transforming this expression back to the lab coordinates by replacing x with $x-v_1 t$ leads to the result for the correlation function in "lab" coordinates:

$$C(|x-v_1t|, r_\perp, |t|) \sim \begin{cases} r_\perp^{2\chi}, & |x-v_1t| \ll r_\perp^{\zeta}, |t| \ll r_\perp^{z}, \\ |x-v_1t|^{\frac{2\chi}{\zeta}}, & |x-v_1t| \gg r_\perp^{\zeta}, |x-v_1t| \gg |t|^{\frac{1}{2}}, \\ |t|^{\frac{2\chi}{z}}, & |t| \gg r_\perp^{z}, |t| \gg |x-v_1t|^2 . \end{cases} \tag{6.3.18}$$

6.4 Loose Ends

I'll present here our argument that, except for the λ term, all of the nonlinear terms in the equation of motion (6.1.5) are irrelevant, in the renormalization group sense [29].

Most of these nonlinearities arise from the $U(v)\mathbf{v}$ term in (6.1.5). It is therefore convenient to solve for $U(v)$ directly in terms of $\mathbf{u}$ and its spatial derivatives. Following [37], we do so by first taking the dot product of both sides of equation (6.1.5) with $\mathbf{v}$ itself. This gives

$$\frac{1}{2}\left[\partial_t + \lambda\left(\mathbf{v}\cdot\nabla\right)\right]|\mathbf{v}|^2 = U(v)|\mathbf{v}|^2 - \mathbf{v}\cdot\nabla P - |\mathbf{v}|^2\mathbf{v}\cdot\nabla P_2 + \mu_\perp \mathbf{v}\cdot\nabla_\perp^2\mathbf{v} + \mu_x\mathbf{v}\cdot\partial_x^2\mathbf{v} + \mathbf{v}\cdot\mathbf{f}. \tag{6.4.1}$$

Using our expression

$$\mathbf{v}(\mathbf{r},t) = (v_0 + u_x(\mathbf{r},t))\hat{\mathbf{x}} + \mathbf{u}_\perp(\mathbf{r},t), \tag{6.4.2}$$

to rewrite this in terms of the fluctuation $\mathbf{u}$, dropping "obviously irrelevant" terms – i.e., terms that differ from others in the equation of motion only by having more powers of the fields u_x or $\mathbf{u}_\perp$ – and solving for $U(v)$, we obtain

$$\begin{aligned} U(v) = {} & \frac{1}{v_0}\partial_x P + \mathbf{v}\cdot\nabla P_2 + \left(\frac{1}{2v_0^2}\right)\left(\partial_t + v_1\partial_x + \lambda\mathbf{u}\cdot\nabla\right)\left(2v_0u_x + |\mathbf{u}_\perp|^2\right) \\ & - \frac{f_x}{v_0} - \left(\frac{1}{v_0^2}\right)\left(\mu_\perp\mathbf{u}_\perp\cdot\nabla_\perp^2\mathbf{u}_\perp + \mu_x\mathbf{u}_\perp\cdot\partial_x^2\mathbf{u}_\perp\right) \\ & - \left(\frac{1}{v_0}\right)\left(\mu_\perp\nabla_\perp^2 u_x + \mu_x\partial_x^2 u_x\right), \end{aligned} \tag{6.4.3}$$

where $v_1 = \lambda v_0$.

Inserting this back into our equation of motion (6.1.5), using our expression (6.1.6) for $\mathbf{v}$, and focusing on the components of that equation perpendicular to the direction of mean motion x, we obtain

$$\partial_t \mathbf{u}_\perp + v_1 \partial_x \mathbf{u}_\perp + \lambda(\mathbf{u}_\perp \cdot \nabla_\perp)\mathbf{u}_\perp = -\nabla_\perp P + \mu_\perp \nabla_\perp^2 \mathbf{u}_\perp + \mu_x \partial_x^2 \mathbf{u}_\perp + \mathbf{f}_\perp$$
$$+ \mathbf{u}_\perp \left[\frac{1}{v_0} \partial_x P + \left(\frac{1}{2v_0^2} \right) \left(\partial_t + v_1 \partial_x + \lambda_1 \mathbf{u}_\perp \cdot \nabla_\perp \right) \left(2v_0 u_x + |\mathbf{u}_\perp|^2 \right) \right.$$
$$\left. - \frac{f_x}{v_0} \right] - \mathbf{u}_\perp \left[\left(\frac{1}{v_0^2} \right) \left(\mu_\perp \mathbf{u}_\perp \cdot \nabla_\perp^2 \mathbf{u}_\perp + \mu_x \mathbf{u}_\perp \cdot \partial_x^2 \mathbf{u}_\perp \right) \right.$$
$$\left. + \left(\frac{1}{v_0} \right) \left(\mu_\perp \nabla_\perp^2 u_x + \mu_x \partial_x^2 u_x \right) \right]. \tag{6.4.4}$$

Note that the P_2 term has been canceled out by a piece of the $U(v)$ term. Note also that we have replaced $(\mathbf{u} \cdot \nabla)$ with $(\mathbf{u}_\perp \cdot \nabla_\perp)$, which is justified since $u_x \ll |\mathbf{u}_\perp|$ in the long-wavelength limit, as discussed in our treatment of the linear theory.

It is now straightforward to show by power counting that every term in the last two lines of equation (6.4.4) – that is, every term arising from the $U(v)$ and P_2 terms in equation (6.1.5) – is irrelevant *in dimensions higher than 2*. This can be seen most easily by comparing them with various similar terms on the first line, as we will demonstrate now.

There are 12 terms in total.

(i) $\mathbf{u}_\perp \partial_x P$: This term is subtle. It is tempting, but misleading, to note that this has the same number of spatial derivatives as the $\nabla_\perp P$ term, but one extra power of $\mathbf{u}_\perp$, and so is apparently irrelevant relative to that $\nabla_\perp$ term. The subtlety is that $\nabla_\perp P$ is purely parallel to $\hat{\mathbf{q}}_\perp$, while $\mathbf{u}_\perp \partial_x P$ has components transverse to $\hat{\mathbf{q}}_\perp$, since $\mathbf{u}_\perp$ itself does. Hence, in order to prove that the $\mathbf{u}_\perp \partial_x P$ term is truly irrelevant, we must show that it is negligible relative to some other term in (6.4.4) that also has a transverse (to $\mathbf{q}_\perp$) component. To do so, we need to obtain the power counting of P itself. This can be done by taking $\nabla_\perp \cdot$ both sides of (6.4.4). Neglecting the "obviously irrelevant" terms we solve the resultant equation for $\nabla_\perp^2 P$:

$$\nabla_\perp^2 P = (\partial_t + v_1 \partial_x)\, \partial_x u_x - \lambda \nabla_\perp \cdot \left[(\mathbf{u}_\perp \cdot \nabla_\perp)\mathbf{u}_\perp \right] + \nabla_\perp \cdot \mathbf{f}_\perp, \tag{6.4.5}$$

where we have used the incompressibility condition $\nabla \cdot \mathbf{v} = \nabla_\perp \cdot \mathbf{u}_\perp + \partial_x u_x = 0$ to rewrite the first two terms on the right-hand side in terms of u_x. Here in order to obtain $\nabla_\perp^2 P$ we have also neglected $\mathbf{u}_\perp \partial_x P / v_0$ on the right-hand side of (6.4.4), which we will show in the following to be irrelevant.

Inspection of this equation reveals that $\partial_x P$ has four terms: the first, coming from the $\partial_t \partial_x u_x$ term on the right-hand side of (6.4.5), power counts like $\partial_t u_x$. This is because $\nabla_\perp^2 P \sim \partial_t \partial_x u_x$ and derivatives in all directions power count in the same way, since scaling is isotropic according to our linear theory. Thus, this piece has the same power counting as the explicit $\partial_t u_x$ term that appears later in (6.4.4). We can thus deal with this piece of $\partial_x P$ at the same time as we deal with that explicit term, as we will in a few paragraphs.

Likewise, the next term on the right-hand side of (6.4.5), which is proportional to $\partial_x^2 u_x$, contributes to $\partial_x P$ a term which power counts like $\partial_x u_x$. Since an explicit term of that form appears later in (6.4.4), we will deal with this piece of $\partial_x P$ at the same time as we deal with that explicit term, as we also will in a few paragraphs.

Similarly, the $\nabla_\perp \cdot \left[(\mathbf{u}_\perp \cdot \nabla_\perp)\mathbf{u}_\perp\right]$ term contributes to $\partial_x P$ a term which power counts exactly like the explicit $\partial_x|\mathbf{u}_\perp|^2$ term that appears later in (6.4.4), so we can deal with it when we deal with that term in a few paragraphs.

Finally, the $\nabla_\perp \cdot \mathbf{f}_\perp$ term contributes to $\partial_x P$ a term that power counts exactly like the explicit (f_x/v_0) term that appears later in (6.4.4), so we can deal with *it* when we deal with *that* term in a few paragraphs.

Now turning to the terms explicitly displayed on the last three lines of (6.4.4).

(ii) $\mathbf{u}_\perp \partial_t u_x$: This term has the same number and type of derivatives as the $\partial_t \mathbf{u}_\perp$ term on the first line, but one extra power of u_x, so it is irrelevant compared with that $\partial_t \mathbf{u}_\perp$ term.

(iii) $\mathbf{u}_\perp \partial_x u_x$: This term has the same number of spatial derivatives as the $(\mathbf{u}_\perp \cdot \nabla_\perp)\mathbf{u}_\perp$ term on the first line, but has a u_x instead of a second $\mathbf{u}_\perp$. Again using the fact that $u_x \ll |\mathbf{u}_\perp|$ in the long-wavelength limit, we see that the $\mathbf{u}_\perp \partial_x u_x$ term is negligible relative to the $\mathbf{u}_\perp \cdot \nabla \mathbf{u}_\perp$ term.

(iv) $\mathbf{u}_\perp(\mathbf{u}_\perp \cdot \nabla_\perp u_x)$: This term differs from the $(\mathbf{u}_\perp \cdot \nabla_\perp)\mathbf{u}_\perp$ term by having one extra power of u_x, so it is irrelevant relative to the $(\mathbf{u}_\perp \cdot \nabla_\perp)\mathbf{u}_\perp$ term.

(v) $\mathbf{u}_\perp \partial_t|\mathbf{u}_\perp|^2$: This term has the same derivatives as the $\partial_t \mathbf{u}_\perp$ term, but two more powers of $\mathbf{u}_\perp$, so it is negligible relative to the $\partial_t \mathbf{u}_\perp$ term.

(vi) $\mathbf{u}_\perp \partial_x|\mathbf{u}_\perp|^2$: This term has the same number of spatial derivatives as the $(\mathbf{u}_\perp \cdot \nabla_\perp)\mathbf{u}_\perp$ term but one more power of $\mathbf{u}_\perp$, so it is negligible relative to the $(\mathbf{u}_\perp \cdot \nabla_\perp)\mathbf{u}_\perp$ term.

(vii) $\mathbf{u}_\perp \mathbf{u}_\perp \cdot \nabla_\perp|\mathbf{u}_\perp|^2$: This term similarly has the same number of spatial derivatives as the $(\mathbf{u}_\perp \cdot \nabla_\perp)\mathbf{u}_\perp$ term, but now with two more powers of $\mathbf{u}_\perp$, so it is even more negligible relative to the $(\mathbf{u}_\perp \cdot \nabla_\perp)\mathbf{u}_\perp$ term.

(viii) $f_x \mathbf{u}_\perp$: This term has the same number of powers of the random force as the $\mathbf{f}_\perp$ term, but one extra power of $\mathbf{u}_\perp$, so it is negligible compared with the $\mathbf{f}_\perp$ term.

(ix) and (x) $\mathbf{u}_\perp\left[\left(\mu_\perp \mathbf{u}_\perp \cdot \nabla_\perp^2 \mathbf{u}_\perp + \mu_x \mathbf{u}_\perp \cdot \partial_x^2 \mathbf{u}_\perp\right)\right]$: These terms have one more spatial derivative, and one more power of $\mathbf{u}_\perp$, than the $(\mathbf{u}_\perp \cdot \nabla_\perp)\mathbf{u}_\perp$ term, and so are doubly negligible in comparison with that term.

(xi) and (xii) $\mathbf{u}_\perp\left(\mu_\perp \nabla_\perp^2 u_x + \mu_x \partial_x^2 u_x\right)$: Finally, these terms have one more spatial derivative than the $(\mathbf{u}_\perp \cdot \nabla_\perp)\mathbf{u}_\perp$ term; they also have a u_x, rather than a $\mathbf{u}_\perp$, and so are doubly negligible in comparison with the $(\mathbf{u}_\perp \cdot \nabla_\perp)\mathbf{u}_\perp$ term, since, as we established in Section 6.2, $u_x \ll |\mathbf{u}_\perp|$ in the long-distance limit.

So we have, rather laboriously, established that all of the terms on the second, third, and fourth lines of (6.4.4) are irrelevant, in the RG sense, at long distances. Hence we can drop them all, leaving our effective long-wavelength model for the fluctuations $\mathbf{u}_\perp$ as:

$$\partial_t \mathbf{u}_\perp + \lambda(\mathbf{u}_\perp \cdot \nabla_\perp)\mathbf{u}_\perp = -\nabla_\perp P + \mu_\perp \nabla_\perp^2 \mathbf{u}_\perp + \mu_x \partial_x^2 \mathbf{u}_\perp + \mathbf{f}_\perp, \tag{6.4.6}$$

where we have eliminated the term $v_1 \partial_x \mathbf{u}_\perp$ on the right-hand side of the equality by making a Galilean transformation to a "pseudo-comoving" coordinate system moving along $\hat{\mathbf{x}}$ with a constant speed v_1. The remainder of the analysis of this nonlinear model was given earlier in this chapter (Section 6.3).

7

Heuristic Argument for the Canonical Exponents ("20-20 Hindsight Hand-Waving Argument")

In this chapter, I will rederive the scaling results I found for incompressible flocks in the previous chapter by an extension of the "blob" derivation of the Mermin–Wagner–Hohenberg theorem given in Chapter 2.

7.1 The Blob Wanders Faster Than it Spreads

Consider the group of birds whose velocities at time t will be well correlated with what some reference bird was doing at $t = 0$. This group will be a "blob" whose center moves along the direction of mean motion of the flock at a speed γ.

I'll start by proving by contradiction that, for spatial dimensions $d \leq 4$, the motion of the flockers implies that the width of this blob can *not* scale like the width of the blob of pointers in Chapter 2. Those of you who've been paying attention will note that the critical dimension of 4 is exactly what we found in the dynamical RG previously.

If we *do* assume that this blob grows like the analogous blob for pointers – that is, diffusively – then it will be essentially isotropic, and have width

$$w(t) \propto t^{1/2}, \tag{7.1.1}$$

as illustrated in Figure 7.1.1.

But *unlike* the pointers, this blob will now be moving, laterally as well as along the mean flock motion. Indeed, its velocity perpendicular to the mean direction of flock motion will be $\delta \mathbf{v}_\perp \sim v_0\theta$, where θ is the deviation of the mean direction of the blob from that of the flock. This means that the blob will wander laterally relative to the rest of the flock.

How far will it wander laterally in time t? Roughly,

$$\delta x \sim \sqrt{\langle v_\perp^2 \rangle t} \sim v_0\sqrt{\langle \theta^2 \rangle t}\,. \tag{7.1.2}$$

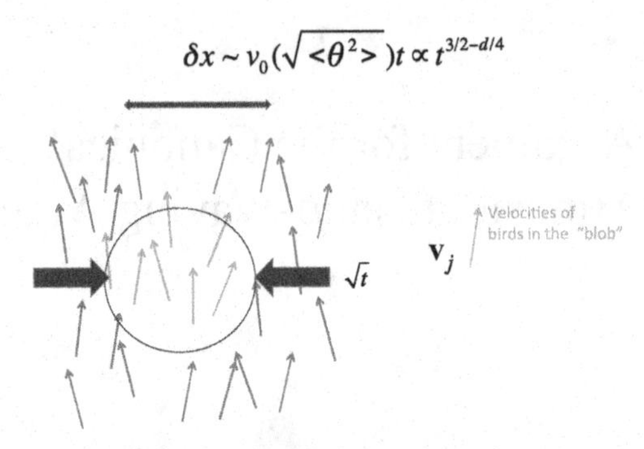

Fig. 7.1.1 Evolution of the "blob" in the *incorrect* (for spatial dimensions $d < 4$) picture that it grows diffusively in all directions. The lateral wandering of the blob δx actually eventually exceeds its diffusive width $w(t)$ for all spatial dimensions $d \leq 4$. Reproduced from [7] by permission of Oxford University Press.

Note that this wandering is purely lateral: Fluctuations in the velocity *along* the mean direction of flock motion go like $v_0[1 - \cos(\theta)] \propto \theta^2$ for small θ, and, so, are much smaller than the lateral velocity fluctuations $\delta \mathbf{v}_\perp$.

If we now assume (and remember, we're going to be showing that this assumption is actually self-contradictory for $d \leq 4$), that the fluctuations of θ of this blob of flockers scale just as those of the blob of pointers in Chapter 2, we have:

$$\sqrt{\langle\theta^2\rangle} \propto r^{1-d/2} \propto t^{1/2-d/4}. \tag{7.1.3}$$

Using this in (7.1.2), I get

$$\delta x \propto t^{3/2-d/4}. \tag{7.1.4}$$

Comparing this to the width $w(t)$ of the blob, I find

$$\frac{\delta x}{w(t)} \propto \frac{t^{3/2-d/4}}{t^{1/2}} \propto t^{1-d/4}. \tag{7.1.5}$$

Note that this ratio diverges as $t \to \infty$ for $d < 4$. Therefore, $d = 4$ is another critical dimension, below which the assumption that information transport is dominated by diffusion breaks down. This is a "breakdown of linearized hydrodynamics." The linear theory, which ignores the convective $\lambda_1 \mathbf{v}_\perp \cdot \nabla \mathbf{v}_\perp$ term – that is, the term that contains the physics of the lateral wandering of the "blob" just calculated – is incorrect: That nonlinear term – i.e., the convective wandering of the blob – actually dominates over the linear diffusive process.

7.2 All Who Wander Are Not Lost: Beating the Mermin–Wagner–Hohenberg Theorem

What happens for $d < 4$? Well, now the blob must become anisotropic, since information is transmitted much more rapidly perpendicular to the mean direction of flock motion, while parallel to the mean direction of flock motion it's still diffusive, since the speed of the flockers does not fluctuate much (it isn't a Goldstone mode), and it doesn't vary much as the direction θ fluctuates ($v_{\parallel} \sim \mathcal{O}(\theta^2)$ versus $\mathbf{v}_{\perp} \sim \mathcal{O}(\theta)$). So the blob actually looks like Figure 7.2.1, where the length of the blob along the mean direction of flock motion still grows diffusively (i.e., like $\sqrt{t}$), while the spatial extent $w(t)$ of the blob in all of the $d-1$ directions perpendicular to $\langle \mathbf{v} \rangle$ is controlled by the fluctuations of $\mathbf{v}_{\perp}$, which I will have to determine self-consistently. I'll do so by the same sort of sloppy hand-waving argument I used in Chapter 2 to "derive" the Mermin–Wagner–Hohenberg theorem. That is, I'll note that

$$\#\,\text{of errors/flocker} \propto t\,. \tag{7.2.1}$$

The number of flockers making errors is

$$N(t) \propto \text{volume of blob} \propto [w(t)]^{d-1}\sqrt{t}\,. \tag{7.2.2}$$

Hence, the total number of errors made inside the blob after a time t is

$$\text{total}\,\#\,\text{of errors} \propto N(t)t\,. \tag{7.2.3}$$

This gives for the rms fluctuations of θ

$$\sqrt{\langle \theta^2 \rangle} \approx \frac{\sqrt{\text{total}\,\#\,\text{of errors}}}{N(t)} \propto \frac{\sqrt{N(t)t}}{N(t)} \propto \sqrt{\frac{t}{N(t)}} \propto t^{1/4} w^{(1-d)/2}\,. \tag{7.2.4}$$

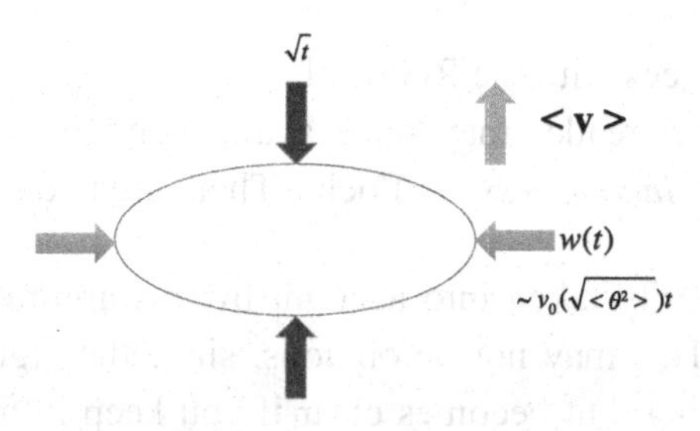

Fig. 7.2.1 Correct picture of the evolving blob. It still grows diffusively in the direction of the mean flock velocity (vertical in this figure), but grows laterally at a rate dominated by motion of the flockers (i.e., convection), rather than diffusion. This convective width $w(t)$ can be determined self-consistently via equation (7.2.5). Reproduced from [7] by permission of Oxford University Press.

As for our earlier estimate of δx, we can estimate the lateral width $w(t)$ as

$$w(t) \sim v_0 \sqrt{\langle\theta^2\rangle} t \propto t^{5/4} w^{(1-d)/2} . \tag{7.2.5}$$

This is the promised self-consistent condition on $w(t)$. It is easily solved to give

$$w(t) \propto t^{\frac{5}{2(d+1)}} , \tag{7.2.6}$$

which can be inverted to give the dynamical exponent z:

$$t(w) \propto w^z , \tag{7.2.7}$$

with

$$z = \frac{2(d+1)}{5} . \tag{7.2.8}$$

Note that this is the same result I got from the dynamical RG argument of Chapter 4! I can get the anisotropy exponent by using the fact that the blob still grows diffusively in the parallel direction:

$$w_{\parallel} \propto t(w)^{1/2} \propto w^{\zeta} , \tag{7.2.9}$$

with

$$\zeta = z/2 = \frac{(d+1)}{5} , \tag{7.2.10}$$

which also agrees with the result of the RG analysis of Section 7.1. (To obtain the second proportionality in (7.2.9), I've used the relation (7.2.7) to relate $t(w)$ to w.)

Finally, using (7.2.7) in (7.2.4), I get

$$\sqrt{\langle\theta^2\rangle} \propto w^{\chi} \tag{7.2.11}$$

with

$$\chi = z/4 + (1-d)/2 = (3-2d)/5 , \tag{7.2.12}$$

which, once again, agrees with the RG result.

The ubiquitous alert reader may once again wonder why the argument just presented is limited to *incompressible* flocks. There are two reasons for this.

(1) The only nonlinearity taken into account by this argument is the convective nonlinearity λ_1. This may not be obvious, since the argument just presented was so hand-waving, but becomes clear if you keep in mind the fact that the convective nonlinearity simply embodies the physics of transport by motion, which was at the heart of the earlier argument. In the compressible model of Chapter 4, however, there are also many nonlinearities involving the density, which can (and do) also contribute to the anomalous hydrodynamics, in ways

that are not so easy (indeed, impossible, at least based on my best efforts!) to capture by a simple hand-waving argument.

In incompressible flocks, of course, the density fluctuations are forbidden, and so the nonlinearities associated with the density drop out of the problem.

(2) The hand-waving argument also implicitly assumes that there is no graphical renormalization of either the noise or the diffusion constant μ_x along the direction of flock motion. Both of these assumptions are true for the incompressible flock, because the λ_1 term, which is the sole relevant nonlinearity in the problem, can be written, *in the incompressible case*, as a total $\perp$ derivative, as shown in equation (6.3.3). In Chapter 9, we study "Malthusian" flocks – that is, flocks with birth and death, so that number is not conserved. For those systems, the λ_1 term is again the only remaining relevant term, but now the velocity field is not divergenceless (i.e., the flow is not incompressible). We then find that *both* the diffusion constant μ_x and the noise D *are* renormalized, which also invalidates the above hand-waving argument, which neglected either possibility. The neglect of the renormalization of μ_x came in when I took the width of the "blob" along the direction of motion to scale like $\sqrt{t}$, while the assumption that the noise strength D was unrenormalized was implicit in the statement that the number of errors per unit time within the blob scaled like the number of flockers within it.

So the hand-waving argument, while appealing in its simplicity, doesn't get everything right, except for the very simple case of an incompressible flock. And even in that case, the argument still doesn't work in two dimensions, as you'll see in Chapter 8.

8

Incompressible Flocks in Spatial Dimensions $d = 2$: Mapping to the KPZ Equation

As I discussed in Chapter 5, before anyone had ever heard of an active fluid, the KPZ equation was a hot (in the sense of fashionable) nonequilibrium system. In this chapter I'll review the discovery [50], by the same suspects (Chiu Fan Lee, Leiming Chen, and me) as in Chapter 6, of a surprising connection between these growing interfaces in *one* dimension and incompressible flocks in *two* ($d = 2$): These two very different systems in fact belong to the same universality class. This universality class also includes two equilibrium systems: two-dimensional smectic liquid crystals, and a peculiar kind of constrained two-dimensional ferromagnet. (Don't worry if you've never heard of a smectic before; I'll describe these fascinating liquid-crystal systems later.) We used these connections to show that two-dimensional incompressible flocks are robust against fluctuations, and exhibit universal long-ranged, anisotropic spatio-temporal correlations of those fluctuations. We thereby determined the exact values of the anisotropy exponent ζ and the roughness exponents $\chi_{x,y}$ that characterize these correlations.

The alert reader might be wondering why we couldn't simply take the results of Chapter 6 for incompressible flocks in $d > 2$ and just set $d = 2$ in those expressions. The reason we can *not* do this, which I also discussed in Chapter 6, is illustrated in Figure 8.0.1, which represents the fluctuations at some particular $\mathbf{q}$ in Fourier space. In order to be a Goldstone mode, the Fourier components $\mathbf{u}(\mathbf{q}, t)$ of the velocity fluctuation $\mathbf{u}(\mathbf{r}, t) \equiv \mathbf{v}(\mathbf{r}, t) - \langle \mathbf{v} \rangle$ must be orthogonal to $\langle \mathbf{v} \rangle$. At the same time, incompressibility requires that $\mathbf{u}(\mathbf{q}, t)$ be orthogonal to $\mathbf{q}$. This is no problem in spatial dimensions $d > 2$, because in those dimensions there are some directions ($d - 2$ of them, to be precise) that *are* orthogonal to *both* $\langle \mathbf{v} \rangle$ and $\mathbf{q}$ (specifically, the direction(s) orthogonal to the plane of the figure in Figure 8.0.1). But in $d = 2$, obviously, there are none ($d - 2 = 0$). That is, in $d = 2$, you're not allowed to point out of the plane of the figure.

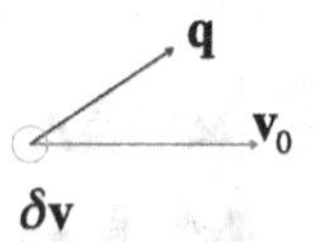

Fig. 8.0.1 Illustration of why fluctuations are actually *smaller* in a two-dimensional incompressible flock than in a three-dimensional one. In both two and three dimensions, Goldstone-mode fluctuations of the velocity at a given wavevector $\mathbf{q}$ must be orthogonal to *both* $\mathbf{q}$ (because of the incompressibility) and $\langle \mathbf{v} \rangle \equiv \mathbf{v}_0$. In three (or more) dimensions, this is easy: The velocity just points perpendicular to the plane spanned by $\mathbf{q}$ and $\mathbf{v}_0$, as illustrated by the small circle in this figure (which indicates an arrow pointing directly out of the page). In two dimensions, however, the velocity is confined to the plane, and so cannot point perpendicular to both $\mathbf{q}$ and $\mathbf{v}_0$, *unless* $\mathbf{q}$ and $\mathbf{v}_0$ are parallel. This restriction on the phase space of Goldstone fluctuations makes the velocity fluctuations much smaller in two dimensions than in three.

To put this in the language of Chapter 6, there is *no* $\mathbf{u}_T$ field in $d = 2$, since that field has $d - 2$ components in general.

Note that there is one exception to this statement: when $\mathbf{q} \parallel \langle \mathbf{v} \rangle$. Then we *can* have Goldstone-mode behavior of the velocity fluctuations. We will see this explicitly when we calculate the velocity fluctuations below. By continuity, if velocity fluctuations become large for $\mathbf{q} \parallel \langle \mathbf{v} \rangle$, they must also be large if $\mathbf{q}$ is *nearly* parallel to $\langle \mathbf{v} \rangle$. We saw this explicitly in Chapter 6, where this was exactly the behavior we found for u_P. In $d > 2$, however, this restriction of the phase space for Goldstone-mode-like behavior of u_P meant that its fluctuations were negligible compared with those of $\mathbf{u}_T$, which was soft for *all* directions of $\mathbf{q}$. However, in $d = 2$, there *is* no $\mathbf{u}_T$, so $\mathbf{u}_P$ becomes the dominant fluctuating field.

This completely changes the nature of the problem. It also means that, paradoxically, and contrary to all of our previous experience in this book (and almost everywhere else), fluctuations are actually *smaller* in $d = 2$ than in $d > 2$. Because of this, the case $d = 2$ requires separate, and special, treatment, which I'll present in this chapter.

Surprisingly, there proves to be enough phase space in this narrow window of wavevector to lead to anomalous hydrodynamics, but not enough to disorder the two-dimensional system, not even in the linear theory. Even more surprisingly, that anomalous hydrodynamics, in a peculiar sense that I'll describe in more detail in the following, proves to be exactly that which we found in Chapter 3 for the *one*-dimensional KPZ equation. Indeed, this problem also connects to an equilibrium problem known as the two-dimensional smectic. This set of connections is illustrated in Figure 8.0.2.

I'll now present the derivation of these results, starting with the formulation of the model.

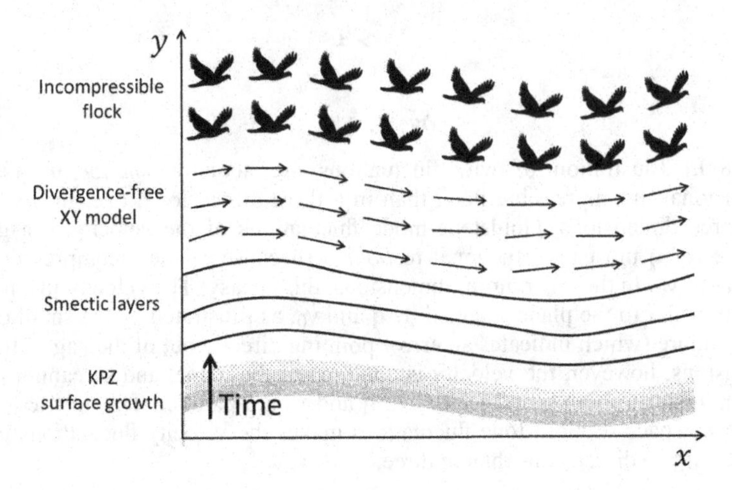

Fig. 8.0.2 The flow lines of the ordered phase of a two-dimensional incompressible polar active fluid, the magnetization lines of the ordered phase of divergence-free two-dimensional XY magnets, dislocation-free two-dimensional smectic layers, and the surfaces of a growing 1 + 1-dimensional crystal (which can be obtained by taking equal-time-interval snap shots), undulate in *exactly* the same way over space; their fluctuations share *exactly* the same asymptotic scaling behavior at large length scales. Note that the vertical axis is time for KPZ surface growth and the y Cartesian coordinate for the other three systems. This figure was conceived and made by Chiu Fan Lee of Imperial College (with some inspiration from M. C. Escher!), and reproduced from [50].

8.1 Model

We started with the hydrodynamic model for incompressible polar active fluids without momentum conservation in *arbitrary* spatial dimensions of Chapter 6:

$$\partial_t \mathbf{v} + \lambda(\mathbf{v}\cdot\nabla)\mathbf{v} = U(v)\mathbf{v} - \nabla P(v) - \mathbf{v}\,(\mathbf{v}\cdot\nabla P_2) + \mu_\perp \nabla_\perp^2 \mathbf{v} + \mu_x \partial_x^2 \mathbf{v} + \mathbf{f}, \quad (8.1.1)$$

along the familiar condition for incompressible flow

$$\nabla\cdot\mathbf{v} = 0. \quad (8.1.2)$$

As always, the U term makes the local $\mathbf{v}$ have a nonzero magnitude v_0 in the ordered phase, by the simple expedient of having $U > 0$ for $v < v_0$, $U = 0$ for $v = v_0$, and $U < 0$ for $v > v_0$. In addition, as before, the $\mathbf{f}$ term is a random driving force representing the noise, and is assumed to be Gaussian with white-noise correlations:

$$\langle f_i(\mathbf{r},t) f_j(\mathbf{r}',t')\rangle = 2D\delta_{ij}\delta^d(\mathbf{r}-\mathbf{r}')\delta(t-t'), \quad (8.1.3)$$

where the "noise strength" D is a constant parameter of the system, and i,j denote Cartesian components.

Keep in mind that all of the parameters in (8.1.1) are functions of the speed $v = \sqrt{|\mathbf{v}|^2}$. This is particularly true of the anisotropic pressure P_2. Using

$$\partial_n P_2(v) = P_2'(v)\partial_n\sqrt{|\mathbf{v}|^2} = \frac{1}{2\sqrt{|\mathbf{v}|^2}}\partial_n(|\mathbf{v}|^2) = \frac{1}{2\sqrt{|\mathbf{v}|^2}}\partial_n(v_\ell v_\ell) = \frac{v_\ell}{v}\partial_n v_\ell\,, \tag{8.1.4}$$

we rewrote (8.1.1) in tensor notation as:

$$\begin{aligned}\partial_t v_m = &-\partial_m P + U(v)v_m - \lambda(v)v_n(\partial_n v_m) - \lambda_4(v)v_m v_n v_\ell(\partial_n v_\ell) + \mu_T(v)\partial_n\partial_n v_m \\ &+ \mu_2(v)v_\ell v_n \partial_\ell\partial_n v_m + f_m,\end{aligned} \tag{8.1.5}$$

where we defined $\lambda_4(v) \equiv \frac{1}{v}\frac{dP_2(v)}{dv}$.

We next implemented the "standard procedure" I've outlined in earlier chapters of this book to this problem. That is, first, we did the linear theory, then showed using the DRG that nonlinearities are relevant, and therefore change the scaling behavior from that predicted by the linear theory. Finally, by a series of mappings onto other problems, we obtained two of the exponents – the roughness exponent χ, and the anisotropy exponent ζ – *exactly*, ultimately by mapping the problem onto the one-dimensional KPZ equation, for which I obtained exact exponents in Chapter 3 (36 years after Kardar, Parisi, and Zhang did the same!).

8.2 Linear Theory

The linear theory of the $d = 2$ incompressible flock is exactly the same as that for $d > 2$, *except* for the fact that there is no $\mathbf{u}_T$. Furthermore, u_P is now just u_y, if we define our coordinates so that x is along the mean direction of motion.

That is, we wrote

$$\mathbf{v} = (v_0 + u_x(\mathbf{r},t))\hat{x} + u_y(\mathbf{r},t)\hat{y} = v_0\hat{\mathbf{x}} + \mathbf{u}(\mathbf{r},t)\,. \tag{8.2.1}$$

Using the results of Chapter 6, we therefore have

$$u_y(\mathbf{q},\omega) = \frac{P_{ym}(\mathbf{q})f_m(\mathbf{q},\omega)}{-\mathrm{i}\left[\omega - c(\hat{\mathbf{q}})q\right] + \Gamma(\mathbf{q}) + 2\alpha\left(\frac{q_y}{q}\right)^2}\,, \tag{8.2.2}$$

where, as in Chapter 6, we've defined the direction-dependent "sound speed"

$$c(\hat{\mathbf{q}}) \equiv \lambda_1 v_0\frac{q_x}{q} + \lambda_4 v_0^3\frac{q_y^2 q_x}{q^3}\,, \tag{8.2.3}$$

and the "longitudinal mass" $\alpha \equiv -\frac{v_0}{2}\left(\frac{dU(v)}{dv}\right)_{v=v_0}$. We've also defined

$$\Gamma(\mathbf{q}) \equiv \mu_\perp q_y^2 + \mu_x q_x^2 \,. \tag{8.2.4}$$

We also have the correlations of u_y:

$$\langle |u_y(\mathbf{q},\omega)|^2\rangle = \frac{2D\left(\frac{q_x^2}{q^2}\right)}{\left[\omega - c(\hat{\mathbf{q}})q\right]^2 + \left[\Gamma(\mathbf{q}) + 2\alpha\left(\frac{q_y}{q}\right)^2\right]^2}\,, \tag{8.2.5}$$

and the equal-time, spatially Fourier-transformed velocity autocorrelation:

$$\left\langle |u_y(\mathbf{q},t)|^2\right\rangle = \frac{Dq_x^2}{2\alpha q_y^2 + \Gamma(\mathbf{q})q^2} \approx \frac{Dq_x^2}{2\alpha q_y^2 + \mu q_x^4}\,, \tag{8.2.6}$$

where the second, approximate equality applies for all $\mathbf{q} \to \mathbf{0}$.

Equation (8.2.6) implies that fluctuations diverge most rapidly as $\mathbf{q} \to \mathbf{0}$ if $\mathbf{q}$ is taken to zero along a locus in the $\mathbf{q}$ plane that obeys $q_y \lesssim q_x^2$; along such a locus, asymptotically, $\langle |u_y(\mathbf{q},t)|^2\rangle \propto \frac{1}{q^2}$. In contrast, along all other locii, i.e., those for which $q_y \gg q_x^2$, we have $\langle |u_y(\mathbf{q},t)|^2\rangle \propto \frac{q_x^2}{q_y^2} \ll \frac{1}{q^2}$. That is, $u_y(\mathbf{q},t)$ looks *massive* for any $\mathbf{q}$'s that are not very nearly parallel to the average velocity (which I remind the reader lies along the x-axis in the coordinate system we chose). This is exactly the behavior for which I argued at the beginning of this chapter.

As we did in the previous chapters, our next step is to determine the scaling of the fluctuations $\mathbf{u}$ of the velocity with length and time scales, *and* to determine the relative scaling of the two Cartesian components x and y of position with each other, and with time t. That is, we seek the "roughness exponents" $\chi_{x,y}$ characterizing the scaling of velocity fluctuations $u_{x,y}$, the anisotropy exponent ζ characterizing the scaling of lengths perpendicular to the direction of mean flock motion (i.e., along y) with length along x, and the dynamical exponent z characterizing the scaling of time t with those lengths. Knowing this scaling (in particular, $\chi_{x,y}$) will allow us to answer the most important question about this system: Is the ordered state actually stable against fluctuations?

We can get the dynamical exponent z predicted by the linear theory by inspection of (8.2.2), although some care is required. The form of the first term in the denominator might suggest $\omega \propto q$, which would imply $z = 1$. However, the propagating $c(\hat{\mathbf{q}})q$ term in this expression does not appear in our final expression (8.2.6) for the fluctuations; rather, these are controlled entirely by the damping term $\Gamma(\mathbf{q}) + 2\alpha\left(\frac{q_y}{q}\right)^2$. Balancing ω against that term in the dominant regime of wavevector $q_y \sim q_x^2$ gives $\omega \propto q_x^2$, which implies $z = 2$.

Now we seek χ_y, which will determine whether or not the ordered state is stable against fluctuations in an arbitrarily large system. This can be obtained by looking at the real-space fluctuations $\langle u_y^2(\mathbf{r},t)\rangle = \int_{q_x \gtrsim \frac{1}{L}} \frac{d^2q}{(2\pi)^2} \langle |u_y(\mathbf{q},t)|^2\rangle$, where L is the lateral extent of the system in the x-direction (its extent in the y-direction is taken for the purposes of this argument to be infinite). Using (8.2.6), this integral is readily seen to converge in the infrared, and, hence, as system size $L \to \infty$. Since the integral is finite, and proportional to the noise strength D, it is clear that, for sufficiently small D, the transverse u_y fluctuations in real space can be made small enough that long-ranged orientational order, and, hence, a nonzero $\langle \mathbf{v}(\mathbf{r},t)\rangle$, will be preserved in the presence of fluctuations; the ordered state *is* stable against fluctuations for sufficiently small noise strength D.

The exponent χ_y can be obtained by looking at the departure δu_y^2 of the u_y fluctuations from their infinite system limit: $\delta u_y^2 \equiv \langle u_y^2(\mathbf{r},t)\rangle|_{L=\infty} - \langle u_y^2(\mathbf{r},t)\rangle|_L = \int_{q_x \lesssim \frac{1}{L}} \frac{d^2q}{(2\pi)^2} \langle |u_y(\mathbf{q},t)|^2\rangle$; we'll define the "roughness exponent" χ_y by the way this quantity scales with system size L: $\delta u_y^2 \propto L^{2\chi_y}$. Note that this definition of χ_y requires $\chi_y < 0$, since it depends on the existence of an ordered state, which necessarily implies that the velocity fluctuations δu_y^2 do not diverge as $L \to \infty$. If $\langle u_y^2(\mathbf{r},t)\rangle|_{L=\infty}$ is not finite, one can obtain χ_y by performing exactly the type of scaling argument outlined here directly on $\langle u_y^2(\mathbf{r},t)\rangle|_L$ itself.

Approximating (8.2.6) for the dominant regime of wavevector $q_y \sim q_x^2$, and changing variables in the integral from $q_{x,y}$ to $Q_{x,y}$ according to $q_x \equiv \frac{Q_x}{L}$, $q_y \equiv \frac{Q_y}{L^2}$ shows that $\delta u_y^2 \propto L^{-1}$, and hence $\chi_y = -\frac{1}{2}$.

Note also that the fluctuations of u_x are much smaller than those of u_y. This can be seen by using the incompressibility condition, which implies, in Fourier space, $u_x = -\frac{q_y u_y}{q_x}$. Using this gives

$$\langle |u_x(\mathbf{q},t)|^2\rangle = \frac{q_y^2}{q_x^2}\langle |u_y(\mathbf{q},t)|^2\rangle = \frac{Dq_y^2}{2\alpha q_y^2 + \Gamma(\mathbf{q})q^2} \approx \frac{Dq_y^2}{2\alpha q_y^2 + \mu q_x^4}, \quad (8.2.7)$$

which is clearly finite as $\mathbf{q} \to \mathbf{0}$ along *any* locus; indeed, it is bounded above by $\frac{D}{2\alpha}$.

We can calculate a roughness exponent χ_x for u_x for the linear theory from this result exactly as we calculated the roughness exponent χ_y for u_y; we find $\chi_x = 1 - \zeta + \chi_y = -\frac{3}{2}$. We shall see in the next section that the first line of this equality also holds in the full nonlinear theory, even though the values of the exponents χ_x, ζ, and χ_y all change.

Since the $\langle |u_y(\mathbf{q},t)|^2\rangle >> \langle |u_x(\mathbf{q},t)|^2\rangle$, the total fluctuations of the velocity $\langle |\mathbf{u}(\mathbf{q},t)|^2\rangle = \langle |u_y(\mathbf{q},t)|^2\rangle + \langle |u_x(\mathbf{q},t)|^2 \approx \langle |u_y(\mathbf{q},t)|^2\rangle$. This implies that the roughness exponent χ for the total fluctuations of the velocity is just $\chi = \chi_y = -1/2$.

Note that this is the same as the value of χ we obtained from the linearized theory for incompressible flocks in $d = 3$: $\chi = (2 - d)/2 = -1/2$ in $d = 3$. This illustrates the paradoxical behavior of the incompressible flock problem described earlier: The fluctuations are no bigger in $d = 2$ than in $d = 3$, at least in the linear theory.

Since u_x has much smaller fluctuations than u_y, we will have to work to higher order in u_y than in u_x when we treat the nonlinear theory, as we will do in Section 8.3.

8.3 Nonlinear Theory

8.3.1 Stripping the Nonlinear Theory Down to Relevant Terms

We next went beyond the linear theory, and expanded the full equation of motion to higher order in $\mathbf{u}$. We obtained:

$$\partial_t u_m = -\partial_m P - 2\alpha u_x \delta_{mx} - \lambda_1 v_0 \partial_x u_m + \mu_T \nabla^2 u_m + \mu_2 v_0^2 \partial_x^2 u_m + f_m \\ - \frac{\alpha}{v_0}\left[\frac{u_y^3}{v_0}\delta_{my} + 2u_x u_y \delta_{my} + u_y^2 \delta_{mx}\right] - \lambda_1 u_y \partial_y u_y \delta_{my}. \tag{8.3.1}$$

We have kept terms that might naively appear to be higher order in the small fluctuations (e.g., the $u_y^3\delta_{my}$ term relative to the $u_x u_y \delta_{my}$ term) because, as we saw in the linearized theory, the two different components $u_{x,y}$ of $\mathbf{u}$ scale differently at long length scales. Hence, it is not immediately obvious, e.g., which of the two terms just mentioned is actually most important at long distances. We have therefore, for now, kept them both. For essentially the same reason, it is not obvious whether $u_y^2\delta_{mx}$ or $u_y^3\delta_{my}$ is more important, so we shall for now keep both of these terms as well.

On the other hand, it *is* immediately obvious that a term like, e.g., $u_x u_y^2 \delta_{mx}$ is less relevant than $u_y^2\delta_{mx}$, since, *whatever* the relative scaling of u_x and u_y, $u_x u_y^2 \delta_{mx}$ is much smaller at large distances than $u_y^2\delta_{mx}$, since u_x is.

Likewise, we have dropped the term $\frac{1}{2}\left(\frac{d\lambda_1}{dv}\right)_{v=v_0} u_y^2 \partial_x u_y \delta_{my}$, since it is manifestly smaller, by one ∂_x, than the $u_y^3\delta_{my}$ term already displayed explicitly in (8.3.1).

This sort of reasoning guides us very quickly to the reduced model (8.3.1). As in the linear theory, acting on both sides of (8.3.1) with the transverse projection operator $P_{lm}(\mathbf{q}) = \delta_{lm} - q_l q_m/q^2$ which projects orthogonal to the spatial wavevector $\mathbf{q}$ eliminates the pressure term. Then taking the $l = y$ component of the resulting equation gives

$$\partial_t u_y(\mathbf{q},t) = -\mathrm{i}v_1 q_x u_y(\mathbf{q},t) - \Gamma(\mathbf{q})u_y(\mathbf{q},t) + P_{yx}(\mathbf{q})\mathcal{F}_{\mathbf{q}}\left[-2\alpha\left(u_x(\mathbf{r},t) + \frac{u_y^2(\mathbf{r},t)}{2v_0}\right)\right]$$
$$+ P_{yy}(\mathbf{q})\mathcal{F}_{\mathbf{q}}\left[-\frac{\alpha}{v_0}\left(\frac{u_y^3}{v_0} + 2u_x u_y\right) - \lambda_1 u_y \partial_y u_y\right] + P_{ym}(\mathbf{q})f_m(\mathbf{q},t), \tag{8.3.2}$$

where $\mathcal{F}_{\mathbf{q}}$ represents the Fourier component at wavevector $\mathbf{q}$, i.e., $\mathcal{F}_{\mathbf{q}}[g(\mathbf{r})] \equiv \int \mathrm{d}^2 r\, g(\mathbf{r})\mathrm{e}^{-\mathrm{i}\mathbf{q}\cdot\mathbf{r}}$; the "bare" value of the speed v_1, before rescaling and renormalization, is $v_1 = \lambda_1 v_0$, and $\Gamma(\mathbf{q})$ is given by equation (8.2.4).

With the full nonlinear theory in hand, we next implemented the RG procedure I've described in the earlier chapters of this book. The first step of that procedure, as always, is to determine the RG rescalings necessary to keep the scale of the fluctuations fixed under RG, ignoring nonlinear effects. Once those scaling exponents are known, we can then use them to decide which nonlinearities (if any) are important. We'll find that some of them are.

So we rescale coordinates (x, y), time t, and the components of the real-space velocity field $u_{x,y}(\mathbf{r},t)$ according to

$$x = bx',\ y = b^{\zeta}y',\ t = b^z t', \tag{8.3.3}$$
$$u_y(\mathbf{r},t) = b^{\chi_y}u_y'(\mathbf{r},t), \tag{8.3.4}$$
$$u_x(\mathbf{r},t) = b^{\chi_x}u_x'(\mathbf{r},t) = b^{(\chi_y+1-\zeta)}u_x(\mathbf{r},t), \tag{8.3.5}$$

where we choose the rescaling exponent χ_x of $u_x(\mathbf{r},t)$ to be related to that of $u_y(\mathbf{r},t)$ via

$$\chi_x = \chi_y + 1 - \zeta \tag{8.3.6}$$

in order to to keep the incompressibility condition in its standard form $\nabla \cdot \mathbf{u} = 0$. Note that our convention for the anisotropy exponent here is exactly the opposite of that used in references [33, 34, 35, 36, 37]; that is, we define ζ by $q_y \sim q_x^{\zeta}$ being the dominant regime of wavevector, while [33, 34, 35, 36, 37] define this regime as $q_x \sim q_y^{\zeta}$.

Upon this rescaling, the form of equation (8.3.2) remains unchanged, but the various coefficients become dependent on the rescaling parameter b.

To do the rescaling, we must also determine how the projection operators P_{yx} and P_{yy} rescale upon the rescalings (8.3.3), (8.3.4), and (8.3.5). Since in the linear theory (see, e.g., the u_y–u_y correlation function (8.2.6)) fluctuations are dominated by the regime $q_y \lesssim q_x^2$, it follows that $P_{yx}(\mathbf{q}) = -\frac{q_x q_y}{q^2} \approx -q_y/q_x \ll 1$ and $P_{yy}(\mathbf{q}) = 1 - \frac{q_y^2}{q^2} \approx 1$. This implies that these rescale according to

$$P_{yx}(\mathbf{q}) = b^{(1-\zeta)}P_{yx}(\mathbf{q}'),\ P_{yy}(\mathbf{q}) = P_{yy}(\mathbf{q}'). \tag{8.3.7}$$

Performing the rescalings (8.3.3), (8.3.4), (8.3.5), and (8.3.7) on the equation of motion (8.3.2), we obtain, from the rescalings of the first three (i.e., the linear) terms on the right-hand side, the following rescalings of the parameters:

$$v_{1R} = b^{(z-1)} v_1 \, , \; \mu_R = b^{(z-2)} \mu ,$$
$$\mu_{TR} = b^{(z-2\zeta)} \mu_T , \tag{8.3.8}$$

and

$$\alpha_R = e^{(z-2\zeta+2)\ell} \alpha . \tag{8.3.9}$$

Note that $\Gamma(\mathbf{q})$ involves *two* parameters (μ and μ_T); hence, we get the rescalings of both of these parameters from this term.

Similarly, looking at the rescaling of the nonlinear terms proportional to u_y^2 and u_y^3, respectively, we obtain the rescalings:

$$\left(\frac{\alpha}{v_0}\right)_R = b^{(z+\chi_y-\zeta+1)} \left(\frac{\alpha}{v_0}\right) , \; \left(\frac{\alpha}{v_0^2}\right)_R = b^{(z+2\chi_y)} \left(\frac{\alpha}{v_0^2}\right) . \tag{8.3.10}$$

We recover the first of these by looking at the rescaling of the nonlinear term proportional to $u_x u_y$ as well.

We note that the two rescalings (8.3.10) are both consistent with (8.3.9) if we rescale v_0 according to

$$v_{0R} = b^{(1-\zeta-\chi_y)} v_0 . \tag{8.3.11}$$

By power counting on the $u_y \partial_y u_y$ term, we obtain the rescaling of λ_1:

$$\lambda_{1R} = b^{(z+\chi_y-\zeta)} \lambda_1 . \tag{8.3.12}$$

Finally, by looking at the rescaling of the noise correlations (8.1.3), we obtain the scaling of the noise strength D:

$$D_R = b^{(z-2\chi_y-\zeta-1)} D . \tag{8.3.13}$$

We next used the standard renormalization group logic that I've been using throughout this book to assess the importance of the nonlinear terms in (8.3.2). This logic is to choose the rescaling exponents z, ζ, and χ_y so as to keep the size of the fluctuations in the field $\mathbf{u}$ fixed upon rescaling. Since, as we saw in our treatment of the linearized theory (in particular, equation (8.2.6)), that size was controlled by three parameters – the "longitudinal mass" α, the damping coefficient μ, and the noise strength D – the choice of z, ζ, and χ_y that keeps these fixed will clearly accomplish this. From the rescalings (8.3.8), (8.3.9), and (8.3.13), this leads to three simple linear equations in the three unknown exponents z, ζ, and χ_y:

$$z - 2\zeta + 2 = 0\,, \tag{8.3.14}$$

$$z - 2 = 0\,, \tag{8.3.15}$$

$$z - 2\chi_y - \zeta - 1 = 0\,. \tag{8.3.16}$$

Solving these, we find the values of these exponents in the linearized theory:

$$\zeta_{lin} = z_{lin} = 2, \ \ \chi_{ylin} = -1/2, \ \ \chi_{xlin} = -3/2\,, \tag{8.3.17}$$

which, unsurprisingly, are the linearized exponents we found earlier.

With these exponents in hand, we can now assess the importance of the nonlinear terms at long length scales, simply by looking at how their coefficients rescale. (We don't have to worry about the size of the actual nonlinear terms themselves changing upon rescaling, because we have chosen the rescalings to keep them constant in the linear theory.) The mass α, of course, is kept fixed. Inserting the linearized exponents (8.3.17) into the rescaling relation (8.3.11) for v_0, we see that

$$v_{0R} = b^{-\frac{1}{2}} v_0. \tag{8.3.18}$$

Since v_0 appears in the *denominator* of all three of the nonlinear terms associated with α, and α itself is fixed, this implies that all three of those terms are "relevant," in the RG sense of growing larger as we go to longer wavelengths (i.e., as ℓ grows). As usual in the RG, this implies that these terms will ultimately alter the scaling behavior of the system at sufficiently long distances. In particular, the exponents z, ζ, and $\chi_{x,y}$ will change from their values (8.3.17) predicted by the linear theory.

The same is *not* true of the λ_1 nonlinearity, however, because it is *irrelevant*; that is, it gets *smaller* upon renormalization. This follows from inserting the linearized exponents (8.3.17) into the rescaling relation (8.3.12) for λ_1, which gives

$$\lambda_{1R} = b^{-\frac{1}{2}} \lambda_1, \tag{8.3.19}$$

which shows clearly that λ_1 vanishes as $\ell \to \infty$; that is, in the long-wavelength limit.

Since λ_1 was the *only* remaining nonlinearity associated with the λ terms in our original equation of motion (8.3.1), we can accurately treat the full, long-distance behavior of this problem by leaving out all of those nonlinear terms.

Doing so reduces the equation of motion (8.3.1) to

$$\begin{aligned}\partial_t u_m = &-\lambda_1 v_0 \partial_x u_m - \partial_m P - 2\alpha\left(u_x + \frac{u_y^2}{2v_0}\right)\delta_{xm} \\ &- \frac{2\alpha}{v_0}\left(u_x + \frac{u_y^2}{2v_0}\right)u_y \delta_{ym} \\ &+ \mu\partial_x^2 u_m + \mu_T \partial_y^2 u_m + f_m. \end{aligned} \tag{8.3.20}$$

Before proceeding to analyze this equation, we note the differences between the structure of this problem and that for the compressible case. In the compressible problem, there is no constraint analogous to the incompressibility condition relating u_x and u_y. Hence, u_x is free to relax quickly (to be precise, on a time scale $\frac{1}{2\alpha}$) its local "optimal" value, which is readily seen to be $-\frac{u_y^2}{2v_0}$. Once this relaxation has occurred, all of the nonlinearities associated with α drop out of that compressible problem [37], leaving the λ nonlinearities as the dominant ones. Here, in the incompressible problem, u_x is, because of the incompressibility constraint, *not* free to relax in such a way as to cancel out the α nonlinearities, which, because they involve no spatial derivatives, wind up dominating the λ nonlinearities, which *do* involve spatial derivatives. In addition, the suppression of fluctuations by the incompressibility condition, which we've already seen in the linear theory, makes the λ nonlinearities not only *less* relevant than the α ones, but actually *irrelevant.* Hence, we can drop them in this incompressible problem, leaving us with equation (8.3.20) as our equation of motion.

As one final simplification, we will make a Galilean transformation to a "pseudo-co-moving" coordinate system moving in the direction $\hat{x}$ of mean flock motion at speed $\lambda_1 v_0$. Note that if the parameter λ_1 had been equal to 1, this *would* be precisely the frame co-moving with the flock. The fact that it is not is a consequence of the lack of Galilean invariance in our problem.

This boost eliminates the "convective" term $\lambda_1 v_0 \partial_x u_m$ from the right-hand side of (8.3.20), leaving us with our final simplified form for the equation of motion:

$$\begin{aligned}\partial_t u_m = & -\partial_m P - 2\alpha\left(u_x + \frac{u_y^2}{2v_0}\right)\delta_{xm} \\ & -\frac{2\alpha}{v_0}\left(u_x + \frac{u_y^2}{2v_0}\right)u_y\delta_{ym} \\ & + \mu\partial_x^2 u_m + \mu_T\partial_y^2 u_m + f_m. \end{aligned} \tag{8.3.21}$$

8.3.2 Mapping to an Equilibrium Divergence-Free XY Model

We will now show that equation (8.3.21) also describes an equilibrium system: our old friend the two-dimensional XY model, in its ordered phase, but now subject to the divergence-free constraint $\nabla \cdot \mathbf{M} = 0$, where $\mathbf{M}$ is the magnetization. This connection will enable us to use purely equilibrium statistical mechanics (in particular, the Boltzmann distribution) to determine the equal-time correlations of two-dimensional incompressible polar active fluids.

Writing the Hamiltonian for the two-dimensional XY model in terms of the magnetization $\mathbf{M}$, instead of the angle field θ (as we did in Chapter 2) gives [61]

$$H_{XY} = \int d^2r \left[V(|\mathbf{M}|) + \frac{1}{2}\mu|\vec{\nabla}\mathbf{M}|^2 \right], \tag{8.3.22}$$

where μ is the "spin wave stiffness." In the ordered phase, the "potential" $V(|\mathbf{M}|)$ has a circle of global minima at a nonzero value of $|\mathbf{M}|$, which we will take to be v_0.

Expanding in small fluctuations about this minimum by writing $\mathbf{M} = (v_0 + u_x)\hat{x} + u_y\hat{y}$, we obtain, keeping only "relevant" terms,

$$H_{XY} = \frac{1}{2}\int d^2r \left[2\alpha(u_x + \frac{u_y^2}{2v_0})^2 + \mu|\nabla\mathbf{u}|^2 \right], \tag{8.3.23}$$

where we've defined the "longitudinal mass" $2\alpha \equiv \left(\frac{\partial^2 V}{\partial|\mathbf{M}|^2}\right)\Big|_{|\mathbf{M}|=v_0}$.

We now add to this model the divergence-free constraint $\nabla\cdot\mathbf{M} = 0$, which obviously implies $\nabla\cdot\mathbf{u} = 0$. To enforce this constraint, we introduce to the Hamiltonian a Lagrange multiplier $P(\mathbf{r})$:

$$H' = H_{XY} - \int d^2r\, P(\mathbf{r})(\nabla\cdot\mathbf{u}). \tag{8.3.24}$$

As I discussed in Chapter 2, the simplest dynamical model that will relax back to the equilibrium Boltzmann distribution $e^{-\beta H'(\mathbf{u})}$ for this Hamiltonian H' is [30, 61] the "time-dependent–Ginzburg–Landau" (TDGL) model $\partial_t u_l = -\delta H'/\delta u_l + f_l$, where $\mathbf{f}$ is the thermal noise whose statistics can also be described by equation (8.1.3) with $D = k_B T = 1/\beta$. This TDGL equation is readily seen to be exactly equation (8.3.21) with $\mu_T = \mu$. Therefore, we conclude that the ordered phase of two-dimensional incompressible polar active fluids has the same static (i.e., equal-time) scaling behaviors as the ordered phase of the two-dimensional XY model subject to the constraint $\nabla\cdot\mathbf{M} = 0$.

Note that those statistics will be *very* different from those we found for the *unconstrained* equilibrium XY model in Chapter 2; the constraint has a *huge* effect. In particular, the "constrained" model, which I'll also sometimes call the "incompressible XY model,"[1] has long-ranged order in $d = 2$, as we'll see in Section 8.3.6.

This mapping between a nonequilibrium active fluid model and a "divergence-free" XY model, which I will sometimes call an "incompressible" XY model, allows us to investigate the fluctuations in our original active fluid model by studying the partition function of the equilibrium model.

[1] By "incompressible XY model," I do *not* mean an XY model living on an incompressible lattice. Indeed, I'm really not talking about true compressibility at all. Instead, I simply mean an XY model subject to the constraint $\nabla\cdot\mathbf{M} = 0$.

That, of course, is still a formidable problem. In Section 8.3.2, we'll tackle this problem by a further mapping of the "incompressible" XY model onto another problem: equilibrium two-dimensional smectic A.

8.3.3 The Streaming Function and Flow Lines

That mapping begins by exploiting an old fluid mechanic's trick for dealing with the incompressibility constraint $\nabla \cdot \mathbf{v} = 0$. Being old fluid mechanics ourselves, we used this trick to rewrite our "incompressible" XY Hamiltonian in a form that enabled us to work with an *unconstrained* variable: a so-called streaming function $h(\mathbf{r})$.

The trick is simply to write the components of the fluctuating part of the velocity field (or, equivalently, our magnetization $\mathbf{M}$) as follows:

$$u_x = -v_0 \partial_y h, \quad u_y = v_0 \partial_x h, \tag{8.3.25}$$

where the streaming function h is completely unconstrained. It is easy to see that this rewriting automatically satisfies the incompressibility constraint $\nabla \cdot \mathbf{u} = 0$, regardless of the functional form of h. This is why h is unconstrained. This implies that the streaming function ϕ for the *full* velocity field $\mathbf{v}(\mathbf{r}) = v_0 \hat{x} + \mathbf{u}$, defined via $v_x = \partial_y \phi$, $v_y = -\partial_x \phi$, is given by

$$\phi = v_0 (y - h(\mathbf{r})) . \tag{8.3.26}$$

As in conventional two-dimensional fluid mechanics, contours of the streaming function ϕ are flow lines. When the system is in its uniform steady state (i.e., $\mathbf{v} = v_0 \hat{x}$), these contour lines, defined via

$$\phi = nC, \quad n = 0, 1, 2, 3 \ldots, \tag{8.3.27}$$

where C is some arbitrary constant, are a set of parallel, uniformly spaced lines given by $y_n = nC/v_0$.

Now let's ask what the flow lines are if there are fluctuations in the velocity field: $\mathbf{v} = v_0 \hat{x} + \mathbf{u}$. Combining our expression for ϕ (8.3.26) and the expression (8.3.27) for the flow lines, we see that the positions of the flow lines are now given by

$$y_n = nC/v_0 + h, \quad n = 0, 1, 2, 3 \ldots, \tag{8.3.28}$$

which shows that $h(\mathbf{r})$ can be interpreted as the local displacement of the flow lines from their positions in the ground state configuration.

This geometry is illustrated in Figure 8.3.1.[2]

This picture of a set of lines that "wants" to be parallel and uniformly spaced, but in the presence of fluctuations is displaced by a distance $h(\mathbf{r})$ from a reference

[2] We thank Pawel Romanczuk for pointing out this pictorial interpretation to us.

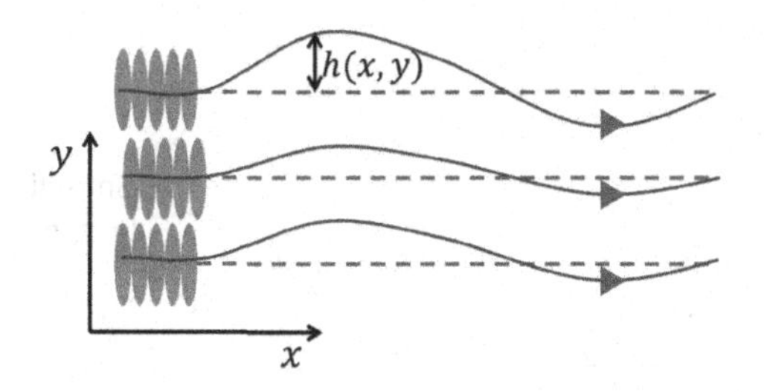

Fig. 8.3.1 In the case of two-dimensional incompressible polar active fluids, the field $h(\mathbf{r})$ is the vertical displacement of the flow lines (i.e., the solid lines) from the set of parallel lines (i.e., the dotted lines) along $\hat{x}$ that would occur in the absence of fluctuations. For a defect-free two-dimensional smectic, it likewise gives the vertical displacement of the smectic layers (i.e., the solid lines) from their reference positions (i.e., the dotted lines) at zero temperature. Reproduced from [50].

set of parallel, uniformly spaced lines, looks very much like a two-dimensional version of my favorite phase of matter: a smectic liquid crystal (i.e., "soap"), for which the layers are actually one-dimensional fluid stripes.

This resemblance between our system and a two-dimensional smectic is not purely visual. Indeed, they can be mapped onto each other exactly. I'll illustrate this in the next two sections. Section 8.3.4 describes the fascinating "smectic A" phase of matter, while Section 8.3.5 presents the mapping between this state and our original two-dimensional incompressible flock via the incompressible XY model.

8.3.4 Smectics A

There are many different types of smectic phases, the names of which are distinguished by a single letter after the word "smectic." I'll talk here about the smectic A phase [24]. This liquid crystalline phase is, as the term "liquid crystal" suggests, a hybrid of a liquid and a crystalline solid. Specifically, a d-dimensional smectic A can be thought of as a one-dimensional stack of $d-1$-dimensional isotropic fluids. In three dimensions, this is a stack of two-dimensional fluid layers, while in two dimensions, it can be, e.g., a stack of one-dimensional fluid layers, or a plane covered by regularly spaced parallel polymers confined to that plane.

These fascinating phases exhibit a number of unique properties, including quasi-long-ranged order – i.e., algebraically decaying translational correlations – in three dimensions [62, 63], anomalous elasticity [64], and a breakdown of linearized hydrodynamics [65, 66].

Any smectic A phase (either equilibrium or active) is characterized by the spontaneous breaking of translational invariance in one direction (in contrast to a crystal, in which translational invariance is broken in all d dimensions, where d

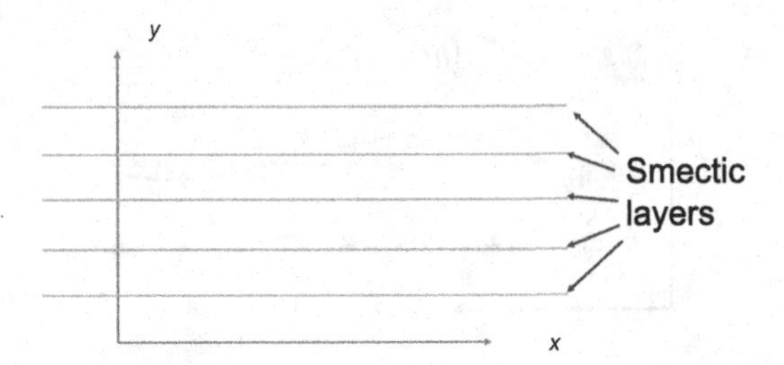

Fig. 8.3.2 Schematic of the ideal smectic state, in which the layers are parallel, and uniformly spaced. We choose our coordinates (x, y) so that the x-axis runs parallel to the layers, as shown.

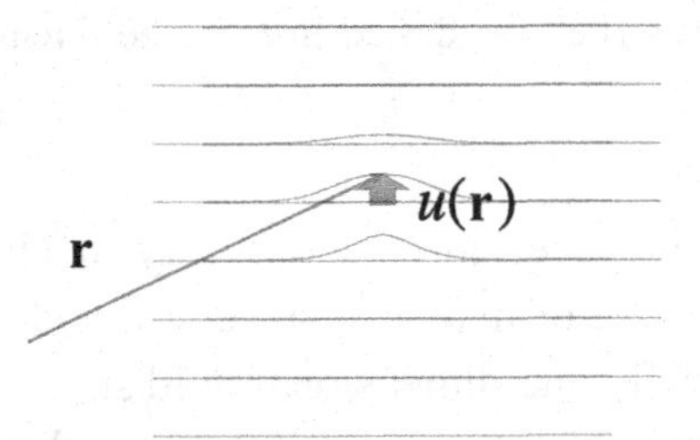

Fig. 8.3.3 Definition of the layer displacement field $u(\mathbf{r}, t)$. The straight parallel lines are the reference positions of the layers, while the curved lines depict a fluctuation in the layer positions. The layer displacement field $u(\mathbf{r}, t)$ is the distance from the reference position to the fluctuating position of the layers in the vicinity of the spatial point $\mathbf{r}$, as illustrated.

is the dimension of space). This is equivalent to saying that the system spontaneously *layers*, with the additional requirement that the layers are "liquid-like," in the sense of being homogeneous along the layers. We will choose our coordinates so that the direction in which translational invariance is broken is y. This means the layers, in the absence of fluctuations, run parallel to x (see Figure 8.3.2).

Since the smectic breaks the continuous translational symmetry of free space, such systems (again whether equilibrium or active) have a "Goldstone mode" associated with the breaking of this symmetry. In smectics, we usually take this Goldstone mode to be the local displacement $u(\mathbf{r}, t)$ of the layers in the vicinity of the spatial point $\mathbf{r}$ away from some reference set of parallel, regularly spaced layer positions (see Figure 8.3.3).

A more mathematical formulation of these ideas is to start by saying that, in the "ideal" (i.e., fluctuation-free) smectic state, the system spontaneously breaks translation invariance by entering a "ground state" in which the density $\rho(\mathbf{r})$ is given by $\rho_{GS}(\mathbf{r})$ ("GS" for "ground state"), which is a periodic function of y only:

$$\rho_{GS}(\mathbf{r}) = F_P(y)\,, \tag{8.3.29}$$

where $F_P(y)$ is a periodic function of y with period a, the layer spacing.

The form of the function $F_P(y)$ is determined by the microscopic interactions in the system which favor the smectic state in the first place. However, because those interactions are rotation and translation invariant, there is, as usual in a system with a spontaneously broken continuous symmetry, an enormous (indeed, infinite and continuous) degeneracy in the ground states.

In particular, *translation* invariance requires that

$$\rho^{u}_{GS}(\mathbf{r}) = F_P(y - u_0)\,, \tag{8.3.30}$$

where the displacement u_0 is *any* constant, must also be a ground state of the system. This state is nothing but the original ground state, translated up (i.e., along y), by an amount u_0. This state can be described as one in which the displacement field $u(\mathbf{r})$ is simply a constant:

$$u(\mathbf{r}) = u_0\,. \tag{8.3.31}$$

This is, of course, almost identical to the *rotation* invariance symmetry of the XY model of Chapter 2, and implies, as it did there, that the Hamiltonian for smectic can involve only *gradients* of the displacement field $u(\mathbf{r})$.

One way we can define $u(\mathbf{r})$ in general is by saying that the density is, in general, for a slowly spatially varying displacement field $u(\mathbf{r})$, given by

$$\rho^{u(\mathbf{r})}_{GS}(\mathbf{r}) = F_P(y - u(\mathbf{r}))\,. \tag{8.3.32}$$

Rotation invariance imposes additional constraints on the Hamiltonian. To see this, note that rotation invariance requires that, if (8.3.29) is a ground-state configuration of the system, then

$$\rho^{\phi}_{GS}(\mathbf{r}) = F_P(y')\,, \tag{8.3.33}$$

where

$$y' = y\cos\phi - x\sin\phi \tag{8.3.34}$$

is the y-coordinate of a Cartesian coordinate system rotated relative to the original one by an angle ϕ, as illustrated in Figure 8.3.4, is also a ground state.

Comparing our general definition (8.3.32) of the smectic displacement field $u(\mathbf{r})$ with our specific expression (8.3.33) enables us to determine the displacement field $u_{pr}(\mathbf{r})$ associated with a pure rotation:

$$y - u_{pr}(\mathbf{r}) = y' = y\cos\phi - x\sin\phi\,. \tag{8.3.35}$$

Solving this equation for $u_{pr}(\mathbf{r})$ gives

$$u_{pr}(\mathbf{r}) = y(1 - \cos\phi) - x\sin\phi\,. \tag{8.3.36}$$

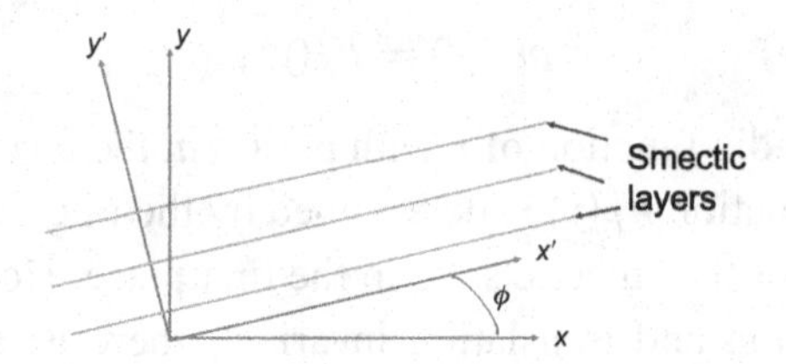

Fig. 8.3.4 A uniform rotation of the smectic ground state, which must itself be another ground state.

For small angle $\phi \ll 1$, this reduces, to linear order in ϕ, to a *nonuniform* displacement field

$$u(\mathbf{r}, t) = \phi x \,. \tag{8.3.37}$$

From this expression, we see that

$$\partial_x u = \phi \,, \tag{8.3.38}$$

that is, the x-derivative of the displacement field u, gives the rotation angle of the layers away from their reference orientation when that angle is small.

The relation (8.3.38) continues to apply for arbitrary layer distortions; that is, the x-derivative of u locally gives the local tilt of the layers away from their reference orientation (provided that tilt is small, of course).

Rotation invariance therefore *forbids* (to quadratic order) the inclusion of terms that depend on $\partial_x u$ in the elastic Hamiltonian, since such terms will be nonzero for the uniform rotation (8.3.37). Therefore, the leading-order term involving x-derivatives of $u(\mathbf{r}, t)$ in H is a term proportional to $(\partial_x^2 u)^2$, which represents the energy cost of *bending* the layers.

There is no such prohibition against terms involving $\partial_y u$. Indeed, a term proportional to $(\partial_y u)^2$ can easily be seen to represent the energy cost of *compressing* the layers closer together (for $\partial_y u < 0$) or stretching them further apart (for $\partial_y u > 0$). It is straightforward to show that

$$\delta a = a \partial_y u \,, \tag{8.3.39}$$

where δa is the departure of the local layer spacing from its energetically optimal value a.

These considerations lead, to quadratic order in u, to the elastic Hamiltonian [24, 61, 62, 63]:

$$H_{\rm sm}^{\rm quad} = \frac{1}{2} \int d^2 r \left[B(\partial_y u)^2 + K(\partial_x^2 u)^2 \right] . \tag{8.3.40}$$

Note that this form implies a two-to-one anisotropy of scaling: One y-derivative is equivalent to two x-derivatives. Let's keep this in mind as we consider higher order terms in the elastic energy.

Up to now, we have only worked to quadratic order in u. What happens at higher order? To answer that, first consider the u field of a pure rotation $u_{pr}(\mathbf{r})$ up to *quadratic* order in the rotation angle ϕ. From equation (8.3.36), we get

$$u_{pr}(\mathbf{r}) = y\phi^2/2 - x\phi + O(\phi^3)\,. \tag{8.3.41}$$

From this expression, it is clear that, to this (higher) order in ϕ, $\partial_y u_{pr}$ does *not* vanish; indeed,

$$\partial_y u_{pr} = \phi^2/2\,. \tag{8.3.42}$$

Hence, the quadratic energy (8.3.40) will *not* vanish upon a pure rotation.

The full *anharmonic* energy, of course, *must* vanish for a pure rotation. Therefore, the anharmonic terms must cancel off the contribution to the energy coming from the harmonic terms.

Let's see if we can construct that anharmonic energy by finding a quantity that reduces to $\partial_y u_{pr}$ to linear order in derivatives of u, but vanishes to $O(\phi^2)$. The form of the result (8.3.42), along with the result $\partial_x u_{pr} = \phi$, which continues to hold to quadratic order, suggests one fairly straightforward quantity $E(\{u(\mathbf{r})\})$ that has these properties:

$$E(\{u(\mathbf{r})\}) = \partial_y u - (\partial_x u)^2/2\,. \tag{8.3.43}$$

Indeed, with a little more playing around, and the use of a few trigonometric identities, one can show that

$$E_2(\{u(\mathbf{r})\}) = \partial_y u - |\nabla u|^2/2 = \partial_y u - (\partial_x u)^2/2 - (\partial_y u)^2/2 \tag{8.3.44}$$

in fact vanishes identically for the pure rotation (8.3.42).

Hence, if we modify our Hamiltonian (8.3.40) by simply replacing $\partial_y u$ with $E_2(\{u(\mathbf{r})\})$ everywhere the former appears, we will have a fully rotation invariant Hamiltonian. This gives

$$H_{\mathrm{sm}} = \frac{1}{2}\int \mathrm{d}^2 r \left[B(\partial_y u - |\nabla u|^2/2)^2 + K(\partial_x^2 u)^2\right]. \tag{8.3.45}$$

However, in light of our earlier observation that each y-derivative power counts like two x derivatives, it is clear that we can ignore the $(\partial_y u)^2$ term implicitly included in the $|\nabla u|^2$ piece of the above expression. Doing so gives our final Hamiltonian:

$$H_{\mathrm{sm}} = \frac{1}{2}\int \mathrm{d}^2 r \left[B(\partial_y u - (\partial_x u)^2/2)^2 + K(\partial_x^2 u)^2\right]. \tag{8.3.46}$$

We will now show that this Hamiltonian also describes our two-dimensional "incompressible" XY model, and hence, two-dimensional incompressible flocks.

8.3.5 Connecting the "Incompressible" XY Model and 2D Smectics

As noted earlier, the resemblance between our system and a two-dimensional smectic is not purely visual. Indeed, making the substitution (8.3.25), the Hamiltonian (8.3.46) becomes (ignoring irrelevant terms like $(\partial_x\partial_y h)^2$, which is irrelevant compared to $(\partial_x^2 h)^2$ because y-derivatives are less relevant than x-derivatives)

$$H_s = \frac{1}{2}\int d^2r\left[B\left(\partial_y h - \frac{(\partial_x h)^2}{2}\right)^2 + K(\partial_x^2 h)^2\right], \tag{8.3.47}$$

where $B = 2\alpha v_0^2$ and $K = \mu v_0^2$.

This Hamiltonian is *exactly* the Hamiltonian for the equilibrium two-dimensional smectic model we just derived with h in equation (8.3.47) playing the role of the displacement field u of the smectic layers. For the equilibrium two-dimensional smectic the Partition function is

$$Z_s = \int D[h]e^{-H_s/k_BT}, \tag{8.3.48}$$

where it should be noted that there is no constraint on the functional integral over $h(\mathbf{r})$ in this expression, since, as noted earlier, $h(\mathbf{r})$ is unconstrained.

Since the variable transformation equation (8.3.25) is linear, the partition functions for the smectic, Z_s (equation (8.3.48)) and the constrained XY model

$$Z_{XY} = \int D[u_x]D[u_y]\delta(\nabla\cdot\mathbf{u} = 0)e^{-H_{XY}/k_BT}, \tag{8.3.49}$$

are the same up to a constant Jacobian factor, which changes none of the statistics.

To summarize what we have learned so far: We have successfully mapped the model for the ordered phase of an incompressible polar active fluid onto the ordered phase of the equilibrium two-dimensional XY model with the constraint $\nabla\cdot\mathbf{u} = 0$, which in turn we have mapped onto the standard equilibrium two-dimensional smectic model [67]. The scaling behaviors of the former can therefore be obtained by studying the latter. Note that the connection between our problem and the dipolar magnet, which was studied in [67], is that the long-ranged dipolar interaction in magnetic systems couples to, and therefore suppresses, the longitudinal component of the magnetization. See [67] for more details.

8.3.6 Mapping the 2D Smectic to the (1+1)-dimensional KPZ Equation

Fortunately, the scaling behaviors of the equilibrium two-dimensional smectic model are known, thanks to an ingenious further mapping [68, 69] of this problem onto the (1+1)-dimensional KPZ equation [31] that I described in hideous detail in Chapter 3. What you need to remember from that chapter is that I was able, by a very simple argument, to get *exact* scaling exponents for that problem. Using this mapping which I'm about to describe, I'll therefore be able to get exact exponents for the "incompressible" magnet, and, hence, for the incompressible two-dimensional flock as well.

In this mapping, which connects the *equal-time* correlation functions of the two-dimensional smectic to those of the (1+1)-dimensional KPZ equation, the y-coordinate in the smectic is mapped onto time t in the (1+1)-dimensional KPZ equation with $h(x, t)$ the height of the "surface" at position x and time t above some reference height. As a result, the *dynamical* exponent z_{KPZ} of the (1+1)-dimensional KPZ equation becomes the anisotropy exponent ζ of the two-dimensional smectic. Since the scaling laws of the (1+1)- dimensional KPZ equation are known exactly [31], those of the equal-time correlations of the two-dimensional smectic can be obtained as well.

This mapping is wonderfully clever, and, once you've been shown the trick by Golubovic and Wang, remarkably simple. I'll review their ingenious argument now.

They started by "solving" the $(1+1)$-dimensional KPZ equation for the random force f:

$$f(x,t) = \partial_t h(x,t) - \nu\partial_x^2 h(x,t) - \frac{\lambda}{2}(\partial_x h(x,t))^2 \,. \tag{8.3.50}$$

Now recall that the probability distribution of f is a zero-mean Gaussian with variance $2D$. This means that the probability of a given configuration $f(x, t)$ is given by the functional integral

$$P(\{f(x,t)\}) = \int Df(\{h(x,t)\}) \exp\left(-\frac{1}{4D}\int \mathrm{d}x\,\mathrm{d}t f^2(x,t)\right)/Z_f \,, \tag{8.3.51}$$

where the "partition function" Z is just a normalization. Substituting the relation (8.3.50) between the noise f and the height field h into this expression gives the probability distribution for the field h:

$$P(\{h(x,t)\}) = \int Dh(\{h(x,t)\}) \exp\left(-\frac{1}{4D}\int \mathrm{d}x\,\mathrm{d}t\left[\partial_t h(x,t) - \nu\partial_x^2 h(x,t) - \frac{\lambda}{2}(\partial_x h(x,t))^2\right]^2\right)/Z_h \,. \tag{8.3.52}$$

I've actually pulled a bit of a fast one on you here. In changing variables of (functional) integration from the noise f to the height field h, I should also pick up a "Jacobean factor" associated with this change of variables of integration.

Had the relation between f and h been linear, this Jacobean would simply have been a constant, and so could have been absorbed into the partition function Z_h. In any event, it would have no effect on the actual probability distribution, since it would be independent of h. Although our relation is actually nonlinear, the dependence of the Jacobean on h proves to be weak enough that we can ignore it, and take the probability distribution of h to be given by (8.3.53).

To make the problem look a little more like my two-dimensional smectic problem, I'm going to change the name of the time variable from t to y, and think of x and y as orthogonal Cartesian components in a two-dimensional plane. Thus we have

$$P(\{h(x,y)\}) = \int Dh(\{h(x,y)\}) \exp\left(-\frac{1}{4D}\int \mathrm{d}x\,\mathrm{d}y\left[\partial_y h(x,y) - \nu\partial_x^2 h(x,y) - \frac{\lambda}{2}(\partial_x h(x,y))^2\right]^2\right)/Z_h\,. \tag{8.3.53}$$

We can now think of this as the Boltzmann weight for a two-dimensional equilibrium system whose fluctuating variable is a height field $h(x,y)$ with its Hamiltonian over $k_B T$ given by

$$\beta H = \frac{1}{4D}\int \mathrm{d}x\,\mathrm{d}y\left[\partial_y h(x,y) - \nu\partial_x^2 h(x,y) - \frac{\lambda}{2}(\partial_x h(x,y))^2\right]^2. \tag{8.3.54}$$

Now expanding out the square of the integrand in the above expression, we get (or, rather, Golubovic and Wang [68, 69] got):

$$\begin{aligned}\left[\partial_y h(x,y) - \nu\partial_x^2 h(x,y) - \frac{\lambda}{2}(\partial_x h(x,y))^2\right]^2 &= \left(\partial_y h(x,y) - \frac{\lambda}{2}(\partial_x h(x,y))^2\right)^2 \\ &\quad + \nu^2(\partial_x^2 h(x,y))^2 \\ &\quad - 2\nu\left[\partial_y h - \lambda(\partial_x h)^2/2\right](\partial_x^2 h).\end{aligned} \tag{8.3.55}$$

The terms on the third line of this expression can be written as a total divergence:

$$-2\nu\left[\partial_y h - \lambda(\partial_x h)^2/2\right](\partial_x^2 h) = \nabla\cdot\mathbf{A}, \tag{8.3.56}$$

where I've defined

$$\mathbf{A} = \nu \left\{ \hat{\mathbf{x}} \left[2h\partial_x \partial_y h + \left(\frac{\lambda}{3} \right) (\partial_x h)^3 \right] - \hat{\mathbf{y}} \left[2h\partial_x^2 h + (\partial_x h)^2 \right] \right\} . \tag{8.3.57}$$

Hence, by the divergence theorem, the integral of the terms in that third line over x and y become surface terms, and so can be ignored.

Doing so, our effective Hamiltonian (or, more precisely, that Hamiltonian divided by $k_B T$) becomes

$$\beta H = \frac{1}{4D} \int \mathrm{d}x \, \mathrm{d}y \left[\left(\partial_y h(x,y) - \frac{\lambda}{2} (\partial_x h(x,y))^2 \right)^2 + \nu^2 \left(\partial_x^2 h(x,y) \right)^2 \right] . \tag{8.3.58}$$

The alert reader will notice that this looks almost *exactly* like our smectic Hamiltonian (8.3.46). Indeed, with a simple linear change of variables,

$$h(x,y) \equiv u(x,y)/\lambda, \tag{8.3.59}$$

it becomes *exactly* the same:

$$\beta H = \frac{1}{4D\lambda^2} \int \mathrm{d}x \, \mathrm{d}y \left[\left(\partial_y u(x,y) - \frac{1}{2} (\partial_x u(x,y))^2 \right)^2 + \nu^2 \left(\partial_x^2 u(x,y) \right)^2 \right] . \tag{8.3.60}$$

Specifically, the alert reader will note that this is just βH for the smectic system described in the last subsection, with its compression modulus B and bend modulus K related to the parameters of the original $1+1$-dimensional KPZ equation via

$$\frac{B}{k_B T} = \frac{1}{2D\lambda^2}, \quad \frac{K}{k_B T} = \frac{\nu^2}{2D\lambda^2} . \tag{8.3.61}$$

Thus, I (or, rather, Golubovic and Wang) have succeeded in mapping the two-dimensional smectic onto the $1+1$-dimensional KPZ equation. Note the two peculiar (and related) features of this mapping:

(1) we have mapped a *two*-dimensional problem (the smectic) onto a *one*-dimensional problem (the $1+1$-dimensional KPZ equation), and
(2) time t in the KPZ equation is identified, in this mapping, with the y-coordinate in the smectic problem, and, hence, in our original two-dimensional incompressible flock problem as well.

Point (2) implies that the *dynamical* exponent z of the $1+1$-dimensional KPZ equation becomes the *anisotropy* exponent ζ of the smectic problem, and, hence, of the original two-dimensional incompressible flock problem as well. Furthermore, the roughness exponent χ of the KPZ problem is also the roughness exponent for the displacement field u of the smectic problem, and, hence, of the *streaming function* h of the incompressible flock problem.

This essentially solves our problem, since we were able in Chapter 3 to determine the exponents z and χ for the $1+1$-dimensional KPZ equation *exactly*. Using those results here gives [68, 69]

$$\zeta = 3/2 \tag{8.3.62}$$

and

$$\chi_h = 1/2 \tag{8.3.63}$$

as the exponents for the two-dimensional smectic, where χ_h gives the scaling of the smectic layer displacement field h with spatial coordinate x. Given the streaming function relation (8.3.25) between h and u, we see that the scaling exponent χ_y for u_y is just

$$\chi_y = \chi_h - 1 = -1/2\,, \tag{8.3.64}$$

and that the scaling exponent χ_x for u_x is just

$$\chi_x = \chi_y + 1 - \zeta = -1\,. \tag{8.3.65}$$

The fact that both of the scaling exponents χ_y and χ_x are less than zero implies that both u_y and u_x fluctuations remain finite as system size $L \to \infty$; this, in turn, implies that the system has long-ranged orientational order. That is, the ordered state is stable against fluctuations, at least for sufficiently small noise D.

Incidentally, although it was not the problem we originally tackled, these negative roughness exponents χ also apply to the "incompressible" XY model that we encountered *en passant* as we were going through our series of mappings. Therefore, as I claimed earlier, that model also has long-ranged order in $d = 2$.

Note that this does not violate the Mermin–Wagner–Hohenberg theorem, because the incompressibility constraint amounts to an effective long-ranged interaction between spins.

With the scaling exponents (8.3.63), (8.3.64), and (8.3.65) in hand, we can immediately write the scaling form for the velocity correlations:

$$\begin{aligned} C_y(\mathbf{r} - \mathbf{r}') &\equiv \left\langle [u_y(\mathbf{r}, t) - u_y(\mathbf{r}', t)]^2 \right\rangle \\ &= B_y |x - x'|^{-1/2} \Psi_y(\kappa) \end{aligned} \tag{8.3.66}$$

and

$$\begin{aligned} C_x(\mathbf{r} - \mathbf{r}') &\equiv \left\langle [u_x(\mathbf{r}, t) - u_x(\mathbf{r}', t)]^2 \right\rangle \\ &= B_x |x - x'|^{-1} \Psi_x(\kappa)\,, \end{aligned} \tag{8.3.67}$$

where $\Psi_{x,y}$ are universal scaling functions, the scaling variable $\kappa \equiv \frac{X}{Y^{2/3}}$, $X = |x - x'|/\xi_x$, and $Y = |y - y'|/\xi_y$, $B_{x,y}$ are nonuniversal overall multiplicative factors extracted from the scaling functions, and the nonuniversal nonlinear lengths $\xi_{x,y}$ are calculated in [50].

8.3.7 Summary

So we have once again been able to get exact scaling exponents for a very nontrivial active system without evaluating any Feynman diagrams. Notice, however, that there are some limits to our knowledge here. To *fully* specify the scaling behavior of active incompressible two-dimensional fluids, we need *four* exponents: the two roughness exponents $\chi_{x,y}$, the anisotropy exponent ζ, and the dynamical exponent z. Unfortunately, because our mapping to the $1+1$-dimensional KPZ equation only related the *equal-time* behavior of the active fluid problem to the time-dependent behavior of the $1+1$-dimensional KPZ equation (with the Cartesian coordinate y of the active problem mapping onto the time t in the KPZ problem), these mappings give us no information about the time dependence of the original active fluid system. That is, we don't get the dynamical exponent z from this argument.

Indeed, we know of no exact argument for z at all. The only way to obtain that exponent is by a full RG, complete with all of the machinery of Feynman diagrams and so on. This calculation has been done [70], and finds $z = 1.67 \pm .05$.

9

"Malthusian" Flocks (Flocks with Birth and Death)

The alert reader will have noticed that, after exhaustively describing the entire elaborate machinery of Feynman diagrams and nonlinear perturbation theory for the RG in Chapter 3, I have not so far actually used any of that machinery to treat any of the problems I've discussed. Instead, for all of them, by one trick or another, I've been able to obtain *exact* scaling exponents *without* evaluating any Feynman diagrams at all.

This may have given many of you the false impression that one never needs any of that machinery, and that, furthermore, one can always get exact exponents by being sufficiently clever.

This is certainly not the case. Indeed, in Chapter 3, we saw the other extreme: No one knows how to obtain the scaling exponents for the two-dimensional KPZ equation, even after doing all of the Feynman diagrams and the full perturbative RG.

Another example of how difficult a problem can be even with all of the tools of the RG in hand is the problem with which I started this book: compressible flocks. While possibly not as bad as the two-dimensional KPZ equation, because, as far as we know, it *might* turn out that this problem has a stable, perturbatively accessible RG fixed point, in practice, the dynamical RG for this problem is simply intractable. What's worse, even if someone did spend the 20+ years I estimated in Chapter 4 that it would take to do a one-loop RG for that problem, it's quite possible – and, indeed, I'd put my money on it – that after doing so, one would find no perturbatively accessible fixed point, as happens in the $2 + 1$-dimensional KPZ equation.

So you might be thinking that Feynman diagrams and the perturbative RG are *never* useful: that every problem is either easy enough that you can get exact exponents without using those techniques – as we've done in the two preceding chapters for compressible flocks in $d = 2$ and $d > 2$, as well as for the $1 + 1$-dimensional KPZ equation in Chapter 3 – or utterly intractable.

In this chapter, I'll describe a counterexample to that discouraging idea: Malthusian flocks [21]. This problem, which I again studied with my usual partners in crime Chiu Fan Lee and Leiming Chen, comes from considering a way to eliminate density fluctuations from a flock *without* making it incompressible: "birth" and "death" of the constituent active particles. Such birth and death imply that the number of those particles is *not* conserved. We like to call such systems "Malthusian flocks" [21, 71], in honor of the great eighteenth-century scientist whose work has, unfortunately, fallen out of fashion recently. As I have done throughout this book, in this study we focus on systems moving through a background medium (e.g., the cytoskeleton, the intracellular matrix in a biological system, or, aerogel in a synthetic system) that can exert frictional forces on the active particles. Therefore, these systems also lack momentum conservation. Since our systems do *not* conserve the number of active particles either, this leaves them with no conservation laws at all.

The only hydrodynamic variables, therefore, are the "Goldstone mode" components $\mathbf{v}_\perp$ of the velocity, which, as in all of the examples treated in previous chapters, are those components perpendicular to the mean velocity. The density, which, you'll recall, was an additional hydrodynamic variable for number-conserving flocks (which, for the balance of this chapter, I'll refer to as "immortal" flocks – although strictly speaking I should call them "immortal and infertile," since they have neither death nor birth). However, here, because of birth and death, ρ becomes a "fast" variable, relaxing back to a local equilibrium value set by the local configuration of the slow variable $\mathbf{v}_\perp$. I'll show this explicitly in Section 9.1.

Some possible experimental realizations of such systems are: growing bacteria colonies and cell tissues [72, 73, 74, 75, 76, 77], and "treadmilling" molecular motor propelled biological macromolecules in a variety of intracellular structures, including the cytoskeleton [78], and mitotic spindles [79], in which molecules are being created and destroyed as they move.

In addition to describing biological and other active systems, the hydrodynamic equation for Malthusian flocks I'll describe here may also be viewed as a generic nonequilibrium d-dimensional d-component spin model, like the XY model of Chapter 2, but with the spin having d components, rather than two, in which the spin vector space $\mathbf{s}(\mathbf{r})$ and the coordinate space $\mathbf{r}$ are treated on an equal footing, and couplings between the two are allowed [80]. In particular, terms like $\mathbf{s} \cdot \nabla \mathbf{s}$ and $(\nabla \cdot \mathbf{s})\mathbf{s}$, will be present in the equation of motion that describes such a generic nonequilibrium system. As a result, the fluctuations in the system can propagate spatially in a spin-direction-dependent manner, but the spins themselves are not moving. Therefore, there are no density fluctuations and the only hydrodynamic variable is the spin field, the equation of motion for which is exactly the same as the one we derive here for a Malthusian flock, with spin playing the role of the velocity field. (Of course, nonequilibrium spin systems in which the spins "live" in real space in the sense described here will *not* map onto Malthusian flocks if those spins live on a lattice, due to the breaking of rotation invariance by the lattice itself. There *are*, however, ways of eliminating these "crystal field" effects.[1])

The remainder of this chapter proceeds along the same lines as for all of the previous problems I've discussed in this book: In Section 9.1, I'll present our derivation of the hydrodynamic equation of motion for Malthusian flocks. In Section 9.2, I'll present the linearized theory of these equations. In Sections 9.3 to 9.7, I'll present our derivation of the dynamical renormalization group (DRG) recursion relations in the $d = 4 - \epsilon$ expansion in a simplifying approximation which proves to be valid to leading order in ϵ.

In Section 9.8, I'll identify the fixed points of those recursion relations; in particular, the stable one that controls the entire ordered phase. With that fixed point in hand, I'll obtain the scaling laws and exponents in Section 9.9. In Section 9.10, I'll present an "uncontrolled" DRG calculation, in which we work to one-loop order in *arbitrary* spatial dimension d, rather than assuming d is close to the critical dimension of 4. This gives us a second set of exponents in $d = 3$, which prove to be quite close to those obtained from the $d = 4 - \epsilon$, thereby increasing our confidence in those numerical values.

In Section 9.11, I'll identify which of our results were artifacts of the lowest order in ϵ-expansion, and are therefore not expected to apply in real systems (fortunately, this is not many of them!).

Section 9.12 summarizes our results and discusses their implications for experiments and simulations. In Section 9.13, I calculate the nonlinear length and time scales at which Malthusian flocks crossover from linear to nonlinear behavior. I calculate a universal amplitude ratio in Section 9.14, and the separatrix between regions of positive and negative density correlation in Section 9.15.

Finally, in Section 9.16, I will present our results for the full one-loop order DRG for this problem, without the aforementioned simplifying approximation. This is by far the most complicated (though logically perfectly straightforward) DRG calculation in this book; indeed, it's the most complicated DRG calculation your humble author has ever done! Fortunately, this calculation shows that the simplifying approximation of Section 9.3 becomes asymptotically exact to leading order in ϵ, or, more generally, in any one-loop calculation.

Throughout this chapter, I'll use "we" when describing work done entirely in collaboration with Leiming Chen and Chiu Fan Lee (which is all of it), and will reserve "I" for my own pedagogical asides (and jokes!).

[1] One could eliminate crystal field effects altogether by allowing the spins to perform diffusive motion (that is uncoupled to the spins' directions) confined to the unit cells around their lattice sites so that the spin spatial distribution approximates a continuum at long time.

9.1 Derivation of the Equation of Motion

We begin by deriving the equations of motion. Our starting equation of motion for the velocity is identical to that of a flock with number conservation, as discussed in Chapter 4, and [33, 34, 35, 36, 37]:

$$\begin{aligned}&\partial_t \mathbf{v} + \lambda_1(\mathbf{v}\cdot\nabla)\mathbf{v} + \lambda_2(\nabla\cdot\mathbf{v})\mathbf{v} + \lambda_3\nabla(|\mathbf{v}|^2)\\ &= U(\rho,|\mathbf{v}|)\mathbf{v} - \nabla P_1 - \mathbf{v}\,(\mathbf{v}\cdot\nabla P_2(\rho,|\mathbf{v}|)) + \mu_B\nabla(\nabla\cdot\mathbf{v}) + \mu_T\nabla^2\mathbf{v}\\ &+ \mu_A(\mathbf{v}\cdot\nabla)^2\mathbf{v} + \mathbf{f}.\end{aligned} \tag{9.1.1}$$

I remind you that $\lambda_i(i = 1 \to 3)$, U, $\mu_{B,T,A}$ and the "pressures" $P_{1,2}(\rho,|\mathbf{v}|)$ are, in general, functions of the flocker number density ρ and the magnitude $|\mathbf{v}|$ of the local velocity. We will expand all of them to the order necessary to include all terms that are "relevant" in the sense of changing the long-distance behavior of the flock.

The fact that this equation is identical to that for "immortal" flocks derived in Chapter 4 is a direct consequence of the fact that the symmetries of our Malthusian flock problem are identical to those of that earlier immortal flock problem. That is, the dynamics is assumed to be rotation invariant, but not Galilean invariant.

Because the symmetries are identical, all of the symmetry arguments that I used in Chapter 4 to derive the equation of motion for the velocity for immortal flocks apply for nonnumber-conserving (i.e., Malthusian) flocks just as well. So the *velocity* equation of motion for Malthusian flocks *must* be identical to that for immortal flocks. (The density equation will change, as you'll see in a few moments.)

Now I'll just remind you here about a few of the features of this equation.

The $U(\rho,|\mathbf{v}|)$ term is responsible for spontaneous flock motion. As discussed in Chapter 4, all we need to assume about $U(\rho,|\mathbf{v}|)$ is that it satisfies $U(\rho_0,|\mathbf{v}| < v_0) > 0$, and $U(\rho_0,|\mathbf{v}| > v_0) < 0$ in the ordered phase. (Here as always ρ_0 is the steady-state density of the system in the absence of fluctuations.) These conditions on $U(\rho,|\mathbf{v}|)$ ensure that in the absence of fluctuations, the flock will move at a speed v_0.

As always (in particular, as in Chapter 4), the **f** term is a random Gaussian white noise, reflecting errors made by the flockers, with correlations:

$$\langle f_i(\mathbf{r},t)f_j(\mathbf{r}',t')\rangle = 2D\delta_{ij}\delta^d(\mathbf{r}-\mathbf{r}')\delta(t-t'), \tag{9.1.2}$$

where the noise strength D is a constant hydrodynamic parameter, and i,j label vector components.

So what *is* the difference between number-conserving and Malthusian flocks? As I said at the outset, it's the existence of birth and death in the latter. While this does *not* affect the velocity equation of motion (9.1.1), it *radically* changes the equation of motion for ρ. In immortal flocks, this is just the usual continuity

Thomas Malthus (1766–1834)
"The power of population is so superior to the power of the earth to produce subsistence for man, that premature death must in some shape or other visit the human race. The vices of mankind are active and able ministers of depopulation. They are the precursors in the great army of destruction, and often finish the dreadful work themselves. But should they fail in this war of extermination, sickly seasons, epidemics, pestilence, and plague advance in terrific array, and sweep off their thousands and tens of thousands. Should success be still incomplete, gigantic inevitable famine stalks in the rear, and with one mighty blow levels the population with the food of the world." —Malthus T.R. 1798.
An essay on the principle of population.
Chapter VII, p61[25]

Fig. 9.1.1 Thomas Malthus describing population dynamics in quintessentially eighteenth-century language. The quote is taken from [71], which is the oldest work cited in this book.

equation of compressible fluid dynamics. For Malthusian flocks, the equation needs an additional term representing the effects of birth and death.

Malthus described the effects of birth and death on human populations beautifully in [71], as illustrated in Figure 9.1.1.

You don't see that sort of language in scientific journals anymore. What Malthus would have been obliged to say had he published today in, say, *Physical Review Letters*, is illustrated in Figure 9.1.2.

Either of these can be summarized as follows: *Any* collection of entities that is reproducing and dying can only reach a nonzero steady-state population density if the death rate exceeds the birth rate for population densities greater than the steady-state density, and the converse for population densities less than the steady-state density [71]. This "Malthusian" condition implies that the net local growth rate of number density, in the absence of motion, which we'll call $\kappa(\rho)$ (which is called $g(\rho)$ in Figure 9.1.2), which is just the local birth rate per unit volume minus the local death rate (also per unit volume), vanishes at some fixed-point density ρ_0, with larger densities decreasing (i.e., $\kappa(\rho > \rho_0) < 0$), and smaller densities increasing (i.e., $\kappa(\rho < \rho_0) > 0$). That is, we expect $\kappa(\rho)$ to look like Figure 9.1.2.

This implies that the equation of motion for the density is now simply

$$\partial_t \rho + \nabla \cdot (\mathbf{v}\rho) = \kappa(\rho). \tag{9.1.3}$$

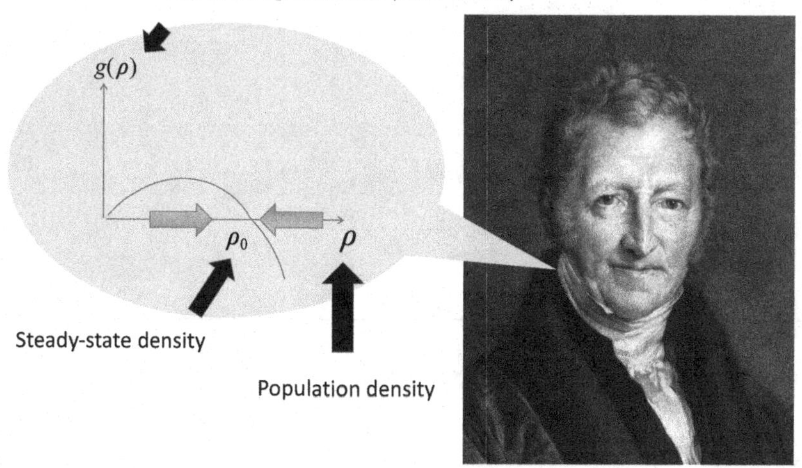

Fig. 9.1.2 The likely *Physical Review Letters* version of Figure 9.1.1. Thick arrows illustrate the direction of population density evolution over time, showing that any initial density will converge, in the absence of motion, to the steady-state density ρ_0.

Note that in the absence of birth and death, $\kappa(\rho) = 0$, and equation (9.1.3) reduces to the usual continuity equation (4.2.40), as it should, since "flocker number" is then conserved.

As you should expect by now after all of the discussion of, and arguments for, universality I've presented in this book, the detailed form of $\kappa(\rho)$ has no effect on the scaling behavior of this system in the long-distance, long-time limit. Indeed, the only features of it that matter at all for the hydrodynamic description are the value of ρ_0, and the slope of $\kappa(\rho)$ as a function of ρ *at* ρ_0, as illustrated in Figure 9.1.3. I'll now explain why this is so.

Since birth and death quickly restore the fixed-point density ρ_0, departures of ρ from ρ_0 are no longer hydrodynamic variables (since a hydrodynamic variable is, by definition, slow). Therefore, like all nonhydrodynamic variables, it can be expressed, at long time scales, as a purely local (in both space *and* time) function of the truly hydrodynamic variables (in our case, the velocity). To show this explicitly, we will do our usual trick of writing $\rho(\mathbf{r}, t) = \rho_0 + \delta\rho(\mathbf{r}, t)$ and expand both sides of equation (4.2.40) to leading order in $\delta\rho$. This gives

$$\partial_t \delta\rho + \rho_0 \nabla \cdot \mathbf{v} \cong \kappa'(\rho_0)\delta\rho \, . \tag{9.1.4}$$

I'll now perform a simple, effective, and powerful hydrodynamic trick on this equation of motion. This trick is what enables us to eliminate *all* nonhydrodynamic (i.e., "fast") variables from *any* problem that we're trying to study hydrodynamically

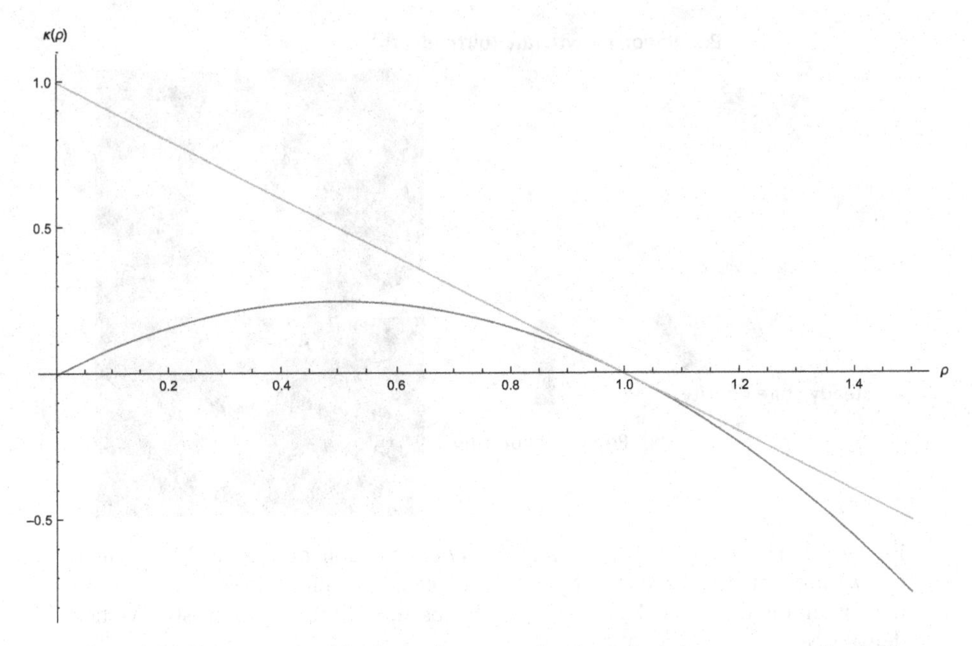

Fig. 9.1.3 The curve in this figure is a generic plot of $\kappa(\rho)$ versus ρ. All $\kappa(\rho)$'s that have this qualitative shape will have the same long-distance, long-time scaling. Furthermore, the only features of $\kappa(\rho)$ that even appear in the hydrodynamic theory are the value ρ_0 of ρ at which $\kappa(\rho)$ vanishes (which is also the steady-state density in the absence of motion), and its slope $\kappa'(\rho = \rho_0)$ (that is, the slope of the tangent to $\kappa(\rho)$ at $\rho = \rho_0$). That tangent is the straight line in this figure.

(that is, in the long-length scale, long-time limit). The trick is simply to drop the $\partial_t \rho$ term relative to the $\kappa'(\rho_0)\delta\rho$ term since we're interested in the hydrodynamic limit, in which the fields evolve extremely slowly. Equivalently, you can think of this an an application of the "gradient expansion" I've discussed earlier in this book; the only difference being that now I'm applying it to time, rather than space, derivatives. Dropping that time-derivative term gives

$$\rho_0 \nabla \cdot \mathbf{v} \cong \kappa'(\rho_0)\delta\rho\,. \tag{9.1.5}$$

The beauty of this trick is that this equation can be readily solved to give the field ρ entirely in terms of derivatives of the velocity field $\mathbf{v}$:

$$\delta\rho \cong \frac{\rho_0 \nabla \cdot \mathbf{v}}{\kappa'(\rho_0)} \equiv -\frac{\Delta\mu_B}{\sigma}(\nabla \cdot \mathbf{v}), \tag{9.1.6}$$

where we've defined $\Delta\mu_B = -\frac{\sigma\rho_0}{\kappa'(\rho_0)}$. (This strange definition is chosen with hindsight to simplify the final resulting equation, as you'll see in a moment.) The parameter $\Delta\mu_B$ is a positive constant, since $\kappa'(\rho_0) < 0$, because $\kappa(\rho > \rho_0) < 0$ and $\kappa(\rho < \rho_0) > 0$), and σ (which must be positive for stability) is the first expansion coefficient for P_1 (i.e., the analog of the inverse compressibility in an equilibrium system).

We can now insert this solution (9.1.6) for $\delta\rho$ in terms of $\mathbf{v}$ into the isotropic pressure P_1; the resulting closed equation of motion for $\mathbf{v}$ is

$$\begin{aligned}\partial_t\mathbf{v} + \lambda_1(\mathbf{v}\cdot\nabla)\mathbf{v} + \lambda_2(\nabla\cdot\mathbf{v})\mathbf{v} + \lambda_3\nabla(|\mathbf{v}|^2) = U(|\mathbf{v}|)\mathbf{v} - \mathbf{v}\,(\mathbf{v}\cdot\nabla P_2(|\mathbf{v}|))\\ + \mu_B'\nabla(\nabla\cdot\mathbf{v})\\ + \mu_T\nabla^2\mathbf{v} + \mu_A(\mathbf{v}\cdot\nabla)^2\mathbf{v} + \mathbf{f},\end{aligned} \tag{9.1.7}$$

where we've defined

$$\mu_B' \equiv \mu_B + \Delta\mu_B\,. \tag{9.1.8}$$

What we've just done is a simple illustration of a general fact that I've often alluded to earlier in this book: "Fast" variables can always be expressed in terms of the "slow" variables in the hydrodynamic (long distance and time) limit, leaving us with closed equations of motion for the slow, hydrodynamic variables. By "closed," I mean that those equations only involve the slow hydrodynamic variables themselves: No other variables need be considered.

You can see this by inspection of (9.1.7): Only $\mathbf{v}$ itself appears in the equation; the density has disappeared.

Note that this did *not* happen because we simply *ignored* the density: That would have been wrong, as can be seen from the fact that the "bulk viscosity" μ_B' is *not* the same as that μ_B appearing in the original equation of motion (9.1.1), but is shifted from that by $\Delta\mu_B$, as shown in (9.1.8). This shift comes entirely from the effects of the density. However, the net result of including the density has been entirely accounted for by a suitable shift (9.1.8) in the parameters.

This will always be the case in *any* problem if one is interested only in the hydrodynamic limit.

The alert reader (the *supernaturally* alert reader!) will have noticed that, to eliminate ρ completely from the problem, I had to drop the ρ dependence of $U(\rho, |\mathbf{v}|)$ from the problem. This is justified by the gradient expansion: From the solution (9.1.6) for $\delta\rho$, we see that $\delta\rho$ is proportional to spatial derivatives of $\mathbf{v}$. These are irrelevant relative to the direct dependence of $U(\rho, |\mathbf{v}|)$ on $|\mathbf{v}|$ itself, and so can be dropped. Thus we need only consider $U(\rho_0, |\mathbf{v}|)$, which I've just called $U(|\mathbf{v}|)$ in (9.1.7).

I've applied the same reasoning to the anisotropic pressure P_2 in (9.1.7).

We're not quite done eliminating fast variables yet. To see this, remember that, as we did for immortal flocks in Chapter 4, we want to study the ordered state (i.e., the state in which $\langle\mathbf{v}(\mathbf{r},t)\rangle = v_0\hat{\mathbf{x}}$, where we've chosen the spontaneously picked direction of mean flock motion as our x-axis). In this state, we can expand the $\mathbf{v}$ equation of motion for small departures $\mathbf{u}(\mathbf{r},t) \equiv u_x\hat{\mathbf{x}} + \mathbf{u}_\perp(\mathbf{r},t)$ of $\mathbf{v}(\mathbf{r},t)$ from uniform motion with speed v_0:

$$\mathbf{v}(\mathbf{r},t) = (v_0 + u_x)\hat{\mathbf{x}} + \mathbf{u}_\perp(\mathbf{r},t), \tag{9.1.9}$$

where, henceforth x and $\perp$ denote components along and perpendicular to the mean velocity, respectively.

In this hydrodynamic approach, we're interested only in fluctuations of $\mathbf{u}(\mathbf{r},t)$ that vary slowly in space and time, which in this case are the Goldstone modes $\mathbf{u}_\perp$. The component u_x of the fluctuation of the velocity *along* the direction of mean motion is *not* such a fluctuation. Rather, like the density fluctuation $\delta\rho$, it is a nonhydrodynamic or "fast" variable of the sort we've seen many times by now. It therefore can be eliminated from the equations of motion in much the same manner as we just eliminated the density fluctuations.

I did this elimination for the problem with number conservation in Chapter 4; here I'll very briefly review the argument, as applied to our equation of motion (9.1.7).

To focus on fluctuations in the magnitude of the velocity (which are, strictly speaking, the fast variable here), we take the dot product of both sides of (9.1.7) with $\mathbf{v}$ itself. This gives

$$\begin{aligned}\frac{1}{2}\left(\partial_t|\mathbf{v}|^2 + (\lambda_1 + 2\lambda_3)(\mathbf{v}\cdot\nabla)|\mathbf{v}|^2\right) + \lambda_2(\nabla\cdot\mathbf{v})|\mathbf{v}|^2 \\ = U(|\mathbf{v}|)|\mathbf{v}|^2 - |\mathbf{v}|^2\mathbf{v}\cdot\nabla P_2 + \mu_B'\mathbf{v}\cdot\nabla(\nabla\cdot\mathbf{v}) \\ + \mu_T\mathbf{v}\cdot\nabla^2\mathbf{v} + \mu_A\mathbf{v}\cdot\left((\mathbf{v}\cdot\nabla)^2\mathbf{v}\right) + \mathbf{v}\cdot\mathbf{f}.\end{aligned} \tag{9.1.10}$$

In this hydrodynamic approach, we're as always interested only in fluctuations $\mathbf{u}_\perp(\mathbf{r},t)$ and $\delta\rho(\mathbf{r},t)$ that vary slowly in space and time. Hence, terms involving spatio-temporal derivatives of $\mathbf{u}_\perp(\mathbf{r},t)$ and $\delta\rho(\mathbf{r},t)$ are always negligible, in the hydrodynamic limit, compared with terms involving the same number of powers of fields with fewer spatio-temporal derivatives. Furthermore, the fluctuations $\mathbf{u}_\perp(\mathbf{r},t)$ and $\delta\rho(\mathbf{r},t)$ can themselves be shown to be small in the long-wavelength limit. Hence, we need only keep terms in (9.1.10) up to linear order in $\mathbf{u}_\perp(\mathbf{r},t)$ and $\delta\rho(\mathbf{r},t)$. The $\mathbf{v}\cdot\mathbf{f}$ term can likewise be dropped.

These observations can be used to eliminate many terms in equation (9.1.10), and solve for the quantity U; we obtain $U = \lambda_2\nabla\cdot\mathbf{v} + \mathbf{v}\cdot\nabla P_2$. Inserting this expression for U back into equation (9.1.7), we find that P_2 and λ_2 cancel out of the $\mathbf{v}$ equation of motion, leaving, ignoring irrelevant terms:

$$\partial_t \mathbf{v} + \lambda_1(\mathbf{v}\cdot\nabla)\mathbf{v} + \lambda_3\nabla(|\mathbf{v}|^2) = \mu_T\nabla^2\mathbf{v} + \mu_B'\nabla(\nabla\cdot\mathbf{v}) + \mu_A(\mathbf{v}\cdot\nabla)^2\mathbf{v} + \mathbf{f}. \tag{9.1.11}$$

This can be made into an equation of motion for $\mathbf{u}_\perp$ involving only $\mathbf{u}_\perp(\mathbf{r},t)$ itself by projecting perpendicular to the direction of mean flock motion $\hat{\mathbf{x}}$, and eliminating u_x using $U = \lambda_2\nabla\cdot\mathbf{v} + \mathbf{v}\cdot\nabla P_2$ and the expansion

$$U \approx -\Gamma_1(|\mathbf{v}| - v_0 \approx -\Gamma_1\left(u_x + \frac{u_\perp^2}{2v_0}\right), \tag{9.1.12}$$

where we've defined $\Gamma_1 \equiv -\left(\frac{\partial U}{\partial|\mathbf{v}|}\right)_{\rho,0}$, with subscripts 0 denoting functions of $|\mathbf{v}|$ evaluated at $|\mathbf{v}| = v_0$. In going from the first to the second approximate equality in (9.1.12), we have used the simple geometrical fact, which follows from the expansion (9.1.9) and the Pythagorean theorem, that $|\mathbf{v}| - v_0 \approx u_x + \frac{u_\perp^2}{2v_0}$, to leading order in u_x and $u_\perp^2$.

Thus eliminating u_x, we obtain:

$$\partial_t\mathbf{u}_\perp + \gamma\partial_x\mathbf{u}_\perp + \lambda(\mathbf{u}_\perp\cdot\nabla_\perp)\mathbf{u}_\perp = \mu_1\nabla_\perp^2\mathbf{u}_\perp + \mu_2\nabla_\perp(\nabla_\perp\cdot\mathbf{u}_\perp) + \mu_x\partial_x^2\mathbf{u}_\perp + \mathbf{f}_\perp, \tag{9.1.13}$$

where we've defined $\lambda \equiv \lambda_1^0$, $\gamma \equiv \lambda_1^0 v_0$, $\mu_2 \equiv \mu_B'^0 + 2v_0\lambda_3^0(\lambda_2^0 - \Gamma_2\Delta\mu_B/\sigma)/\Gamma_1$, $\mu_x \equiv \mu_T^0 + \mu_A^0 v_0^2$, and $\mu_1 \equiv \mu_T^0$, where the superscripts 0 denote coefficients evaluated at $\rho = \rho_0$ and $|\mathbf{v}| = v_0$. In writing (9.1.13) we have ignored irrelevant terms which come from the higher-order expansion of the coefficients in $\delta\rho$ and $u_x + \frac{u_\perp^2}{2v_0}$ than the zeroth order.

Changing coordinates to a new Galilean frame $\mathbf{r}'$ moving with respect to our original frame (which, we remind the reader, is that of the fixed background medium through which the flock moves) in the direction $\hat{\mathbf{x}}$ of mean flock motion at speed γ – i.e.,

$$\mathbf{r}' \equiv \mathbf{r} - \gamma t\hat{\mathbf{x}}, \tag{9.1.14}$$

we obtain

$$\partial_t\mathbf{u}_\perp + \lambda(\mathbf{u}_\perp\cdot\nabla_\perp)\mathbf{u}_\perp = \mu_1\nabla_\perp^2\mathbf{u}_\perp + \mu_2\nabla_\perp(\nabla_\perp\cdot\mathbf{u}_\perp) + \mu_x\partial_x^2\mathbf{u}_\perp + \mathbf{f}_\perp, \tag{9.1.15}$$

where we have dropped the prime in $\mathbf{r}$.

This equation will be the basis of our remaining theoretical analysis. Note that to obtain correlations in the original (unboosted) coordinate system, we need to take into account the boost (9.1.14).

9.2 Linear Theory

9.2.1 Propagators

In this section we treat the linear approximation to the model (9.1.15). The procedure is very similar to that applied to immortal flocks in Chapter 4, but simpler, since we no longer have that pesky density field ρ to worry us. Keeping only the linear terms in (9.1.15), and writing the resultant equation of motion in Fourier space, we obtain

$$-\mathrm{i}\omega \mathbf{u}_\perp(\tilde{\mathbf{k}}) = -\mu_1 k_\perp^2 \mathbf{u}_\perp(\tilde{\mathbf{k}}) - \mu_2 \mathbf{k}_\perp \left(\mathbf{k}_\perp \cdot \mathbf{u}_\perp(\tilde{\mathbf{k}})\right) - \mu_x k_x^2 \mathbf{u}_\perp(\tilde{\mathbf{k}}) + \mathbf{f}_\perp(\tilde{\mathbf{k}}), \tag{9.2.1}$$

where $\tilde{\mathbf{k}} \equiv (\mathbf{k}, \omega)$, and

$$\mathbf{u}_\perp(\tilde{\mathbf{k}}) = \frac{1}{\left(\sqrt{2\pi}\right)^{d+1}} \int \mathrm{d}t\, \mathrm{d}^d r\, \mathbf{u}_\perp(\mathbf{r}, t) \mathrm{e}^{\mathrm{i}(\omega t - \mathbf{k}\cdot\mathbf{r})}. \tag{9.2.2}$$

As I did in Chapter 4, we can solve the linear equation (9.2.1) by separating $\mathbf{u}_\perp$ into its "longitudinal" component along $\mathbf{k}_\perp$ and its remaining $d-2$ "transverse" components perpendicular to $\mathbf{k}_\perp$. (I remind the reader that $\mathbf{u}_\perp$ has only $d-1$ independent components, since it is by definition orthogonal to the mean direction of flock motion $\hat{\mathbf{x}}$.)

That is, we write:

$$\mathbf{u}_\perp(\tilde{\mathbf{k}}) = u_L(\tilde{\mathbf{k}})\hat{\mathbf{k}}_\perp + \mathbf{u}_T(\tilde{\mathbf{k}}), \tag{9.2.3}$$

with $\mathbf{k}_\perp \cdot \mathbf{u}_T = 0$ by definition. These components u_L and $\mathbf{u}_T$ can be computed using

$$u_L(\tilde{\mathbf{k}}) = \hat{\mathbf{k}}_\perp \cdot \mathbf{u}_\perp(\tilde{\mathbf{k}}) \tag{9.2.4}$$

and

$$u_i^T(\tilde{\mathbf{k}}) = P_{ij}^\perp(\mathbf{k}) u_j^\perp(\tilde{\mathbf{k}}), \tag{9.2.5}$$

where you've seen the "transverse projection operator"

$$P_{ij}^\perp(\mathbf{k}) \equiv \delta_{ij}^\perp - \frac{k_i^\perp k_j^\perp}{k_\perp^2}, \tag{9.2.6}$$

which projects any vector into the $(d-2)$-dimensional space orthogonal to both the direction of mean flock motion $\hat{\mathbf{x}}$ and $\mathbf{k}_\perp$, before. As in Chapter 4, we decompose the random force $\mathbf{f}_\perp$ in exactly the same way.

We can now easily rewrite the equation of motion (9.2.1) for $\mathbf{u}_\perp$ as decoupled equations for u_L and $\mathbf{u}_T$. To obtain the former, we take the dot product of $\hat{\mathbf{k}}_\perp$ with (9.2.1); this gives a closed equation of motion for u_L:

$$-\mathrm{i}\omega u_L(\tilde{\mathbf{k}}) = -\mu_L k_\perp^2 u_L(\tilde{\mathbf{k}}) - \mu_x k_x^2 u_L(\tilde{\mathbf{k}}) + f_L(\tilde{\mathbf{k}}), \tag{9.2.7}$$

where we have defined

$$\mu_L \equiv \mu_1 + \mu_2. \tag{9.2.8}$$

Likewise, acting on both sides of (9.2.1) with the transverse projection operator (9.2.22) gives a closed equation of motion for $\mathbf{u}_T$:

$$-\mathrm{i}\omega \mathbf{u}_T(\tilde{\mathbf{k}}) = -\mu_1 k_\perp^2 \mathbf{u}_T(\tilde{\mathbf{k}}) - \mu_x k_x^2 \mathbf{u}_T(\tilde{\mathbf{k}}) + \mathbf{f}_T(\tilde{\mathbf{k}}). \tag{9.2.9}$$

Before proceeding to solve these two simple linear equations for u_L and $\mathbf{u}_T$ in terms of the forces f_L and $\mathbf{f}_T$, it is informative first to determine the eigenfrequencies $\omega(\mathbf{k})$ of the normal modes of this system. These are clearly just

$$\omega_L(\mathbf{k}) = -\mathrm{i}\left(\mu_L k_\perp^2 + \mu_x k_x^2\right) \tag{9.2.10}$$

for the longitudinal mode, and

$$\omega_T(\mathbf{k}) = -\mathrm{i}\left(\mu_1 k_\perp^2 + \mu_x k_x^2\right) \tag{9.2.11}$$

for the transverse mode. In order for the system to be stable, we must have the imaginary part $I_{L,T}(\omega(\mathbf{k})) < 0$ for both modes; this requires that

$$\mu_{L,1,x} > 0. \tag{9.2.12}$$

Note that this condition (9.2.12) does *not* require $\mu_2 > 0$; using the definition (9.2.8) of μ_L in (9.2.12), we see that the stability requirement is weaker:

$$\mu_2 > -\mu_1, \tag{9.2.13}$$

or, equivalently,

$$\frac{\mu_2}{\mu_1} > -1. \tag{9.2.14}$$

Now we turn to the solutions of the equations of motion (9.2.7) and (9.2.9). These can be immediately read off:

$$u_L(\tilde{\mathbf{k}}) = G_L(\tilde{\mathbf{k}}) f_L(\tilde{\mathbf{k}}), \tag{9.2.15}$$

$$\mathbf{u}_T(\tilde{\mathbf{k}}) = G_T(\tilde{\mathbf{k}}) \mathbf{f}_T(\tilde{\mathbf{k}}), \tag{9.2.16}$$

where we've defined the longitudinal and transverse "propagators"

$$G_L(\tilde{\mathbf{k}}) = \frac{1}{-\mathrm{i}\omega + \mu_L k_\perp^2 + \mu_x k_x^2}, \tag{9.2.17}$$

$$G_T(\tilde{\mathbf{k}}) = \frac{1}{-\mathrm{i}\omega + \mu_1 k_\perp^2 + \mu_x k_x^2}. \tag{9.2.18}$$

These propagators will also have an important role to play in our DRG analysis later.

The solutions (9.2.15) and (9.2.16) for u_L and $\mathbf{u}_T$ can be summarized in a single equation using the relations (9.2.4) and (9.2.5) between $\mathbf{u}_\perp$ and its components u_L and $\mathbf{u}_T$, along with the analogous relations between $\mathbf{f}_\perp$ and f_L and $\mathbf{f}_T$; we obtain

$$u_i^\perp(\tilde{\mathbf{k}}) = G_{ij}(\tilde{\mathbf{k}}) f_j^\perp(\tilde{\mathbf{k}}), \tag{9.2.19}$$

where

$$G_{ij}(\tilde{\mathbf{k}}) \equiv L_{ij}^\perp(\mathbf{k}_\perp) G_L(\tilde{\mathbf{k}}) + P_{ij}^\perp(\mathbf{k}_\perp) G_T(\tilde{\mathbf{k}}), \tag{9.2.20}$$

with, as in Chapter 4,

$$L_{ij}^\perp(\mathbf{k}_\perp) \equiv \frac{k_i^\perp k_j^\perp}{k_\perp^2} \tag{9.2.21}$$

being the "longitudinal projection operator," which projects any vector along $\mathbf{k}_\perp$, and

$$P_{ij}^\perp(\mathbf{k}_\perp) \equiv \delta_{ij}^\perp \frac{k_i^\perp k_j^\perp}{k_\perp^2} \tag{9.2.22}$$

being the "transverse projection operator," which projects any vector into the plane perpendicular to the mean direction of flock motion (what we're calling the "$\perp$-plane") and then orthogonal to $\mathbf{k}_\perp$ within the $\perp$-plane.

Note that equation (9.2.19) is logically exactly the same as our formal solution (3.4.3) for the height field in h in the KPZ equation. There are two important differences to note, however: one is that we now have more fields, so the propagator G_{ij} is a tensor. The second, which is going to cause us a great deal of misery when we have to expand graphs in powers of external momentum, is that the projection operators $L_{ij}(\mathbf{k}_\perp)$ and $P_{ij}(\mathbf{k}_\perp)$ also depend on momentum, and sometimes have to be expanded as well. This is nuisancesome, but manageable, as you'll see below.

9.2.2 Velocity Correlation Functions

Using (9.2.19), we obtain the autocorrelations:

$$\left\langle u_i^\perp(\tilde{\mathbf{k}}) u_j^\perp(\tilde{\mathbf{k}}')\right\rangle = G_{im}(\tilde{\mathbf{k}}) G_{jn}(\tilde{\mathbf{k}}') \left\langle f_m^\perp(\tilde{\mathbf{k}}) f_n^\perp(\tilde{\mathbf{k}}')\right\rangle = 2DC_{ij}(\tilde{\mathbf{k}})\delta(\mathbf{k}+\mathbf{k}')\delta(\omega+\omega'), \tag{9.2.23}$$

where in the second equality we have used the correlations of the noise in Fourier space:

$$\left\langle f_m^\perp(\tilde{\mathbf{k}}) f_n^\perp(\tilde{\mathbf{k}}')\right\rangle = 2D\delta_{mn}\delta(\mathbf{k}+\mathbf{k}')\delta(\omega+\omega'), \tag{9.2.24}$$

and we've defined

$$C_{ij}(\tilde{\mathbf{k}}) \equiv L_{ij}^{\perp}(\mathbf{k})|G_L(\tilde{\mathbf{k}})|^2 + P_{ij}^{\perp}(\mathbf{k})|G_T(\tilde{\mathbf{k}})|^2 . \tag{9.2.25}$$

In writing (9.2.23), we have, being liberals, made liberal use of the property shared by both projection operators $L_{ij}^{\perp}$ and $P_{ij}^{\perp}$ that their squares are themselves.

Fourier transforming the above correlation function gives the velocity correlations in real space and time. First, let's calculate the equal-time correlation function:

$$\begin{aligned}
\langle \mathbf{u}_{\perp}(\mathbf{r},t)\cdot\mathbf{u}_{\perp}(\mathbf{0},0)\rangle &= \frac{1}{(2\pi)^{d+1}}\int \mathrm{d}\omega \mathrm{d}\omega' \mathrm{d}^d k \mathrm{d}^d k' \left\langle \mathbf{u}_{\perp}(\tilde{\mathbf{k}})\cdot\mathbf{u}_{\perp}(\tilde{\mathbf{k}}')\right\rangle \mathrm{e}^{\mathrm{i}\mathbf{k}\cdot\mathbf{r}} \\
&= \frac{2D}{(2\pi)^{d+1}}\int \mathrm{d}\omega \mathrm{d}^d k\, \mathrm{e}^{\mathrm{i}\mathbf{k}\cdot\mathbf{r}} \left[\frac{1}{\omega^2 + \left(\mu_L k_{\perp}^2 + \mu_x k_x^2\right)^2}\right. \\
&\qquad \left. + \frac{d-2}{\omega^2 + \left(\mu_1 k_{\perp}^2 + \mu_x k_x^2\right)^2}\right] \\
&= D[U_L(\mathbf{r}) + (d-2)U_T(\mathbf{r})],
\end{aligned} \tag{9.2.26}$$

where in the last step we evaluated the two simple integrals over frequency ω, and we've defined

$$U_L(\mathbf{r}) = \frac{1}{(2\pi)^d}\int \mathrm{d}^d k \left[\frac{\mathrm{e}^{\mathrm{i}\mathbf{k}\cdot\mathbf{r}}}{\mu_L k_{\perp}^2 + \mu_x k_x^2}\right], \quad U_T(\mathbf{r}) = \frac{1}{(2\pi)^d}\int \mathrm{d}^d k \left[\frac{\mathrm{e}^{\mathrm{i}\mathbf{k}\cdot\mathbf{r}}}{\mu_1 k_{\perp}^2 + \mu_x k_x^2}\right]. \tag{9.2.27}$$

Clearly, $U_{L,T}(\mathbf{r})$ satisfy the anisotropic Poisson equations:

$$\left(\mu_L \nabla_{\perp}^2 + \mu_x \partial_x^2\right) U_L(\mathbf{r}) = -\delta^d(\mathbf{r}), \quad \left(\mu_1 \nabla_{\perp}^2 + \mu_x \partial_x^2\right) U_T(\mathbf{r}) = -\delta^d(\mathbf{r}). \tag{9.2.28}$$

The solutions to the above equations are, for $d > 2$,

$$U_L(\mathbf{r}) = \frac{\left(\frac{\mu_L}{\mu_x}x^2 + r_{\perp}^2\right)^{(2-d)/2}}{S_d(d-2)\sqrt{\mu_x\mu_L}}, \quad U_T(\mathbf{r}) = \frac{\left(\frac{\mu_1}{\mu_x}x^2 + r_{\perp}^2\right)^{(2-d)/2}}{S_d(d-2)\sqrt{\mu_x\mu_1}}, \tag{9.2.29}$$

where S_d is the surface area of a d-dimensional unit sphere.

You can see this most easily by rescaling lengths so that equation (9.2.28) becomes the *isotropic* Poisson's equation, and then solving that using your first-year E&M techniques.

Inserting the above results into Eq. (9.2.26), we get

$$\langle \mathbf{u}_{\perp}(\mathbf{r},t)\cdot\mathbf{u}_{\perp}(\mathbf{0},t)\rangle \propto r^{-(d-2)}, \tag{9.2.30}$$

where the "constant" of proportionality is direction dependent, as can be seen from the explicit forms (9.2.29).

Now we calculate the time-dependent correlations. Setting the spatial separation to zero in the correlation function to get

$$\langle \mathbf{u}_\perp(\mathbf{0},t)\cdot\mathbf{u}_\perp(\mathbf{0},0)\rangle = \frac{1}{(2\pi)^{d+1}}\int \mathrm{d}\omega\mathrm{d}\omega'\mathrm{d}^d k\mathrm{d}^d k' \left\langle \mathbf{u}_\perp(\tilde{\mathbf{k}})\cdot\mathbf{u}_\perp(\tilde{\mathbf{k}}')\right\rangle \mathrm{e}^{-\mathrm{i}\omega t}$$

$$= \frac{2D}{(2\pi)^{d+1}}\int \mathrm{d}\omega\mathrm{d}^d k\,\mathrm{e}^{-\mathrm{i}\omega t}\left[\frac{1}{\omega^2+\left(\mu_L k_\perp^2+\mu_x k_x^2\right)^2}+\frac{d-2}{\omega^2+\left(\mu_1 k_\perp^2+\mu_x k_x^2\right)^2}\right]$$

$$= \frac{D}{(2\pi)^d}\int \mathrm{d}^d k\left[\frac{\mathrm{e}^{-(\mu_L k_\perp^2+\mu_x k_x^2)|t|}}{\mu_L k_\perp^2+\mu_x k_x^2}+\frac{(d-2)\mathrm{e}^{-(\mu_1 k_\perp^2+\mu_x k_x^2)|t|}}{\mu_1 k_\perp^2+\mu_x k_x^2}\right]$$

$$= |t|^{-\left(\frac{d-2}{2}\right)}D\int\frac{\mathrm{d}^d q}{(2\pi)^d}\left[\frac{\mathrm{e}^{-(\mu_L q_\perp^2+\mu_x q_x^2)}}{\mu_L q_\perp^2+\mu_x q_x^2}+\frac{(d-2)\mathrm{e}^{-(\mu_1 q_\perp^2+\mu_x q_x^2)}}{\mu_1 q_\perp^2+\mu_x q_x^2}\right]$$

$$\propto |t|^{-\left(\frac{d-2}{2}\right)}, \tag{9.2.31}$$

where in the penultimate equality we have made the change of vectorial variable, $\mathbf{q} = |t|^{\frac{1}{2}}\mathbf{k}$, while in the ultimate proportionality we have used the fact that the integral over $\mathbf{q}$ is a finite constant (i.e., independent of time t), for spatial dimensions $d > 2$.

We can easily generalize these results to arbitrary spatio-temporal separations. We start with

$$\langle \mathbf{u}_\perp(\mathbf{r},t)\cdot\mathbf{u}_\perp(\mathbf{0},0)\rangle = \frac{1}{(2\pi)^{d+1}}\int \mathrm{d}\omega\mathrm{d}\omega'\mathrm{d}^d k\mathrm{d}^d k' \left\langle \mathbf{u}_\perp(\tilde{\mathbf{k}})\cdot\mathbf{u}_\perp(\tilde{\mathbf{k}}')\right\rangle \mathrm{e}^{\mathrm{i}(\mathbf{k}\cdot\mathbf{r}-\omega t)}$$

$$= \frac{2D}{(2\pi)^{d+1}}\int \mathrm{d}\omega\mathrm{d}^d k\,\mathrm{e}^{\mathrm{i}(\mathbf{k}\cdot\mathbf{r}-\omega t)}\left[\frac{1}{\omega^2+\left(\mu_L k_\perp^2+\mu_x k_x^2\right)^2}+\frac{d-2}{\omega^2+\left(\mu_1 k_\perp^2+\mu_x k_x^2\right)^2}\right]. \tag{9.2.32}$$

Changing the variables of integration from $\mathbf{k}$ and ω to $\mathbf{Q}$ and Υ:

$$\mathbf{k}_\perp = \mathbf{Q}_\perp/r_\perp,\quad k_x = Q_x/r_\perp,\quad \omega = \Upsilon/r_\perp^2, \tag{9.2.33}$$

we obtain

$$\langle \mathbf{u}_\perp(\mathbf{r},t)\cdot\mathbf{u}_\perp(\mathbf{0},0)\rangle = r_\perp^{-(d-2)}H_u\left(\frac{x}{r_\perp},\frac{t}{r_\perp^2}\right)\propto\begin{cases} r^{-(d-2)}, & r\gg|t|^{\frac{1}{2}},\\ |t|^{-\frac{(d-2)}{2}}, & |t|\gg r_\perp^2,\end{cases} \tag{9.2.34}$$

where we've defined the scaling function

$$H_u(a,b) \equiv \frac{2D}{(2\pi)^{d+1}} \int \mathrm{d}\Upsilon \mathrm{d}^d Q \, \mathrm{e}^{\mathrm{i}[\mathbf{Q}_\perp \cdot \hat{\mathbf{r}}_\perp + Q_x a - \Upsilon b]} \left[\frac{1}{\Upsilon^2 + \left(\mu_L Q_\perp^2 + \mu_x Q_x^2\right)^2} + \frac{d-2}{\Upsilon^2 + \left(\mu_1 Q_\perp^2 + \mu_x Q_x^2\right)^2} \right]. \quad (9.2.35)$$

9.2.3 Density Correlations

Although it is not a "soft mode" of Malthusian flocks, since it is not conserved in these systems, the density ρ nonetheless exhibits long-ranged spatio-temporal correlations by virtue of being "enslaved" to the slow u_L field via (9.1.6). This mechanism is very similar to the "generic scale invariance" found in driven anisotropic diffusion [81].

Using the relation (9.1.6) in Fourier space, we obtain

$$\left\langle \delta\rho(\tilde{\mathbf{k}})\delta\rho(\tilde{\mathbf{k}}') \right\rangle = \frac{2D' k_\perp^2 \delta(\mathbf{k}+\mathbf{k}')\delta(\omega+\omega')}{\omega^2 + (\mu_L k_\perp^2 + \mu_x k_x^2)^2}, \quad (9.2.36)$$

where we've defined

$$D' \equiv D\left(\frac{\rho_0}{\kappa'(\rho_0)}\right)^2. \quad (9.2.37)$$

The spatio-temporal correlations can be calculated by Fourier transforming (9.2.36) back to real space and time. In particular, the equal time correlation function is

$$\begin{aligned} \langle \delta\rho(\mathbf{r},0)\delta\rho(\mathbf{0},0)\rangle &= \frac{1}{(2\pi)^{d+1}} \int \mathrm{d}\omega \mathrm{d}\omega' \mathrm{d}^d k \mathrm{d}^d k' \left\langle \delta\rho(\tilde{\mathbf{k}})\delta\rho(\tilde{\mathbf{k}}') \right\rangle \mathrm{e}^{\mathrm{i}\mathbf{k}\cdot\mathbf{r}} \\ &= \frac{1}{(2\pi)^d} \int \mathrm{d}^d k \, \frac{D' k_\perp^2 \mathrm{e}^{\mathrm{i}\mathbf{k}\cdot\mathbf{r}}}{\mu_L k_\perp^2 + \mu_x k_x^2}, \end{aligned} \quad (9.2.38)$$

where in the last equality we have used (9.2.36). To calculate this correlation function, we write

$$\langle \delta\rho(\mathbf{r},0)\delta\rho(t,0)\rangle = -D'\nabla_\perp^2 U_L(\mathbf{r}), \quad (9.2.39)$$

where $U_L(\mathbf{r})$ is given in (9.2.29). Inserting (9.2.29) into the above expression gives

$$\langle \delta\rho(\mathbf{r},t)\delta\rho(\mathbf{0},t)\rangle = \left(\frac{\rho_0}{\kappa'(\rho_0)}\right)^2 \frac{D}{S_d} \sqrt{\frac{\mu_x^{d-1}}{\mu_L}} \left[\frac{\mu_L(d-1)x^2 - \mu_x r_\perp^2}{\left(\mu_L x^2 + \mu_x r_\perp^2\right)^{(2+d)/2}} \right] \propto r^{-d}.$$

In particular, for $d = 3$, we have

$$\langle \delta\rho(\mathbf{r},t)\delta\rho(\mathbf{0},t)\rangle \sim r^{-3}. \quad (9.2.40)$$

It is clear from (9.2.40) that the equal time correlation function of the density fluctuation $\delta\rho$ vanishes on the surface

$$x = \pm\left(\sqrt{\frac{\mu_x}{\mu_L(d-1)}}\right)r_\perp, \tag{9.2.41}$$

which, in $d = 3$, is a cone. For $|x| > \sqrt{\mu_x/\mu_L(d-1)}r_\perp$, $\langle\delta\rho(\mathbf{r},t)\delta\rho(\mathbf{0},t)\rangle$ is positive; otherwise, the correlation is negative.

The qualitative shape of the regions of positive and negative density correlations can be understood heuristically as follows. We first recall that in the hydrodynamic limit, we can ignore velocity fluctuations in the x-direction. Hence, equation (9.1.6) implies that $\delta\rho \propto -\nabla_\perp \cdot \mathbf{u}_\perp$. That is, a negative $\delta\rho(\mathbf{0},t)$ at the origin results from a positive divergence of $\mathbf{u}_\perp$ at the origin. Therefore, a negative $\delta\rho(\mathbf{0},t)$ will occur if, e.g., $u_y(-\epsilon\hat{\mathbf{y}},t) > u_y(\epsilon\hat{\mathbf{y}},t)$, where ϵ is a small distance. On the other hand, since we know that the equal-time correlation of $\mathbf{u}_\perp$ is always positive, we expect $u_y(-\epsilon\hat{\mathbf{y}},t) > u_y(\epsilon\hat{\mathbf{y}},t)$ will lead to $u_y(A\hat{x} - \epsilon\hat{\mathbf{y}},t) > u_y(A\hat{x} + \epsilon\hat{\mathbf{y}},t)$, where A is any positive or negative number. This leads further to $\delta\rho(A\hat{\mathbf{x}},t) < 0$. Therefore, we expect that, more often than not, $\delta\rho(A\hat{\mathbf{x}},t) < 0$ if $\delta\rho(\mathbf{0},t) < 0$. Thus, this case will make a positive contribution to $\langle\delta\rho(\mathbf{0},t)\delta\rho(A\hat{\mathbf{x}},t)\rangle$.

One can make a similar argument for the case in which $\delta\rho(\mathbf{0},t) > 0$, and conclude that usually $\delta\rho(A\hat{\mathbf{x}},t) > 0$ if $\delta\rho(\mathbf{0},t) > 0$. Thus, this case will also make a positive contribution to $\langle\delta\rho(\mathbf{0},t)\delta\rho(A\hat{\mathbf{x}},t)\rangle$.

This explains the positive region of $\langle\delta\rho(\mathbf{0},t)\delta\rho(\mathbf{r},t)\rangle$ close to the $\hat{x}$-axis (i.e., $r_\perp \ll x$) in the $\mathbf{r}$-space, which implies the density correlations along the direction of mean flock motion are positive. Now, as the equal-time density correlation function is the Laplacian of a function (9.2.39), the overall spatial integral of the correlation function must be zero. Therefore, there must be a negative region of $\langle\delta\rho(\mathbf{0},t)\delta\rho(\mathbf{r},t)\rangle$ in $\mathbf{r}$-space. This must be the region close to the $\perp$ plane (i.e., $x \ll r_\perp$). The separatrix that separates these positive and negative regions must therefore lie between the two, which means it must run at an angle to the $\hat{x}$-axis both forward and back. This is, of course, exactly where the cone we found in our more detailed calculation above lies.

In Section 9.12, we will show that the shape of this separatrix will be modified if we go beyond the linear theory.

Now we turn to the temporal correlations:

$$\begin{aligned}\langle\delta\rho(\mathbf{0},t)\delta\rho(\mathbf{0},0)\rangle &= \frac{1}{(2\pi)^{d+1}}\int \mathrm{d}\omega\mathrm{d}\omega'\mathrm{d}^dk\mathrm{d}^dk'\left\langle\delta\rho(\tilde{\mathbf{k}})\delta\rho(\tilde{\mathbf{k}}')\right\rangle \mathrm{e}^{-\mathrm{i}\omega t}\\ &= \frac{1}{(2\pi)^{d+1}}\int \mathrm{d}\omega\mathrm{d}^dk\,\frac{2D'k_\perp^2\mathrm{e}^{-\mathrm{i}\omega t}}{\omega^2+\left(\mu_Lk_\perp^2+\mu_xk_x^2\right)^2}\end{aligned}$$

$$
\begin{aligned}
&= \int \frac{\mathrm{d}^d\mathbf{k}}{(2\pi)^d} \frac{D' k_\perp^2 \mathrm{e}^{-(\mu_L k_\perp^2 + \mu_x k_x^2)|t|}}{\mu_L k_\perp^2 + \mu_x k_x^2} \\
&= |t|^{-d/2} \int \frac{\mathrm{d}^d\mathbf{q}}{(2\pi)^d} \frac{D' q_\perp^2 \mathrm{e}^{-(\mu_L q_\perp^2 + \mu_x q_x^2)}}{\mu_L q_\perp^2 + \mu_x q_x^2} \propto |t|^{-d/2} ,
\end{aligned}
$$

where, again, in the penultimate equality we have made the change of variable, $\mathbf{q} = |t|^{\frac{1}{2}}\mathbf{k}$, and in the ultimate proportionality we have used the fact that the integral over $\mathbf{q}$ is a finite constant (i.e., independent of time t).

For arbitrary spatio-temporal separations, the correlation function is given by

$$
\begin{aligned}
\langle \delta\rho(\mathbf{r}, t)\delta\rho(\mathbf{0}, 0)\rangle &= \frac{1}{(2\pi)^{d+1}} \int \mathrm{d}\omega \mathrm{d}\omega' \mathrm{d}^d k \mathrm{d}^d k' \left\langle \delta\rho(\tilde{\mathbf{k}})\delta\rho(\tilde{\mathbf{k}}')\right\rangle \mathrm{e}^{\mathrm{i}(\mathbf{k}\cdot\mathbf{r}-\omega t)} \\
&= \frac{1}{(2\pi)^{d+1}} \int \mathrm{d}\omega \mathrm{d}^d k \frac{2D' k_\perp^2 \mathrm{e}^{\mathrm{i}(\mathbf{k}\cdot\mathbf{r}-\omega t)}}{\omega^2 + \left(\mu_L k_\perp^2 + \mu_x k_x^2\right)^2} . \qquad (9.2.42)
\end{aligned}
$$

Making the changes of variables of integration prescribed by (9.2.33), we obtain

$$
\langle \delta\rho(\mathbf{r}, t)\delta\rho(\mathbf{0}, 0)\rangle = r_\perp^{-d} H_\rho\left(\frac{x}{r_\perp}, \frac{t}{r_\perp^2}\right) \propto \begin{cases} r^{-d}, & r \gg |t|^{\frac{1}{2}} , \\ |t|^{-\frac{d}{2}}, & |t| \gg r^2 , \end{cases}
$$

where we've defined the scaling function

$$
H_\rho(u, v) \equiv \frac{2D'}{(2\pi)^{d+1}} \int \mathrm{d}\Upsilon \mathrm{d}^d Q \frac{Q_\perp^2 \mathrm{e}^{\mathrm{i}[\mathbf{Q}_\perp\cdot\hat{\mathbf{r}}_\perp + Q_x u - \Upsilon v]}}{\Upsilon^2 + \left(\mu_L Q_\perp^2 + \mu_x Q_x^2\right)^2} . \qquad (9.2.43)
$$

In any spatial dimension d, these correlations decay too rapidly to give rise to giant number fluctuations (GNF) [44, 45, 82]; that is, they are *not* sufficiently long-ranged to make the rms number fluctuations $\delta N \equiv \sqrt{\langle (N - \langle N \rangle)^2 \rangle}$ in a large region grow more rapidly than the square root of the mean number $\sqrt{\langle N \rangle}$. However, they are sufficiently long-ranged to make δN depend on the shape of the region in which the particle number N is being counted [47].

Unfortunately, as we will see in the next section, these scaling laws, in particular the power law with which correlations decay with distance r, are changed by nonlinear effects, leading to a more rapid decay which eliminates this shape dependence. Nonetheless, the strange power law dependence of density correlations persists (albeit with different exponents than found here in the linear theory), and still displays universal exponents which can be readily measured in experiments and simulations.

9.3 One-Loop RG Calculation with μ_2 Set to Zero

Now we use the standard DRG approach to analyze the effect of nonlinearities. First we need to write the full model (9.1.15) in Fourier space in tensor form:

$$-\mathrm{i}\omega u_i^{\perp}(\tilde{\mathbf{k}}) = -\left(\mu_1 k_\perp^2 + \mu_x k_x^2\right) u_i^{\perp}(\tilde{\mathbf{k}}) + f_i^{\perp}(\tilde{\mathbf{k}}) - \mu_2 k_i^{\perp}\left(\mathbf{k}_\perp \cdot \mathbf{u}_\perp(\tilde{\mathbf{k}})\right) - \frac{\mathrm{i}\lambda}{\left(\sqrt{2\pi}\right)^{d+1}} \int_{\tilde{\mathbf{q}}} \left[\mathbf{u}_\perp(\tilde{\mathbf{k}} - \tilde{\mathbf{q}}) \cdot \mathbf{q}_\perp\right] u_i^{\perp}(\tilde{\mathbf{q}}), \tag{9.3.1}$$

where $\tilde{\mathbf{q}} \equiv (\mathbf{q}, \Omega)$ and $\int_{\tilde{\mathbf{q}}} \equiv \int \mathrm{d}\Omega\, \mathrm{d}^d q$. Going through essentially the same calculation as the one which leads to (9.2.19), we get

$$u_i^{\perp}(\tilde{\mathbf{k}}) = G_{ij}(\tilde{\mathbf{k}})\left\{ f_j^{\perp}(\tilde{\mathbf{k}}) - \frac{\mathrm{i}\lambda}{\left(\sqrt{2\pi}\right)^{d+1}} \int_{\tilde{\mathbf{q}}} [\mathbf{u}_\perp(\mathbf{k}_\perp - \mathbf{q}_\perp) \cdot \mathbf{q}_\perp] u_j^{\perp}(\tilde{\mathbf{q}}) \right\}. \tag{9.3.2}$$

To probe what happens for $d < 4$, we use a DRG analysis together with the ϵ-expansion method to one-loop level [29]. At last you'll get to see the full-blown DRG in action!

As always, first we decompose the Fourier modes $\mathbf{u}_\perp(\tilde{\mathbf{k}})$ into a rapidly varying part $\mathbf{u}_\perp^{>}(\tilde{\mathbf{k}})$ and a slowly varying part $\mathbf{u}_\perp^{<}(\tilde{\mathbf{k}})$ in (9.3.1). Because the system is anisotropic, we'll use a convenient anisotropic Brillouin zone. Specifically, the rapidly varying part is supported in the momentum shell $-\infty < k_x < \infty$, $\Lambda \mathrm{e}^{-\mathrm{d}\ell} < k_\perp < \Lambda$, where $\mathrm{d}\ell$ is an infinitesimal and Λ is the ultraviolet cutoff. The slowly varying part is supported in $-\infty < k_x < \infty$, $0 < k_\perp < \Lambda \mathrm{e}^{-\mathrm{d}\ell}$.

We then perform the usual two-step DRG procedure. In step 1, we eliminate $\mathbf{u}_\perp^{>}(\tilde{\mathbf{k}})$ from (9.3.1). We do this by perturbatively solving (9.3.2) iteratively for $\mathbf{u}_\perp^{>}(\tilde{\mathbf{k}})$. Here, this solution is a power series in λ which depends on $\mathbf{u}_\perp^{<}(\tilde{\mathbf{k}})$. We substitute this solution into (9.3.1) and average over the short-wavelength components $\mathbf{f}^{>}(\tilde{\mathbf{k}})$ of the noise $\mathbf{f}$, which gives a closed equation of motion for $\mathbf{u}_\perp^{<}(\tilde{\mathbf{k}})$.

Step 2 then consists of rescaling the length and time. The length rescaling is done anisotropically, since our problem has become anisotropic in the broken symmetry state. So, as in previous chapters, we rescale according to

$$\mathbf{r}_\perp = b\mathbf{r}'_\perp,\; x = b^{\zeta} x',\; t = b^{z} t',\; \mathbf{u}_\perp = b^{\chi} \mathbf{u}'_\perp, \tag{9.3.3}$$

which restores the ultraviolet cutoff back to Λ. We then reorganize the resultant equation of motion so that it has the same form as (9.3.1), but with various coefficients renormalized.

The calculation of the renormalization of the coefficients arising from the process of eliminating $\mathbf{u}_\perp^{>}(\tilde{\mathbf{k}})$ can, as always, be represented by graphs. The basic rules for the graphical representation will be illustrated later in Figure 9.4.1. Again, this should all be familiar from our earlier discussion of the KPZ equation, with the only new wrinkle being that the propagators and correlation functions (9.2.20) and (9.2.23) now carry tensor indices.

Having done step 2 for a variety of models by now, including in particular the very similar immortal flock model, doing it here is completely straightforward, so I won't go through the details here. Its net effect, as always, is to give us the rescaling parts of the recursion relations for the various parameters. Knowing these tells us that the recursion relations for our five parameters must take the form:

$$\frac{1}{D}\frac{dD}{d\ell} = z - 2\chi - d + 1 - \zeta + \text{graphs}, \tag{9.3.4}$$

$$\frac{1}{\lambda}\frac{d\lambda}{d\ell} = z + \chi - 1, \tag{9.3.5}$$

$$\frac{1}{\mu_x}\frac{d\mu_x}{d\ell} = z - 2\zeta + \text{graphs}, \tag{9.3.6}$$

$$\frac{1}{\mu_1}\frac{d\mu_1}{d\ell} = z - 2 + \text{graphs}, \tag{9.3.7}$$

$$\frac{1}{\mu_2}\frac{d\mu_2}{d\ell} = z - 2 + \text{graphs}. \tag{9.3.8}$$

Note that I have not included graphical corrections for λ. This is because it gets none, because the "pseudo-Galilean invariance" argument I presented in Chapter 6's treatment of incompressible flocks applies just as well for Malthusian flocks, since the velocity-dependent terms in the equation of motion are the same, and the incompressibility constraint did not enter the argument.

Our motivation for calculating the rescaling parts first is, as always, to allow us to decide whether or not we need to do the hard work of calculating the graphical corrections in (9.3.4), (9.3.6), (9.3.7), and (9.3.8) at all. We'll do this by the usual RG logic: We'll assume that the graphical corrections are, at least in the early stages of the RG, negligible. This amounts to assuming that the bare λ is small. We'll then choose the rescaling exponents z, ζ, and χ to keep the parameters that control the scale of the fluctuations in the linear theory fixed in the absence of the neglected graphical corrections. As we saw from the linear theory of Section 9.2, those parameters are the noise strength D and the three viscosities $\mu_{x,1,2}$. Keeping those four parameters fixed when the graphical corrections are negligible forces us to choose z, ζ, and χ to satisfy three independent linear equations (not four, because μ_1 and μ_2 have the same rescaling), which we can simply read off from (9.3.4), (9.3.6), and either (9.3.7) or (9.3.8) (which are identical):

$$z - 2\chi - d + 1 - \zeta = 0, \quad z - 2\zeta = 0, \quad z - 2 = 0, \tag{9.3.9}$$

whose easy solutions are:

$$z = 2, \quad \zeta = 1, \quad \chi = \frac{2-d}{2}. \tag{9.3.10}$$

These are, of course, just the scaling exponents predicted by the linear theory of Section 9.2.

Continuing with the standard RG procedure, we now insert these exponents into the recursion relation (9.3.5) for the sole relevant nonlinear coupling λ. This gives

$$\frac{1}{\lambda}\frac{d\lambda}{d\ell} = \frac{4-d}{2}. \tag{9.3.11}$$

Obviously, if $d < 4$, an initially small λ (specifically, one small enough that the graphical corrections can initially be ignored) will grow (exponentially) upon renormalization, meaning that eventually the graphical corrections will no longer be negligible. (I remind the reader that the increase in the effect of the nonlinear term due to the growth of λ can *not* be offset by a reduction in the size of the fluctuations upon renormalization, since we chose z, ζ, and χ to keep the size of the fluctuations fixed, by keeping the noise strength D and the three viscosities $\mu_{x,1,2}$, which control them, fixed.)

So in the real world, where $d < 4$, we'll need to calculate the graphical corrections.

Before doing that calculation in all its hideous complexity, let's first consider an *immensely* simplifying limit: a model in which one of the viscosities in the original equation of motion (9.1.15) vanishes; specifically, consider $\mu_2 = 0$.

This restriction *tremendously* simplifies the calculation, by making the propagators G_{ij} and correlation function C_{ij} diagonal. This also has the advantage of eliminating the need to expand the transverse and longitudinal projection operators $P_{ij}(\mathbf{q}_\perp)$ and $L_{ij}(\mathbf{q}_\perp)$ in powers of external wavevector, which proves to be an enormous simplification.

It also proves to be sufficient to explore this region, since it contains the only stable fixed point in the problem, which we can find, and the exponents of which we can calculate, using this restricted approach. However, to demonstrate the stability of this fixed point against nonzero μ_2, and to show that it is the *only* stable fixed point (and, indeed, the only fixed point other than the unstable Gaussian fixed point), it is necessary to extend these calculations to nonzero μ_2, which we do in Section 9.16.

I'll start by showing that, when $\mu_2 = 0$, both the propagator and the correlation function (which I remind you are tensors) become diagonal.

First, note from its definition (9.2.8) that $\mu_L = \mu_1$ when $\mu_2 = 0$. It then follows from our expressions (9.2.17) and (9.2.18) for the longitudinal and transverse propagators $G_{L,T}$ that $G_L = G_T \equiv G$, where I'm using

$$G(\tilde{\mathbf{k}}) = \frac{1}{-i\omega + \mu_1 k_\perp^2 + \mu_x k_x^2} \tag{9.3.12}$$

to denote the common value of G_L and G_T. Likewise, for the remainder of this discussion of the $\mu_2 = 0$ case, I'll use

$$\Gamma(\mathbf{k}) = \mu_1 k_\perp^2 + \mu_x k_x^2 \tag{9.3.13}$$

to denote the common value of $\Gamma_L(\mathbf{k})$ and $\Gamma_T(\mathbf{k})$.

It then follows from the expression (9.2.20) for the tensor G_{ij} that

$$G_{ij}(\tilde{\mathbf{k}}) \equiv L_{ij}^\perp(\mathbf{k}_\perp)G_L(\tilde{\mathbf{k}}) + P_{ij}^\perp(\mathbf{k}_\perp)G_T(\tilde{\mathbf{k}}) = G(\tilde{\mathbf{k}})\left[L_{ij}^\perp(\mathbf{k}_\perp) + P_{ij}^\perp(\mathbf{k}_\perp)\right] = G(\tilde{\mathbf{k}})\delta_{ij}^\perp , \tag{9.3.14}$$

where I've used the fact that $L_{ij}^\perp(\mathbf{k}_\perp) + P_{ij}^\perp(\mathbf{k}_\perp) = \delta_{ij}^\perp$.

The demonstration that C_{ij} is also diagonal is almost identical, so I won't present it in detail here. The result is

$$C_{ij}(\tilde{\mathbf{k}}) = 2D\delta_{ij}^\perp |G(\tilde{\mathbf{k}})|^2 = \frac{2D\delta_{ij}^\perp}{\omega^2 + (\mu_1 k_\perp^2 + \mu_x k_x^2)^2} . \tag{9.3.15}$$

For $\mu_2 = 0$, the equation of motion simplifies to:

$$-i\omega \mathbf{u}_i^\perp = -\left(\mu_1 k_\perp^2 + \mu_x k_x^2\right)\mathbf{u}_i^\perp - \frac{i\lambda}{\left(\sqrt{2\pi}\right)^{d+1}} \int_{\tilde{\mathbf{q}}} [\mathbf{u}_\perp(\tilde{\mathbf{q}}) \cdot (\mathbf{k}_\perp - \mathbf{q}_\perp)]\, \mathbf{u}_i^\perp(\tilde{\mathbf{k}} - \tilde{\mathbf{q}}) + \mathbf{f}_i^\perp . \tag{9.3.16}$$

We can formally solve this equation for $\mathbf{u}$, which gives

$$u_i^\perp(\tilde{\mathbf{k}}) = G_{ij}(\tilde{\mathbf{k}})\left\{f_j^\perp(\tilde{\mathbf{k}}) - \frac{i\lambda}{\left(\sqrt{2\pi}\right)^{d+1}} \int_{\tilde{\mathbf{q}}} [\mathbf{u}_\perp(\mathbf{k}_\perp - \mathbf{q}_\perp) \cdot \mathbf{q}_\perp]\, u_j^\perp(\tilde{\mathbf{q}})\right\}. \tag{9.3.17}$$

We'll now use this simplified equation to calculate the graphical corrections.

9.4 Graphical Rules

The graphical elements for this problem are shown in Figure 9.6.1. As always, the line with an arrow (see Figure 9.4.1(a)) represents the propagator $G_{ij}(\tilde{\mathbf{k}})$, while lines

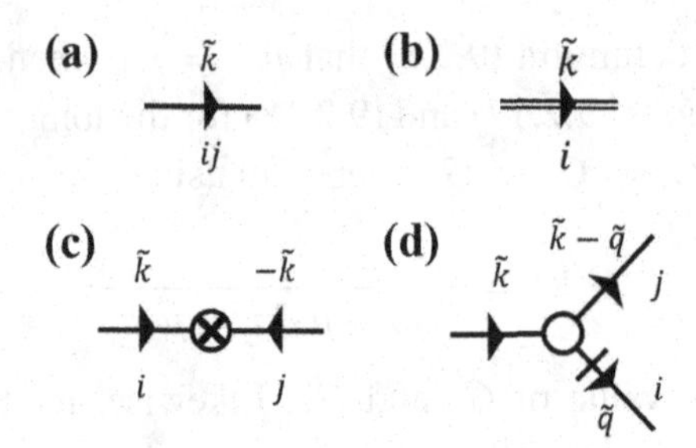

Fig. 9.4.1 Rules of graphical representation: (a) = $G_{ij}(\tilde{\mathbf{k}})$; (b) = $u_i^{\perp}(\tilde{\mathbf{k}})$; (c) = $2DC_{ij}(\tilde{\mathbf{k}})$; (d) the nonlinear term proportional to = $-\mathrm{i}\lambda_1$; the slash represents a factor $q_j^{\perp}$. Reproduced from [23].

with two arrows and a crossed circle represent the correlation function $C_{ij}(\tilde{\mathbf{k}})$. This is all exactly as we did for the KPZ equation in Chapter 3, with the sole exception that now, the lines carry Cartesian indices ij. We will use the Einstein summation convention on these graphs, so that repeated indices are summed over.

The (quadratic) nonlinearity λ in the equation of motion (9.1.15) is represented by a three-pronged vertex, as per the usual Feynman diagram rules, with, again, the only new wrinkle being that each line carries a Cartesian index.

We begin by perturbatively calculating the corrections to the equation of motion (9.1.15). These are represented by the Feynman diagrams shown in Figure 9.6.1. These will give rise to corrections to the viscosities $\mu_{1,2}$. One might have expected them to correct the other viscosity μ_x as well, but that correction proves to vanish to this order in perturbation theory.

So hang on to your hats, and let's start.

9.5 Noise Renormalization

The graphs in Figure 9.5.1 represent the following two corrections to the noise correlator $\left\langle f_\ell(\tilde{\mathbf{k}})f_u(-\tilde{\mathbf{k}})\right\rangle$:

$$\Delta\left\langle f_\ell(\tilde{\mathbf{k}})f_u(-\tilde{\mathbf{k}})\right\rangle_{D,a} = \frac{2\lambda^2D^2}{(2\pi)^{d+1}}\int_{\tilde{\mathbf{q}}}^{>} q_i^{\perp}q_m^{\perp}C_{im}(\tilde{\mathbf{k}}-\tilde{\mathbf{q}})C_{\ell u}(\tilde{\mathbf{q}}), \tag{9.5.1}$$

$$\Delta\left\langle f_\ell(\tilde{\mathbf{k}})f_u(-\tilde{\mathbf{k}})\right\rangle_{D,b} = \frac{2\lambda^2D^2}{(2\pi)^{d+1}}\int_{\tilde{\mathbf{q}}}^{>} q_i^{\perp}(k_m^{\perp}-q_m^{\perp})C_{iu}(\tilde{\mathbf{k}}-\tilde{\mathbf{q}})C_{\ell m}(\tilde{\mathbf{q}}). \tag{9.5.2}$$

Since the noise strength D is the value of this correlation at $\tilde{\mathbf{k}}=\mathbf{0}$, we set $\tilde{\mathbf{k}}=0$ in (9.5.1) and (9.5.2) to get

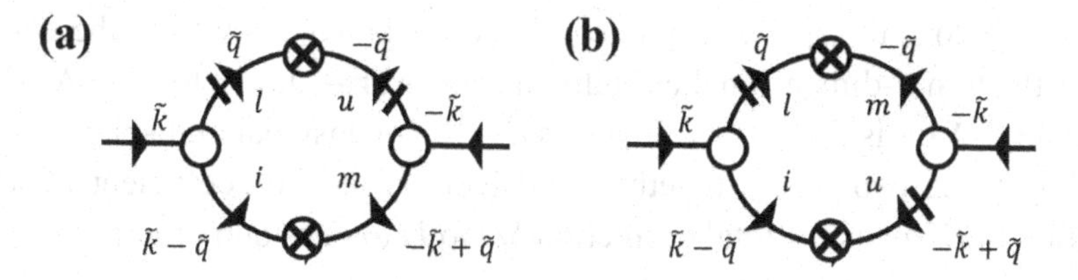

Fig. 9.5.1 Graphical representation of the correction to the noise correlator $\left\langle f_\ell(\tilde{\mathbf{k}})f_u(-\tilde{\mathbf{k}})\right\rangle$. Reproduced from [23].

$$\Delta\left\langle f_\ell(\tilde{\mathbf{k}})f_u(-\tilde{\mathbf{k}})\right\rangle_{D,a} = \frac{2\lambda^2 D^2}{(2\pi)^{d+1}}\int_{\tilde{\mathbf{q}}}^{>} q_i^{\perp}q_m^{\perp}C_{im}(-\tilde{\mathbf{q}})C_{\ell u}(\tilde{\mathbf{q}})$$
$$= \frac{2\lambda^2 D^2\delta_{\ell u}}{(2\pi)^{d+1}}\int_{\tilde{\mathbf{q}}}^{>} q_\perp^2|G(\tilde{\mathbf{k}})|^4\,, \tag{9.5.3}$$

$$\Delta\left\langle f_\ell(\tilde{\mathbf{k}})f_u(-\tilde{\mathbf{k}})\right\rangle_{D,b} = \frac{2\lambda^2 D^2}{(2\pi)^{d+1}}\int_{\tilde{\mathbf{q}}}^{>} q_i^{\perp}(-q_m^{\perp})C_{iu}(-\tilde{\mathbf{q}})C_{\ell m}(\tilde{\mathbf{q}})$$
$$= -\frac{2\lambda^2 D^2}{(2\pi)^{d+1}}\int_{\tilde{\mathbf{q}}}^{>} q_u^{\perp}q_l^{\perp}|G(\tilde{\mathbf{k}})|^4\,, \tag{9.5.4}$$

where in the second equality of both of these expressions we've used equation (9.3.15) for the correlation tensor C_{ij}, and done some obvious contractions of indices.

Doing the angular integral in (9.5.4), using the trick I explained in Chapter 3 when we were discussing the KPZ equation, shows that

$$\int_{\tilde{\mathbf{q}}}^{>} q_u^{\perp}q_l^{\perp}|G(\tilde{\mathbf{k}})|^4 = \frac{\delta_{\ell u}}{d-1}\int_{\tilde{\mathbf{q}}}^{>} q_\perp^2|G(\tilde{\mathbf{k}})|^4\,, \tag{9.5.5}$$

where we have a factor of $d-1$ in the denominator rather than the d we had in Chapter 3 because the vector $q_\perp$ is $d-1$-dimensional, rather than d-dimensional.

Inserting our expression (9.3.12) for the propagator $G(\tilde{\mathbf{k}})$ into the above two formulae, we get

$$\Delta\left\langle f_\ell(\tilde{\mathbf{k}})f_u(-\tilde{\mathbf{k}})\right\rangle_{D,a} = \frac{2\lambda^2 D^2}{(2\pi)^{d+1}}\delta_{\ell u}^{\perp}\int_{\tilde{\mathbf{q}}}^{>} q_\perp^2 \mid G_T(\tilde{\mathbf{q}})\mid^4$$
$$= \frac{3}{32}\frac{D^2\lambda^2}{\sqrt{\mu_x\mu_1^5}}\frac{S_{d-1}}{(2\pi)^{d-1}}\Lambda^{d-4}\mathrm{d}\ell\delta_{\ell u}^{\perp}\,, \tag{9.5.6}$$

$$\Delta\left\langle f_\ell(\tilde{\mathbf{k}})f_u(-\tilde{\mathbf{k}})\right\rangle_{D,b} = -\frac{2\lambda^2 D^2}{(2\pi)^{d+1}}\frac{\delta_{\ell u}^{\perp}}{d-1}\int_{\tilde{\mathbf{q}}}^{>} q_\perp^2 \mid G_T(\tilde{\mathbf{q}})\mid^4$$
$$= -\frac{3}{32(d-1)}\frac{D^2\lambda^2}{\sqrt{\mu_x\mu_1^5}}\frac{S_{d-1}}{(2\pi)^{d-1}}\Lambda^{d-4}\mathrm{d}\ell\delta_{\ell u}^{\perp}, \tag{9.5.7}$$

where we've done the elementary integrals over ω and q_x, as well as the integral of $\mathbf{q}_\perp$ over the thin d-dimensional cylindrical shell $\Lambda(1-\mathrm{d}\ell) < |\mathbf{q}_\perp| < \Lambda$. Here, as in Chapter 3, S_{d-1} is the surface area of a $d-1$-dimensional unit sphere.

Adding these two pieces together, and identifying the coefficient of $\delta^\perp_{\ell u}$ as a correction to D, gives the total correction δD to D to one-loop order:

$$\delta D = \frac{3}{32}\left(1-\frac{1}{d-1}\right)\frac{D^2\lambda^2}{\sqrt{\mu_x\mu_1^5}}\frac{S_{d-1}}{(2\pi)^{d-1}}\Lambda^{d-4}\mathrm{d}\ell = \frac{3}{32}\left(1-\frac{1}{d-1}\right)g_1 D\mathrm{d}\ell\,, \tag{9.5.8}$$

where in the second equality we've defined the dimensionless coupling

$$g_1 \equiv \frac{D\lambda^2}{\sqrt{\mu_x\mu_1^5}}\frac{S_{d-1}}{(2\pi)^{d-1}}\Lambda^{d-4}\,, \tag{9.5.9}$$

which will prove to be ubiquitous in this calculation (as these dimensionless couplings usually do in RG calculations).

9.6 Renormalization of the μ's

9.6.1 *Graph in Figure 9.6.1(a)*

The graph in Figure 9.6.1(a) gives a contribution which we'll call $\Delta(\partial_t u_j^<)_{\mu,a}$ to the equation of motion for $u_j^<(\tilde{\mathbf{k}})$:

$$\begin{aligned}\Delta(\partial_t u_j^<(\tilde{\mathbf{k}})_{\mu,a}) &= -\frac{2D\lambda^2 k_u^\perp u_c^\perp(\tilde{\mathbf{k}})}{(2\pi)^{d+1}}\int_{\tilde{q}}^{>}(k_i^\perp - q_i^\perp)C_{iu}(\tilde{\mathbf{q}})G_{jc}(\tilde{\mathbf{k}}-\tilde{\mathbf{q}})\\ &\equiv -2D\lambda^2 k_u^\perp u_c^\perp(\tilde{\mathbf{k}})\left[(I_1^{\mu,a})_{cju}(\tilde{\mathbf{k}}) - (I_2^{\mu,a})_{cju}(\tilde{\mathbf{k}})\right],\end{aligned} \tag{9.6.1}$$

where

$$\int_{\tilde{\mathbf{q}}}^{>} \equiv \int_{\Lambda>|\mathbf{q}_\perp|>\Lambda e^{-\mathrm{d}\ell}}\mathrm{d}^{d-1}q_\perp\int_{-\infty}^{\infty}\mathrm{d}\Omega\int_{-\infty}^{\infty}\mathrm{d}q_x\,, \tag{9.6.2}$$

$$\begin{aligned}(I_1^{\mu,a})_{cju}(\tilde{\mathbf{k}}) &\equiv \frac{k_i^\perp}{(2\pi)^{d+1}}\int_{\tilde{q}}^{>}C_{iu}(\tilde{\mathbf{q}})G_{jc}(\tilde{\mathbf{k}}-\tilde{\mathbf{q}})\\ &= \frac{k_u^\perp\delta_{jc}^\perp}{(2\pi)^{d+1}}\int_{\tilde{q}}^{>}|G(\tilde{\mathbf{q}})|^2G(\tilde{\mathbf{k}}-\tilde{\mathbf{q}})\,,\end{aligned} \tag{9.6.3}$$

$$\begin{aligned}(I_2^{\mu,a})_{cju}(\tilde{\mathbf{k}}) &\equiv \frac{1}{(2\pi)^{d+1}}\int_{\tilde{q}}^{>}q_i^\perp C_{iu}(\tilde{\mathbf{q}})G_{jc}(\tilde{\mathbf{k}}-\tilde{\mathbf{q}})\\ &= \frac{\delta_{jc}^\perp}{(2\pi)^{d+1}}\int_{\tilde{q}}^{>}q_u^\perp|G(\tilde{\mathbf{q}}|^2)G(\tilde{\mathbf{k}}-\tilde{\mathbf{q}})\,.\end{aligned} \tag{9.6.4}$$

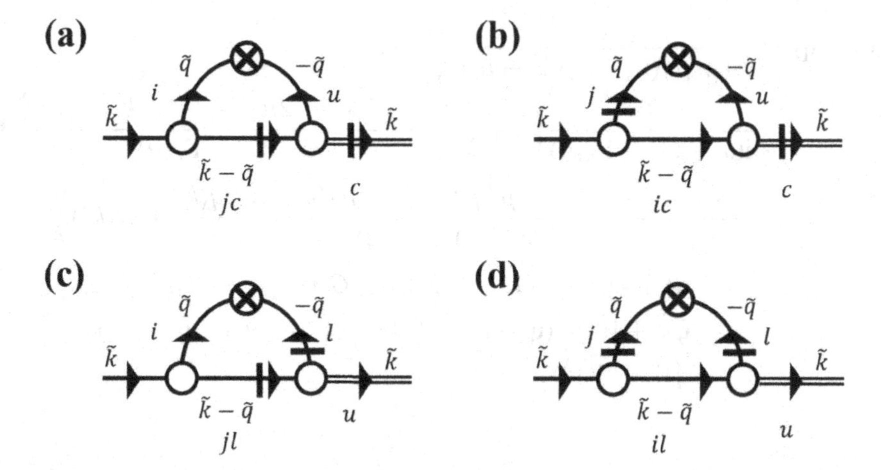

Fig. 9.6.1 Graphical representation of the correction to the linear terms in the equation of motion (9.3.1). Reproduced from [23].

Because we are interested in terms only up to $O(k^2)$, since that is the order of the μ terms in the equation of motion, and (9.6.1) already has an explicit factor $k_u^\perp$, we need only expand these integrals $(I_{1,2}^{5a})_{cju}$ up to linear order in $\mathbf{k}$. Furthermore, since our expression (9.6.3) for the integral $(I_1^{5a})_{cju}$ already has an explicit factor of $k_u^\perp$ out in front, we can, to quadratic order in $\mathbf{k}_\perp$ for the full equation of motion correction, set $\tilde{\mathbf{k}} = 0$ inside the integral in the second equality of (9.6.3).

Doing so, we obtain to this order

$$(I_1^{\mu,a})_{cju}(\tilde{\mathbf{k}}) = \frac{k_u^\perp \delta_{jc}^\perp}{(2\pi)^{d+1}} \int_{\tilde{\mathbf{q}}}^{>} \mid G_T(\tilde{\mathbf{q}}) \mid^2 G_T(-\tilde{\mathbf{q}}) = \frac{k_u^\perp}{16} \frac{D\lambda^2}{\sqrt{\mu_x \mu_1^3}} \frac{S_{d-1}}{(2\pi)^{d-1}} \Lambda^{d-4} \mathrm{d}\ell \,, \tag{9.6.5}$$

where the second equality just follows from doing the perfectly straightforward integrals over ω, and q_x can be done either by simple complex contour techniques, or by even simpler trigonometric substitutions. The integral over $\mathbf{q}_\perp$ then simply gives an additional factor of $\frac{S_{d-1}}{(2\pi)^{d-1}} \Lambda^{d-4} \mathrm{d}\ell$, which is just the hypervolume of the Brillouin zone in $\mathbf{q}_\perp$ space.

We're not quite so lucky with the second integral $(I_2^{\mu,a})_{cju}(\tilde{\mathbf{k}})$: Here we must expand the integrand to linear order in $\mathbf{k}_\perp$ to get a nonzero answer. You can see this by noting that if we just set $\tilde{\mathbf{k}} = 0$ in the integral, the integrand is odd in $\mathbf{q}_\perp$ (due to the explicit factor of $q_u^\perp$) and so vanishes.

The expansion of $G(\tilde{\mathbf{k}} - \tilde{\mathbf{q}})$ in powers of $\mathbf{q}_\perp$ is straightforward (indeed, if you can remember the series for $\frac{1}{1+x}$, you can almost just read it off), and is

$$
\begin{aligned}
G(\tilde{\mathbf{k}} - \tilde{\mathbf{q}}) &= \frac{1}{\mathrm{i}\omega + \mu_1(\mathbf{k}_\perp - \mathbf{q}_\perp)^2 + \mu_x(k_x - q_x)^2} \\
&= \frac{1}{\mathrm{i}(\omega - \Omega) + \Gamma_L(\mathbf{q})}\left(1 + \left(\frac{2\mu_1\mathbf{q}_\perp \cdot \mathbf{k}_\perp + 2\mu_x q_x k_x - \mu_1 k_\perp^2 - \mu_x k_x^2}{\mathrm{i}(\omega - \Omega) + \Gamma_L(\mathbf{q})}\right)\right. \\
&\quad \left. + \frac{4\mu_1^2(\mathbf{q}_\perp \cdot \mathbf{k}_\perp)^2 + 4\mu_x^2 q_x^2 k_x^2 + 4\mu_x\mu_1(\mathbf{q}_\perp \cdot \mathbf{k}_\perp)q_x k_x}{(\mathrm{i}(\omega - \Omega) + \Gamma_L(\mathbf{q}))^2} + \mathcal{O}(k^3)\right) \\
&= G_L(-\tilde{\mathbf{q}}) + (2\mu_1\mathbf{q}_\perp \cdot \mathbf{k}_\perp + 2\mu_x q_x k_x)\, G_L(-\tilde{\mathbf{q}})^2 - \left(\mu_1 k_\perp^2 + \mu_x k_x^2\right) \\
&\quad G_L(-\tilde{\mathbf{q}})^2 + \left[(4\mu_1^2(\mathbf{q}_\perp \cdot \mathbf{k}_\perp)^2 + 4\mu_x^2 q_x^2 k_x^2 + 4\mu_x\mu_1(\mathbf{q}_\perp \cdot \mathbf{k}_\perp)q_x k_x\right] \\
&\quad G_L(-\tilde{\mathbf{q}})^3 + \mathcal{O}(k^3).
\end{aligned} \tag{9.6.6}
$$

As just noted, the integral of the first term gives zero. Since we only need this integral to $O(\mathbf{q}_\perp)$, we can drop the $O(\mathbf{q}_\perp^2)$ terms. This leaves us with

$$
\begin{aligned}
(I_2^{\mu,a})_{cju}(\tilde{\mathbf{k}}) &= \frac{\delta_{jc}^\perp}{(2\pi)^{d+1}} \int_{\tilde{\mathbf{q}}}^{>} q_u^\perp \mid G_T(\tilde{\mathbf{q}}) \mid^2 G_T(\tilde{\mathbf{k}} - \tilde{\mathbf{q}}) \\
&= \frac{\delta_{jc}^\perp}{(2\pi)^{d+1}} \int_{\tilde{\mathbf{q}}}^{>} q_u^\perp \mid G_T(\tilde{\mathbf{q}}) \mid^2 [G_T(-\tilde{\mathbf{q}})]^2 \,(2\mu_1\mathbf{q}_\perp \cdot \mathbf{k}_\perp + 2\mu_x q_x k_x) \\
&= \frac{3k_u^\perp \delta_{jc}^\perp}{64(d-1)} \frac{D\lambda^2}{\sqrt{\mu_x\mu_1^3}} \frac{S_{d-1}}{(2\pi)^{d-1}} \Lambda^{d-4} \mathrm{d}\ell\,,
\end{aligned} \tag{9.6.7}
$$

where we have only kept terms up to $O(k)$. In the last equality in (9.6.7), we have used the fact that the second ($2\mu_x$) term in the parenthesis is odd in q_x, and so integrates to zero. We have also evaluated the first term by using the angular integral trick from Chapter 3 again.

Inserting (9.6.5) and (9.6.7) into (9.6.1), we find:

$$
\begin{aligned}
\Delta(\partial_t u_j^<)_{\mu,a} &= -\left(\frac{1}{32} \frac{D\lambda^2}{\sqrt{\mu_x\mu_1^3}} \frac{S_{d-1}}{(2\pi)^{d-1}} \Lambda^{d-4} \mathrm{d}\ell \left[4 - \frac{3}{(d-1)}\right]\right) \\
&= -\left[\frac{1}{8} - \frac{3}{32(d-1)}\right] (g_1\mu_1 \mathrm{d}\ell) k^2 u_j^<.
\end{aligned} \tag{9.6.8}
$$

Since this contribution to $\partial_t u_j^<$ has the same form as the μ_1 term already present, we can absorb it into a renormalization of μ_1:

$$
(\delta\mu_1)_{\mu,a} = \frac{1}{32} \frac{D\lambda^2}{\sqrt{\mu_x\mu_1^3}} \frac{S_{d-1}}{(2\pi)^{d-1}} \Lambda^{d-4} \mathrm{d}\ell \left[4 - \frac{3}{(d-1)}\right] = \left[\frac{1}{8} - \frac{3}{32(d-1)}\right] g_1\mu_1 \mathrm{d}\ell. \tag{9.6.9}
$$

9.6.2 Graph in Figure 9.6.1(b)

The graph in Figure 9.6.1(b) gives a contribution $\Delta(\partial_t u_j^<)_{\mu,b}$ to the equation of motion for $u_j^<(\tilde{\mathbf{k}})$:

$$\Delta(\partial_t u_j^<)_{\mu,b} = -\frac{2\lambda^2 D k_u^\perp u_c^\perp(\tilde{\mathbf{k}})}{(2\pi)^{d+1}} \int_{\tilde{\mathbf{q}}}^> q_i^\perp C_{ju}(\tilde{\mathbf{q}}) G_{ic}(\tilde{\mathbf{k}} - \tilde{\mathbf{q}})$$
$$= -2\lambda^2 D k_u^\perp u_b^\perp(\tilde{\mathbf{k}}) (I^{\mu,b})_{cju}(\tilde{\mathbf{k}}), \tag{9.6.10}$$

where the steps leading to the second equality are almost identical to those just presented in the evaluation of the graph in Figure 9.6.1(b). I therefore won't go through those steps again here, or anywhere further on in this section, but, rather, will just quote the answers. In (9.6.10), we've defined

$$(I^{\mu,b})_{cju}(\tilde{\mathbf{k}}) \equiv \frac{1}{(2\pi)^{d+1}} \int_{\tilde{\mathbf{q}}}^> q_i^\perp C_{ju}(\tilde{\mathbf{q}}) G_{ic}(\tilde{\mathbf{k}} - \tilde{\mathbf{q}}). \tag{9.6.11}$$

Inserting (9.3.17) and the expansion (9.6.6) into (9.6.11), we get

$$(I^{\mu,b})_{cju}(\tilde{\mathbf{k}}) = \frac{\delta_{uj}^\perp}{(2\pi)^{d+1}} \int_{\tilde{\mathbf{q}}}^> q_c^\perp \mid G_T(\tilde{\mathbf{q}}) \mid^2 G_T(\tilde{\mathbf{k}} - \tilde{\mathbf{q}})$$
$$= \frac{\delta_{uj}^\perp}{(2\pi)^{d+1}} \int_{\tilde{\mathbf{q}}}^> q_c^\perp \mid G_T(\tilde{\mathbf{q}}) \mid^2 [G(-\tilde{\mathbf{q}})]^2 (2\mu_1 \mathbf{q}_\perp \cdot \mathbf{k}_\perp + 2\mu_x q_x k_x) \tag{9.6.12}$$

$$= \frac{3\delta_{uj}^\perp k_c^\perp}{64(d-1)} \frac{D\lambda^2}{\sqrt{\mu_x \mu_1^3}} \frac{S_{d-1}}{(2\pi)^{d-1}} \Lambda^{d-4} d\ell, \tag{9.6.13}$$

where we have only kept terms up to $O(k)$, and we have again used the angle average trick to evaluate the angular integrals. We've also thrown out odd terms that integrate to zero.

Inserting (9.6.13) into (9.6.10), we obtain a modification to the equation of motion for $u_j^<$:

$$\Delta(\partial_t u_j^<)_{\mu,b} = -\left(\frac{3}{32(d-1)} \frac{D\lambda^2}{\sqrt{\mu_x \mu_1^3}} \frac{S_{d-1}}{(2\pi)^{d-1}} \Lambda^{d-4} d\ell \right) k_j^\perp k_c^\perp u_c^\perp . \tag{9.6.14}$$

From the form of this correction (namely, the fact that it is proportional to $k_j^\perp k_b^\perp u_b^\perp$), we can identify this as a correction to μ_2 (which we remind the reader is the parameter whose bare value we took to be zero):

$$(\delta\mu_2)_{\mu,b} = \frac{3}{32(d-1)} \frac{D\lambda^2}{\sqrt{\mu_x \mu_1^3}} \frac{S_{d-1}}{(2\pi)^{d-1}} \Lambda^{d-4} d\ell . \tag{9.6.15}$$

Thus, it would appear at this point that, even starting as we have with a model in which $\mu_2 = 0$, we generate a nonzero μ_2 upon renormalization. This proves to *not* be the case, at least to one-loop order. Instead, to this order, (9.6.15) is exactly canceled by other graphs, as we will now see.

9.6.3 Graph in Figure 9.6.1(c)

The graph in Figure 9.6.1(c) gives a contribution $\Delta(\partial_t u_j^<)_{\mu,c}$ to the equation of motion for $u_j^<(\tilde{\mathbf{k}})$:

$$\Delta(\partial_t u_j^<)_{\mu,c} = \frac{2\lambda^2 D u_u^\perp(\tilde{\mathbf{k}})}{(2\pi)^{d+1}} \int_{\tilde{\mathbf{q}}}^{>} (k_i^\perp - q_i^\perp) q_u^\perp C_{i\ell}(\tilde{\mathbf{q}}) G_{j\ell}(\tilde{\mathbf{k}} - \tilde{\mathbf{q}})$$
$$\equiv 2\lambda^2 D u_u^\perp(\tilde{\mathbf{k}}) \left[(I_1^{\mu,c})_{ju}(\tilde{\mathbf{k}}) + (I_2^{\mu,c})_{ju}(\tilde{\mathbf{k}}) \right], \tag{9.6.16}$$

where

$$(I_1^{\mu,c})_{ju}(\tilde{\mathbf{k}}) \equiv \frac{k_i^\perp}{(2\pi)^{d+1}} \int_{\tilde{\mathbf{q}}}^{>} q_u^\perp C_{i\ell}(\tilde{\mathbf{q}}) G_{j\ell}(\tilde{\mathbf{k}} - \tilde{\mathbf{q}}) \tag{9.6.17}$$

$$(I_2^{\mu,c})_{ju}(\tilde{\mathbf{k}}) \equiv -\frac{1}{(2\pi)^{d+1}} \int_{\tilde{\mathbf{q}}}^{>} q_i^\perp q_u^\perp C_{i\ell}(\tilde{\mathbf{q}}) G_{j\ell}(\tilde{\mathbf{k}} - \tilde{\mathbf{q}}). \tag{9.6.18}$$

Inserting our expressions (9.3.14) and (9.3.15) for the correlation functions and propagators into (9.6.17) leads to

$$(I_1^{\mu,c})_{ju}(\tilde{\mathbf{k}}) = \frac{k_j^\perp}{(2\pi)^{d+1}} \int_{\tilde{\mathbf{q}}}^{>} q_u^\perp \mid G_T(\tilde{\mathbf{q}}) \mid^2 G_T(\tilde{\mathbf{k}} - \tilde{\mathbf{q}})$$
$$= \frac{k_j^\perp}{(2\pi)^{d+1}} \int_{\tilde{\mathbf{q}}}^{>} q_u^\perp \mid G_T(\tilde{\mathbf{q}}) \mid^2 [G_T(-\tilde{\mathbf{q}})]^2 \left(2\mu_1 \mathbf{q}^\perp \cdot \mathbf{k}^\perp + 2\mu_x q_x k_x\right)$$
$$= \frac{3 k_j^\perp k_u^\perp}{64(d-1)} \frac{D\lambda^2}{\sqrt{\mu_x \mu_1^3}} \frac{S_{d-1}}{(2\pi)^{d-1}} \Lambda^{d-4} d\ell. \tag{9.6.19}$$

We deliberately leave $(I_2^{\mu,c})_{ju}(\tilde{\mathbf{k}})$ untouched since we will show in the next section that this piece is canceled out by that from Figure 9.6.1(d).

Inserting only this $(I_1^{\mu,c})_{ju}$ contribution (9.6.19) into (9.6.16) leads to a term in the equation of motion for $u_j^<$:

$$\Delta(\partial_t u_j^<)_{\mu,c} = \left(\frac{3}{32(d-1)} \frac{D\lambda^2}{\sqrt{\mu_x \mu_1^3}} \frac{S_{d-1}}{(2\pi)^{d-1}} \Lambda^{d-4} d\ell \right) k_j^\perp k_u^\perp u_u^\perp, \tag{9.6.20}$$

which, as before, can be interpreted as a correction to μ_2:

$$\delta\mu_2 = -\frac{3}{32(d-1)}\frac{D\lambda^2}{\sqrt{\mu_x\mu_1^3}}\frac{S_{d-1}}{(2\pi)^{d-1}}\Lambda^{d-4}d\ell. \tag{9.6.21}$$

Note that this exactly cancels the contribution to μ_2 from Figure 9.6.1(b) that we just calculated.

9.6.4 Graph in Figure 9.6.1(d)

The graph in Figure 9.6.1(d) gives a contribution $\Delta(\partial_t u_j^<)_{\mu,d}$ to the equation of motion for $u_j^<(\tilde{\mathbf{k}})$:

$$\Delta(\partial_t u_j^<)_{\mu,d} = \frac{2\lambda^2 D u_u^\perp(\tilde{\mathbf{k}})}{(2\pi)^{d+1}}\int_{\tilde{\mathbf{q}}}^{>} q_i^\perp q_u^\perp C_{j\ell}(\tilde{\mathbf{q}})G_{i\ell}(\tilde{\mathbf{k}}-\tilde{\mathbf{q}}) \equiv 2\lambda^2 D u_u^\perp(\tilde{\mathbf{k}})(I^{\mu,d})_{uj}(\tilde{\mathbf{k}}), \tag{9.6.22}$$

where

$$(I^{\mu,d})_{uj}(\tilde{\mathbf{k}}) \equiv \frac{1}{(2\pi)^{d+1}}\int_{\tilde{\mathbf{q}}}^{>} q_i^\perp q_u^\perp C_{j\ell}(\tilde{\mathbf{q}})G_{i\ell}(\tilde{\mathbf{k}}-\tilde{\mathbf{q}}). \tag{9.6.23}$$

This contribution cancels out the $I_2^{\mu,c}$ contribution from Figure 9.6.1(c) above exactly, leaving no correction to μ_2 at all, to one-loop order. Thus, to this order, $\mu_2 = 0$ is a fixed point.

Note also that we get no contributions to the other viscosity (μ_x). As I'll discuss later, we believe this is an artifact of the one-loop calculation, and that μ_x *will* be renormalized at higher-loop order. However, we expect such renormalizations to be small, since they only enter at higher order in perturbation theory. I'll discuss this more later.

9.7 (Non)renormalization of λ

The ever-present (I hope!) alert reader will have noticed that we have not looked at any of the potential graphical corrections to λ. This is because we know that there are none – or, more precisely, that all such corrections must cancel exactly, to all orders. This is a consequence of the fact that λ is "protected" by a pseudo-Galilean symmetry, just as in the incompressible problem treated in Chapter 6. That is, the equation of motion is invariant under the substitutions: $\mathbf{x}_\perp \mapsto \mathbf{x}_\perp + t\lambda\mathbf{w}$ and $\mathbf{u}_\perp \mapsto \mathbf{u}_\perp + \mathbf{w}$ for some arbitrary constant vector $\mathbf{w}$ perpendicular to the mean velocity $\langle\mathbf{u}\rangle$. Since this *exact* symmetry involves λ, and the renormalization group preserves the underlying symmetries of the problem, it follows that λ can *not* be renormalized (except trivially by rescaling): Its graphical corrections *must* vanish in *any* dimension d.

9.8 Summary of All Corrections to One-Loop Order in the $\mu_2 = 0$ Limit

Adding up the results obtained in previous sections gives the total one-loop graphical corrections to the various parameters when $\mu_2 = 0$:

$$\delta\mu_1 = \left[\frac{1}{8} - \frac{3}{32(d-1)}\right] g_1 \mu_1 \mathrm{d}\ell \,, \tag{9.8.1}$$

$$\delta D = \frac{3}{32}\left(1 - \frac{1}{(d-1)}\right) g_1 D \mathrm{d}\ell \,, \tag{9.8.2}$$

$$\delta\mu_2 = 0 \,, \tag{9.8.3}$$

$$\delta\mu_x = 0 \,, \tag{9.8.4}$$

where we've defined the dimensionless coupling

$$g_1 \equiv \frac{D\lambda^2}{\sqrt{\mu_x \mu_1^5}} \frac{S_{d-1}}{(2\pi)^{d-1}} \Lambda^{d-4} \,, \tag{9.8.5}$$

where S_{d-1} is the surface area of a $d-1$-dimensional unit sphere,

Dividing both sides of each of these equations by $\mathrm{d}\ell$, we obtain the graphical contributions to the recursion relations for the parameters of our model, in the special case $\mu_2 = 0$:

$$\left(\frac{\mathrm{d}\mu_1}{\mathrm{d}\ell}\right)_{\text{graph}} = \left[\frac{1}{8} - \frac{3}{32(d-1)}\right] g_1 \mu_1 \,, \tag{9.8.6}$$

$$\left(\frac{\mathrm{d}D}{\mathrm{d}\ell}\right)_{\text{graph}} = \frac{3}{32}\left(\frac{d-2}{d-1}\right) g_1 D \,, \tag{9.8.7}$$

$$\left(\frac{\mathrm{d}\mu_2}{\mathrm{d}\ell}\right)_{\text{graph}} = 0 \,, \tag{9.8.8}$$

$$\left(\frac{\mathrm{d}\mu_x}{\mathrm{d}\ell}\right)_{\text{graph}} = 0 \,. \tag{9.8.9}$$

The vanishing of the correction to μ_2 at this one-loop order, in the restricted model in which the bare $\mu_2 = 0$, tells us that at one-loop order there is a fixed point at $\mu_2 = 0$. It requires analysis at nonzero μ_2, which we perform in Section 9.9, to show that this $\mu_2 = 0$ fixed point is actually stable, and furthermore, is the only fixed point in the problem.

Now we can get the full DRG recursion relations of our restricted $\mu_2 = 0$ problem by performing the rescaling step described earlier. This gives

$$\frac{\mathrm{d}D}{\mathrm{d}\ell} = \left[z - 2\chi - d + 1 - \zeta + \frac{3}{32}\left(\frac{d-2}{d-1}\right) g_1\right] D \,, \tag{9.8.10}$$

$$\frac{\mathrm{d}\lambda}{\mathrm{d}\ell} = (z + \chi - 1)\lambda \,, \tag{9.8.11}$$

$$\frac{d\mu_x}{d\ell} = (z - 2\zeta)\mu_x\,, \tag{9.8.12}$$

$$\frac{d\mu_1}{d\ell} = \left[z - 2 + \left[\frac{1}{8} - \frac{3}{32(d-1)}\right] g_1\right]\mu_1\,, \tag{9.8.13}$$

$$\frac{d\mu_2}{d\ell} = (z - 2)\mu_2 + O(\mu_2), \tag{9.8.14}$$

where the $O(\mu_2)$ term represents the potential nonzero graphical corrections to μ_2 that may (and in fact, do, as we'll see later) occur when $\mu_2 \neq 0$.

As can be seen from (9.8.14), $\mu_2 = 0$ is a fixed point, even when the coupling constant $g_1 \neq 0$. To see if there are fixed points at nonzero g_1, we construct the recursion relation for g_1. As we've found in all of the other problems we've studied, the recursion relation for the dimensionless coupling g_1 is independent of the arbitrary rescaling exponents z, ζ, and χ. To see this, we'll do our usual trick of calculating $\frac{d\ln g_1}{d\ell}$, which is easily related to the derivatives of the logarithms of the parameters D, λ, μ_1, and μ_x by taking the logarithm of the definition (9.8.5) of g_1. Doing so, and taking a derivative with respect to RG time ℓ, gives

$$\frac{1}{g_1}\frac{dg_1}{d\ell} = \frac{1}{D}\frac{dD}{d\ell} + \frac{2}{\lambda}\frac{d\lambda}{d\ell} - \frac{5}{2\mu_1}\frac{d\mu_1}{d\ell} - \frac{1}{2\mu_x}\frac{d\mu_x}{d\ell} = \epsilon + \left(\frac{23 - 14d}{64(d-1)}\right) g_1, \tag{9.8.15}$$

where we've defined $\epsilon = 4 - d$.

The recursion relation (9.8.15) can (obviously) be rewritten as

$$\frac{dg_1}{d\ell} = \epsilon g_1 + \left(\frac{23 - 14d}{64(d-1)}\right) g_1^2. \tag{9.8.16}$$

Finding the values of g_1 at which $\frac{dg_1}{d\ell} = 0$ (a value that we'll refer to as g_1^*) gives two possibilities: $g_1^* = 0$, which corresponds to the unstable fixed point ($g_1^* = 0$) we identified earlier, and a single stable non-Gaussian fixed point at:

$$g_1^* = \frac{64}{11}\epsilon + \mathcal{O}(\epsilon^2), \quad \mu_2^* = 0\,. \tag{9.8.17}$$

Note that the value of g_1 at this non-Gaussian fixed point is $\mathcal{O}(\epsilon)$, so our perturbation theory, which is valid for small g_1, should be accurate for small ϵ; i.e., for spatial dimensions near the upper critical dimension $d = 4$. This validity for small ϵ is, of course, a standard feature of all ϵ expansions.

The fact that (9.8.17) is a fixed point of our recursion relations does not actually prove that this fixed point controls the long distance behavior of any real system. To prove *that*, we need to show that this fixed point is stable. We do that in Section 9.16. For now, we'll just assume that it's stable, and calculate the scaling exponents z, ζ, and χ that follow from this assumption.

9.9 Scaling Exponents

With the location of the fixed point (see equation (9.10.1)) in hand, we can now easily find the universal scaling exponents governing the behavior of all properties (in particular, correlation functions) of Malthusian flocks.

The most direct way to do this is to choose the heretofore arbitrary RG rescaling exponents – that is, the dynamical exponent z, the anisotropy exponent ζ, and the roughness exponent χ – to keep all of the other important parameters (i.e., the noise strength D, the diffusion constants $\mu_{x,1}$, and the convective nonlinearity λ) fixed.[2,3]

Keeping the noise strength D fixed leads, via (9.8.10), to the condition

$$z - 2\chi - d + 1 - \zeta + \frac{g_1^*}{16} = 0\,, \tag{9.9.1}$$

where we've replaced d with 4 everywhere in (9.8.10) except in g_1^*, since the errors thereby introduced are $O(\epsilon^2)$. Inserting the fixed-point value (9.8.17) of g_1^*, and $d = 4 - \epsilon$ into (9.9.1) gives

$$z - 2\chi - \zeta = 3 - \frac{15}{11}\epsilon\,. \tag{9.9.2}$$

We can obtain two more linear conditions on our three exponents z, ζ, and χ, by requiring that μ_x and λ remain fixed. The former condition leads to

$$z = 2\zeta\,, \tag{9.9.3}$$

while the latter implies

$$z + \chi = 1\,. \tag{9.9.4}$$

The three linear equations (9.9.2), (9.9.3), and (9.9.4) are easily solved to give:

$$z = 2 - \frac{6\epsilon}{11} + \mathcal{O}(\epsilon^2), \tag{9.9.5}$$

$$\chi = -1 + \frac{6\epsilon}{11} + \mathcal{O}(\epsilon^2), \tag{9.9.6}$$

$$\zeta = 1 - \frac{3\epsilon}{11} + \mathcal{O}(\epsilon^2)\,. \tag{9.9.7}$$

To the best of our knowledge, the above fixed point and the associated scaling exponents characterize a previously undiscovered universality class (see Figure 9.9.1).

[2] The diffusion coefficient μ_2 is kept fixed simply by flowing to zero, as it can readily be seen to do for the choice of z that we are making here.

[3] We only need to keep any three of μ_1, μ_x, λ, and D fixed. This is because g_1 flows to a nonzero fixed point under the renormalization group. Since $g_1 = \frac{D\lambda^2}{\sqrt{\mu_x \mu_1^5}} \times$ constant and g_1 flows to a nonzero fixed point, keeping any three of μ_1, μ_x, λ, and D fixed will automatically hold the remaining one fixed as well. Choosing μ_x and λ to be fixed is the easiest option since their RG flow equations are the simplest.

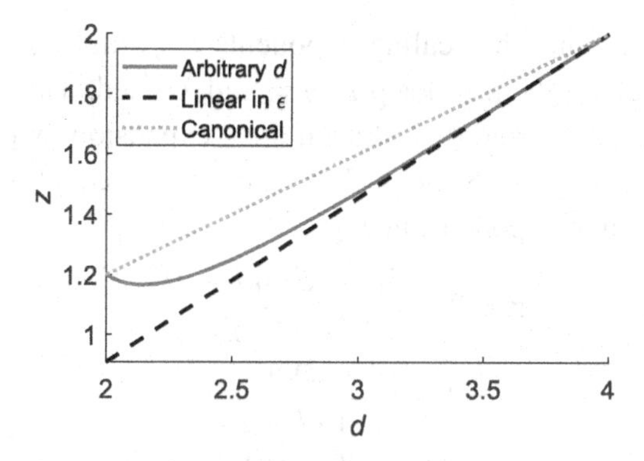

Fig. 9.9.1 Graphical summary of our results for the dynamic exponent z as a function of the spatial dimension d. The result (9.9.5) based on the ϵ-expansion method to $\mathcal{O}(\epsilon)$ is shown by the broken line, while the extrapolation to arbitrary d based on our one-loop result is shown in the solid line (9.10.2). The dashed line is the "canonical" value $z = \frac{2(d+1)}{5}$ that we found for *incompressible* flocks in $d > 2$ in Chapter 6, and is shown here just for purposes of comparison. Reproduced from [22] and [23].

9.10 Beyond Linear Order in ϵ

Our results so far are based on a one-loop calculation, which is exact to linear order in ϵ. However, since all of our expressions for G_{D,μ_1,μ_2} are evaluated for general d, one can potentially extrapolate our results to arbitrary d based on our one-loop calculation, ignoring higher-loop graphs. We must emphasize that this is an uncontrolled approximation, since the higher-loop graphs are of higher order in g_1, but g_1 is *not* small at the fixed point once d is far from 4. Nonetheless, this approach will automatically recover exactly the one-loop $4 - \epsilon$ expansion results we've just obtained.

This truncated one-loop calculation for general d now proceeds in much the same way as our small-ϵ approach.

We start by noting that once again, as for d near 4, $\mu_2 = 0$ is a fixed point of this uncontrolled one-loop approximation. Hence, we again have an unstable Gaussian fixed line at $g_1 = 0$, and a single stable (at least when $\mu_2 = 0$) non-Gaussian fixed point at $\mu_2 = 0$ and a nonzero value of g_1, which we'll calculate in a moment. (We will do a more thorough analysis of the stability of this fixed point for general d later.)

We'll again assume, and verify in Section 9.16, that the fixed point at $\mu_2 = 0$ is stable in this approximation.

Using the recursion relation (9.8.16) in arbitrary spatial dimension d, and setting $\frac{dg_1}{d\ell} = 0$, we get the fixed-point value g_1^* of g_1:

$$g_1^* = \frac{64(4-d)(d-1)}{14d-23}. \tag{9.10.1}$$

We can now determine the scaling exponents z, ζ, and χ, as we did in the ϵ expansion, by choosing them to keep D, μ_x, and λ fixed. This leads to the same conditions (9.9.1), (9.9.3), and (9.9.4) as in the ϵ expansion, but now in (9.9.1) we use the value (9.10.1) for g_1^*. Solving these three linear equations (9.9.1), (9.9.3), and (9.9.4) for the three exponents now gives

$$z = 2 - \frac{2(4-d)(4d-7)}{14d-23}, \tag{9.10.2}$$

$$\zeta = 1 - \frac{(4-d)(4d-7)}{14d-23}, \tag{9.10.3}$$

$$\chi = -1 + \frac{2(4-d)(4d-7)}{14d-23}. \tag{9.10.4}$$

Note that the predicted values of the scaling exponents in three dimensions obtained from these two approaches (ϵ expansion and one loop in arbitrary d) are in fact very quantitatively similar (Figure 9.9.1). For example, the value of z obtained from the ϵ expansion in $d = 3$, obtained from equation (9.9.5) by setting $\epsilon = 1$, is $z_\epsilon = \frac{16}{11}$, while that obtained from our uncontrolled one-loop approximation is $z_u = \frac{28}{19}$. The difference between these is $z_u - z_\epsilon = \frac{4}{209}$, which is only $\frac{1}{77}$ of z_u. The other exponents are comparably close. We thereby conclude that the values given by the uncontrolled approximation in $d = 3$, namely

$$z = \frac{28}{19} \approx 1.47, \tag{9.10.5}$$

$$\zeta = \frac{14}{19} \approx 0.74, \tag{9.10.6}$$

$$\chi = -\frac{9}{19} \approx -0.47, \tag{9.10.7}$$

are likely accurate to ± 1 percent. This implies that the digits shown after the approximate equalities above are probably all correct.

9.11 Beyond One-Loop Order

In this section, we discuss what features of the above results are artifacts of the one-loop truncation. Aside from small quantitative corrections to the precise values of the exponents, which we have just argued are small, there are two more significant changes that we expect will occur in a higher-order calculation (which, we should emphasize, we have *not* done!).

The first of these is that the diffusion constant μ_x will no longer be unrenormalized at higher order. We expect this to be the case because there is no symmetry that "protects" μ_x from renormalizing. Its failure to renormalize at one-loop order is therefore to some extent simply a coincidence, and almost certainly an artifact of the one-loop approximation. In this respect, its failure to renormalize is very similar to the result in ϵ-expansions for ϕ^4 theories of phase transitions [30] that the critical exponent η is zero to $\mathcal{O}(\epsilon)$. As is well known, η becomes nonzero at $\mathcal{O}(\epsilon^2)$; or, equivalently, at two-loop order. We are quite confident that the same thing is true of renormalization of μ_x.

The most important qualitative consequence of this is that the scaling relation

$$z = 2\zeta, \quad \text{(one-loop)}, \tag{9.11.1}$$

which emerges at one-loop order from the requirement that μ_x remain fixed upon renormalization, will no longer hold, since μ_x will now get graphical corrections.

However, the analogy just noted with critical phenomena strongly suggests that the corrections to (9.11.1) will be very small in $d = 3$. The exponent η in ϕ^4 theories is typically [30] of order $\eta \sim 0.01{-}0.02$, so it seems reasonable to expect the corrections to (9.11.1), which also arise only at two-loop order in a problem with a critical dimension of 4, to be comparable in magnitude. So, although it is an artifact of the one-loop approximation, equation (9.11.1) probably holds to within a few percent. But as a matter of principle, (9.11.1) is *not* an exact scaling relation.

The second change that will occur at higher-loop order is that the fixed point will no longer be at $\mu_2 \neq 0$. This is because, as for the renormalization of μ_x, there is no symmetry that prevents a nonzero μ_2 from being generated, even when the initial (bare) $\mu_2 = 0$.

To treat the effects of a nonzero μ_2, it is useful to introduce an additional dimensionless parameter

$$\alpha \equiv \frac{\mu_2}{\mu_1}. \tag{9.11.2}$$

With this definition, the statement that $\mu_2 = 0$ is a fixed point implies that $\alpha = 0$ is a fixed point as well. Therefore, the recursion relation for α near $g_1 = 0$ must, at two-loop order, take the form

$$\frac{d\alpha}{d\ell} = g_1 G_\alpha(\alpha)\alpha + g_1^2 H(\alpha), \tag{9.11.3}$$

where $G_\alpha(\alpha)$ and $H(\alpha)$ are a functions of α that remain finite as $\alpha \to 0$. We will calculate $G_\alpha(\alpha)$ in Section 9.16; as you'll see there, it is quite a complicated function. The function $H(\alpha)$ will presumably be even more formidable. More importantly, it will be nonzero at $\alpha = 0$. Expanding the right-hand side of (9.11.3) for small α and ϵ gives

$$\frac{d\alpha}{d\ell} = g_1 \alpha G_\alpha(\alpha = 0) + g_1^2 H(\alpha = 0), \tag{9.11.4}$$

where the alert reader will recognize the first term on the right-hand side from our linearized recursion relation (9.16.54) to one-loop order. Solving for the fixed-point value α^* of α by setting $\frac{d\alpha}{d\ell} = 0$ and $g_1 = g_1^*$ gives

$$\alpha^* = -\frac{H(\alpha = 0)}{G_\alpha(\alpha = 0)} g_1^* = \mathcal{O}(\epsilon), \tag{9.11.5}$$

where in the last equality we have used the fact that $g_1^* = \mathcal{O}(\epsilon)$. Thus α^* is nonzero, and $\mathcal{O}(\epsilon)$, once higher-loop corrections are taken into account. Unfortunately, it is impossible to say much more about the *value* of α^*, other than that it is nonzero, without actually doing the two-loop calculation necessary to determine the function $H(\alpha)$ in equation (9.11.3). We have not attempted this formidable calculation, and so can say nothing beyond the statement that α^* will be nonzero. (We cannot even determine its sign!)

This has experimental consequences, because, as we'll show in Section 9.12 below, the value of α^* determines a universal amplitude ratio in the velocity correlation function.

9.12 Experimental Consequences

9.12.1 Scaling Laws for Velocity and Density Correlations

The scaling exponents z, ζ, and χ just determined control the scaling properties of velocity and density correlations, as embodied in equations (9.17.10) and (9.17.11) for the velocity and density autocorrelations, respectively. This can be seen by using the "trajectory integral matching formalism" [32], which, as I described in Chapter 3, is simply a fancy way of describing the process of undoing all of the variable and coordinate rescaling done in the renormalization group process. This implies, for example, that the velocity autocorrelation function

$$C_u\Big(r_\perp, x - \gamma t, t; \{D_0, \mu_{x0}, \mu_{10}, \mu_{20}, \lambda_0\}\Big) \equiv \langle \mathbf{u}_\perp(\mathbf{r}, t) \cdot \mathbf{u}_\perp(\mathbf{0}, 0)\rangle \tag{9.12.1}$$

of the original system (whose parameters – the "bare" parameters – are denoted by the subscript 0) can be related to that of the system after a renormalization group time ℓ has elapsed via

$$C_u\Big(r_\perp, x - \gamma t, t; \big\{D_0, \mu_{x0}, \mu_{10}, \mu_{20}, \lambda_0\big\}\Big) = \exp\left[2\int_0^\ell \chi(\ell')\mathrm{d}\ell'\right]$$
$$\times\, C_u\left(r_\perp \mathrm{e}^{-\ell}, (x - \gamma t)\exp\left(-\int_0^\ell \zeta(\ell')\mathrm{d}\ell'\right), t\exp\left(-\int_0^\ell z(\ell')\mathrm{d}\ell'\right); \text{parameters}\right), \tag{9.12.2}$$

where "parameters" stands for $D(\ell), \mu_x(\ell), \mu_1(\ell), \mu_2(\ell), \lambda(\ell)$.

In this relation the combination $x - \gamma t$ appears rather than x due to the boost (9.1.14) we performed to obtain the model equation (9.1.15), which we actually used for the renormalization group.

The relation (9.12.2) holds for an *arbitrary* choice of the rescaling exponents $\chi(\ell)$, $\zeta(\ell)$, and $z(\ell)$; they need not be the special choice (9.9.5), (9.9.6), and (9.9.7) that we made earlier to produce fixed points. Indeed, as our notation suggests, we can even choose different values for these exponents at different renormalization group times ℓ. We will take advantage of this freedom to use (9.12.2) to derive the scaling relation (9.17.10). We will do so by choosing the rescaling exponents $\chi(\ell)$, $\zeta(\ell)$, and $z(\ell)$ according to the following scheme: For $\ell < \ell^*$, where ℓ^* is the renormalization group time at which g_1 and α get close to their fixed point values, we will choose these exponents so that *at* $\ell = \ell^*$, the parameters $D(\ell^*)$, $\mu_x(\ell^*)$, and $\mu_1(\ell^*)$ take on the values $D(\ell^*) = D_{\text{ref}}$, $\mu_x(\ell^*) = \mu_1(\ell^*) = \mu_{\text{ref}}$, where D_{ref} and μ_{ref} are some reference values that we are free to choose. Note that we have deliberately chosen to make $\mu_x(\ell^*) = \mu_1(\ell^*)$.

Since there are three free scaling exponents at our disposal, and an equal number (three) of parameters that we wish to force to take on values of our choosing, we can always find a choice of the rescaling exponents $\chi(\ell)$, $\zeta(\ell)$, and $z(\ell)$ (even with the assumption that these exponents are constant for $\ell > \ell^*$) that will achieve the target values D_{ref} and μ_{ref}.

The precise values of D_{ref} and μ_{ref} that we choose are unimportant; what is important is that we choose the same values *no matter what the initial (bare) values* D_0, μ_{x0}, μ_{10}, μ_{20}, and λ_0 of the parameters were in our original model. The only "memory" of these original parameters on the right-hand side of (9.12.2) will therefore be contained in the value of ℓ^*.[4]

Once we have fixed $D(\ell^*)$, $\mu_x(\ell^*)$, and $\mu_1(\ell^*)$, the parameters $\mu_2(\ell^*)$ and $\lambda(\ell^*)$ are also determined, the former by the relation $\alpha = \frac{\mu_2}{\mu_1}$, the latter by the definition (9.8.5) of g_1. Combining this fact with $g_1(\ell^*) = g_1^*$ and $\alpha(\ell^*) = \alpha^*$ (which follows our definition of ℓ^* as the renormalization group time at which we get close to the fixed point), we have that

$$\mu_2(\ell^*) = \alpha^* \mu_{\text{ref}}, \quad \lambda(\ell^*) = \sqrt{\frac{g_1^* \mu_{\text{ref}}^3 (2\pi)^{d-1} \Lambda^{4-d}}{D_{\text{ref}} S_{d-1}}}. \tag{9.12.3}$$

Hereafter we also refer to $\lambda(\ell^*)$ as λ_{ref}.

Note finally that the value of ℓ^* at which we get close to the fixed point is unaffected by our arbitrary choice of the rescaling exponents $\chi(\ell)$, $\zeta(\ell)$, and $z(\ell)$, since these do not enter the recursion relations for g_1 and α.

For $\ell > \ell^*$, we will choose the rescaling exponents $\chi(\ell)$, $\zeta(\ell)$, and $z(\ell)$ to take on the values (9.9.5), (9.9.6), and (9.9.7) that we showed earlier keep all of the parameters fixed once g_1 and α have flowed to their fixed-point values. For the remainder of this section, we will refer to these values of χ, ζ, and z as the fixed-point values χ_{FP}, ζ_{FP}, and z_{FP}.

[4] This argument can easily be extended to include any one of the infinite number of *irrelevant* parameters that one could add to our starting model (e.g., terms involving, say, $\nabla^4 \mathbf{u}$). The precise value of ℓ^* could also depend on these parameters; all other dependence on these parameters would drop out of the problem, since they would renormalize to zero on the right-hand side of (9.12.2).

This choice will, for all $\ell > \ell^*$, keep all of the parameters fixed at the "reference" values we have just described.

With this choice of $\chi(\ell)$, $\zeta(\ell)$, and $z(\ell)$, we can rewrite equation (9.12.2) as

$$C_u\Big(r_\perp, x-\gamma t, t; \{D_0, \mu_{x0}, \mu_{10}, \mu_{20}, \lambda_0\}\Big) = \exp\left[2\int_0^{\ell_*} \chi(\ell')\mathrm{d}\ell' + 2\chi_{FP}(\ell-\ell_*)\right]$$
$$\times\, C_u\Bigg(r_\perp \mathrm{e}^{-\ell}, (x-\gamma t)\exp\left(-\int_0^{\ell_*}\zeta(\ell')\mathrm{d}\ell' - \zeta_{FP}(\ell-\ell_*)\right),$$
$$t\exp\left(-\int_0^{\ell_*} z(\ell')\mathrm{d}\ell' - z_{FP}(\ell-\ell_*)\right); \left\{D_{\mathrm{ref}}, \mu_{\mathrm{ref}}, \mu_{\mathrm{ref}}, \alpha^*\mu_{\mathrm{ref}}, \lambda_{\mathrm{ref}}\right\}\Bigg). \quad (9.12.4)$$

To derive our scaling law (9.17.10) for C_u, we simply apply this relation (9.12.4) at particular value of ℓ, which we'll call $\ell(r_\perp)$, determined by

$$\mathrm{e}^{-\ell(r_\perp)} r_\perp = a \equiv \frac{1}{\Lambda}. \quad (9.12.5)$$

Setting $\ell = \ell(r_\perp)$ on the right-hand side of (9.12.4) gives

$$C_u\Big(r_\perp, x-\gamma t, t; \big\{D_0, \mu_{x0}, \mu_{10}, \lambda_0\big\}\Big)$$
$$= A r_\perp^{2\chi_{FP}} C_u\left(a, a\frac{(|x-\gamma t|/\xi_x)}{(r_\perp/\xi_\perp)^{\zeta_{FP}}}, \frac{(t/\tau)\tau_0}{(r_\perp/\xi_\perp)^{z_{FP}}}; \left\{D_{\mathrm{ref}}, \mu_{\mathrm{ref}}, \mu_{\mathrm{ref}}, \alpha^*\mu_{\mathrm{ref}}, \lambda_{\mathrm{ref}}\right\}\right)$$
$$\equiv r_\perp^{2\chi_{FP}} F_u\left(\frac{(|x-\gamma t|/\xi_x)}{(r_\perp/\xi_\perp)^{\zeta_{FP}}}, \frac{(t/\tau)}{(r_\perp/\xi_\perp)^{z_{FP}}}\right), \quad (9.12.6)$$

where we've defined scaling function

$$F_u \equiv A C_u\left(a, a\frac{(|x-\gamma t|/\xi_x)}{(r_\perp/\xi_\perp)^{\zeta_{FP}}}, \frac{(t/\tau)\tau_0}{(r_\perp/\xi_\perp)^{z_{FP}}}; \left\{D_{\mathrm{ref}}, \mu_{\mathrm{ref}}, \mu_{\mathrm{ref}}, \alpha^*\mu_{\mathrm{ref}}, \lambda_{\mathrm{ref}}\right\}\right), \quad (9.12.7)$$

the constant

$$A \equiv a^{-2\chi_{FP}} \exp\left[2\int_0^{\ell_*} (\chi(\ell') - \chi_{FP})\mathrm{d}\ell'\right], \quad (9.12.8)$$

and the nonuniversal "nonlinear lengths" $\xi_{\perp,x}$ to satisfy

$$\xi_\perp = \mathrm{e}^{\ell^*} a, \quad (9.12.9)$$

and

$$\frac{\xi_\perp^{\zeta_{FP}}}{\xi_x} = \exp\left[\int_0^{\ell^*} (\zeta_{FP} - \zeta(\ell))\mathrm{d}\ell\right] a^{\zeta_{FP}-1} \quad (9.12.10)$$

and the nonuniversal "nonlinear time" τ to satisfy

$$\frac{\xi_{\perp}^{z_{FP}}}{\tau}\tau_0 = \exp\left[\int_0^{\ell^*}(z_{FP} - z(\ell))d\ell\right]a^{z_{FP}}. \qquad (9.12.11)$$

Here the value of the characteristic time τ_0 is not arbitrary, but set by the cutoff length a and the μ_1 of the rescaled system, namely μ_{ref}. Specifically it is given by

$$\tau_0 = \frac{a^2}{\mu_{\text{ref}}}. \qquad (9.12.12)$$

Note that, like the reference values of other parameters, this characteristic time is the same for all systems, regardless of the bare values of the parameters. The nonlinear lengths $\xi_{x,\perp}$ and the nonlinear time τ are the crossover scales beyond which nonlinear effects become important. We will calculate them in the following section.

Because the parameters appearing in C_u on the right-hand side of (9.12.7) (namely, a, τ_0, D_{ref}, μ_{ref}, $\alpha^*\mu_{\text{ref}}$, and λ_{ref}) are all independent of the initial system under consideration, the scaling function F_u is, as claimed in the introduction, a universal function of its arguments $\frac{(|x-\gamma t|/\xi_x)}{(r_\perp/\xi_\perp)^{\zeta_{FP}}}$ and $\frac{(t/\tau)}{(r_\perp/\xi_\perp)^{z_{FP}}}$, up to the nonuniversal multiplicative factor A, which is given by (9.12.8).

Dropping the subscript "*FP*" on all of the exponents, and recognizing that their values are those given by equation (9.17.9), we can summarize this result for the velocity autocorrelation function as follows:

$$C_u(\mathbf{r},t) \equiv \langle \mathbf{u}_\perp(\mathbf{r},t)\cdot\mathbf{u}_\perp(\mathbf{0},0)\rangle = r_\perp^{2\chi}F_u\left(\frac{(|x-\gamma t|/\xi_x)}{(r_\perp/\xi_\perp)^{\zeta}}, \frac{(t/\tau)}{(r_\perp/\xi_\perp)^{z}}\right)$$

$$\propto \begin{cases} r_\perp^{2\chi}, & (r_\perp/\xi_\perp)^{\zeta} \gg |x-\gamma t|/\xi_x, \quad (r_\perp/\xi_\perp)^{z} \gg (|t|/\tau), \\ |x-\gamma t|^{\frac{2\chi}{\zeta}}, & |x-\gamma t|/\xi_x \gg (r_\perp/\xi_\perp)^{\zeta}, \quad |x-\gamma t|/\xi_x \gg (|t|/\tau)^{\frac{\zeta}{z}}, \\ |t|^{\frac{2\chi}{z}}, & (|t|/\tau) \gg (r_\perp/\xi_\perp)^{z}, \quad (|t|/\tau) \gg (|x-\gamma t|/\xi_x)^{\frac{z}{\zeta}}, \end{cases} \qquad (9.12.13)$$

with, as I said above, the values of the universal exponents z, ζ, and χ given by (9.17.9).

This concludes our derivation of the scaling law for velocity correlations. The derivation of the density correlations is almost identical. The only difference lies in the field rescaling. Since $\delta\rho$ is enslaved to $\mathbf{u}_\perp$ by the relation (9.1.6), the rescaling exponent for $\delta\rho$ is $\chi - 1$ instead of χ, the rescaling exponent of $\mathbf{u}_\perp$. Therefore, in analogy to (9.12.2), we get the following relation between density correlations in the original system and the rescaled system:

$$C_\rho\Big(r_\perp, x-\gamma t, t; \{D_0, \mu_{x0}, \mu_{10}, \mu_{20}, \lambda_0\}\Big) = \exp\left[2\int_0^\ell \chi(\ell') - 1 \mathrm{d}\ell'\right]$$
$$\times C_\rho\left(r_\perp \mathrm{e}^{-\ell}, (x-\gamma t)\exp\left(-\int_0^\ell \zeta(\ell')\mathrm{d}\ell'\right), t\exp\left(-\int_0^\ell z(\ell')\mathrm{d}\ell'\right);\right.$$
$$\left.\{D(\ell), \mu_x(\ell), \mu_1(\ell), \mu_2(\ell), \lambda(\ell)\}\right). \tag{9.12.14}$$

From here on the derivation is virtually identical to that of the velocity correlations, which we will not repeat. The final result is given by

$$C_\rho(\mathbf{r}, t) \equiv \langle \delta\rho(\mathbf{r}, t)\delta\rho(\mathbf{0}, 0)\rangle = r_\perp^{2(\chi-1)} F_\rho\left(\frac{(|x-\gamma t|/\xi_x)}{(r_\perp/\xi_\perp)^\zeta}, \frac{(t/\tau)}{(r_\perp/\xi_\perp)^z}\right)$$
$$\propto \begin{cases} r_\perp^{2(\chi-1)}, & (r_\perp/\xi_\perp)^\zeta \gg |x-\gamma t|/\xi_x, \quad (r_\perp/\xi_\perp)^z \gg (|t|/\tau), \\ |x-\gamma t|^{\frac{2(\chi-1)}{\zeta}}, & |x-\gamma t|/\xi_x \gg (r_\perp/\xi_\perp)^\zeta, \quad |x-\gamma t|/\xi_x \gg (|t|/\tau)^{\frac{\zeta}{z}}, \\ |t|^{\frac{2(\chi-1)}{z}}, & (|t|/\tau) \gg (r_\perp/\xi_\perp)^z, \quad (|t|/\tau) \gg (|x-\gamma t|/\xi_x)^{\frac{z}{\zeta}}. \end{cases} \tag{9.12.15}$$

9.13 Calculation of the Nonlinear Lengths and Times

There are two independent ways of calculating the nonlinear lengths and times appearing in the scaling functions (9.12.6) and (9.12.15) just derived. One way is to continue with the RG approach just presented. We take this approach in the next subsection. An alternative approach, which we present as a check on the RG approach, is to calculate perturbative corrections to the linear theory and calculate the length and time scales on which they become appreciable. These length and time scales prove to be precisely the lengths $\xi_\perp$, and ξ_x, and the time τ.

We'll begin here with the RG calculation; then, in the Section 9.13.2, we'll present the perturbation theory approach.

9.13.1 RG Calculation

The conditions (9.12.10) and (9.12.11) can be solved for the nonlinear length $\xi_\perp$ and nonlinear time τ, giving

$$\xi_x = a\left(\frac{\xi_\perp}{a}\right)^{\zeta_{FP}} \exp\left[-\zeta_{FP}\ell^* + \int_0^{\ell^*} \zeta(\ell)\mathrm{d}\ell\right] \tag{9.13.1}$$

and

$$\tau = \tau_0\left(\frac{\xi_\perp}{a}\right)^{z_{FP}} \exp\left[-z_{FP}\ell^* + \int_0^{\ell^*} z(\ell)\mathrm{d}\ell\right]. \tag{9.13.2}$$

Using our expression (9.12.9) for $\xi_\perp$ in these expressions simplifies them to

$$\xi_x = a \exp\left[\int_0^{\ell^*} \zeta(\ell)\mathrm{d}\ell\right] \tag{9.13.3}$$

and

$$\tau = \tau_0 \exp\left[\int_0^{\ell^*} z(\ell)\mathrm{d}\ell\right]. \tag{9.13.4}$$

The alert reader may be alarmed by the apparent dependence of $\xi_\perp$ and τ in the arbitrary choices of $\zeta(\ell)$ and $z(\ell)$. But those choices are not *completely* arbitrary, since they must lead to the parameters D and μ_{1x} flowing to their reference values. This requirement proves to constrain the very integrals that appear in (9.13.3) and (9.13.4) to (nonuniversal) values that are determined entirely by the bare parameters of the model. Likewise, the nonuniversal overall scale factor A in the correlation function (9.12.7), while apparently dependent on our arbitrary choice of the velocity rescaling exponent $\chi(\ell)$, in fact is not, and is, instead, also determined solely by the nonuniversal values of the bare parameters of the model, as we'll show now.

The requirement that $\mu_x(\ell)$ and $\mu_1(\ell)$ reach equality at $\ell = \ell^*$ constrains the integral of $\zeta(\ell)$ in (9.13.3). To see this, consider the recursion relations for μ_x and μ_1. In complete generality, to arbitrary order in perturbation theory, these can be written:

$$\frac{1}{\mu_x}\frac{\mathrm{d}\mu_x}{\mathrm{d}\ell} = z - 2\zeta(\ell) + Y_x(g_1(\ell), \alpha(\ell)), \tag{9.13.5}$$

$$\frac{1}{\mu_1}\frac{\mathrm{d}\mu_1}{\mathrm{d}\ell} = z - 2 + Y_1(g_1(\ell), \alpha(\ell)). \tag{9.13.6}$$

To one-loop order, $Y_x = 0$ and $Y_1 = g_1 G_{\mu_1}(\alpha)$; here we'll use this more-general form to demonstrate that our conclusion is *not* an artifact of the one-loop approximation, or, indeed, any approximation at all.

The recursion relations (9.13.5) and (9.13.6) taken together imply that the logarithm of the ratio $\frac{\mu_x}{\mu_1}$ obeys the recursion relation

$$\frac{\mathrm{d}}{\mathrm{d}\ell}\ln\left(\frac{\mu_x}{\mu_1}\right) = \frac{1}{\mu_x}\frac{\mathrm{d}\mu_x}{\mathrm{d}\ell} - \frac{1}{\mu_1}\frac{\mathrm{d}\mu_1}{\mathrm{d}\ell} = 2(1-\zeta(\ell)) + Y_x(g_1(\ell), \alpha(\ell)) - Y_1(g_1(\ell), \alpha(\ell)). \tag{9.13.7}$$

The solution of this is

$$\ln\left[\left(\frac{\mu_x(\ell)}{\mu_1(\ell)}\right)\left(\frac{\mu_{10}}{\mu_{x0}}\right)\right] = 2\ell - 2\int_0^\ell \zeta(\ell')\mathrm{d}\ell' + \int_0^\ell [Y_x(g_1(\ell'), \alpha(\ell')) - Y_1(g_1(\ell'), \alpha(\ell'))]\mathrm{d}\ell'. \tag{9.13.8}$$

Evaluating both sides of this expression at $\ell = \ell^*$, and recalling that, by construction, $\mu_x(\ell^*) = \mu_1(\ell^*)$, gives

$$\ln\left(\frac{\mu_{10}}{\mu_{x0}}\right) = 2\ell^* - 2\int_0^{\ell^*} \zeta(\ell)\mathrm{d}\ell + 2\Phi(g_{10}, \alpha_0), \tag{9.13.9}$$

where we've defined

$$\Phi(g_{10}, \alpha_0) \equiv \frac{1}{2}\int_0^{\ell^*} [Y_x(g_1(\ell'), \alpha(\ell')) - Y_1(g_1(\ell'), \alpha(\ell'))]\mathrm{d}\ell'. \tag{9.13.10}$$

Note that, as our notation suggests, Φ is completely determined by the bare values g_{10} and α_0 of g_1 and α; in particular, it is *independent* of the arbitrary choice of the rescaling exponents $\chi(\ell)$, $\zeta(\ell)$, and $z(\ell)$. This is because the recursion relations for g_1 and α are independent of those exponents; so their solutions $g_1(\ell)$ and $\alpha(\ell)$ are determined entirely by the initial conditions $g_1(\ell = 0) = g_{10}$ and $\alpha(\ell = 0) = \alpha_0$. Once those solutions are determined, the integrand in (9.13.10) is also fully determined (since it depends only on $g_1(\ell)$ and $\alpha(\ell)$). Furthermore, the limits on the integral are completely determined by g_{10} and α_0 as well, since ℓ^* is. Hence, Φ is completely determined by g_{10} and α_0, as claimed.

The condition (9.13.9) can be rewritten as

$$\int_0^{\ell^*} \zeta(\ell)\mathrm{d}\ell = \frac{1}{2}\ln\left(\frac{\mu_{x0}}{\mu_{10}}\right) + \ell^* + \Phi(g_{10}, \alpha_0). \tag{9.13.11}$$

Using this in (9.13.3) gives

$$\xi_x = a e^{\ell^*} e^{\Phi}\sqrt{\frac{\mu_{x0}}{\mu_{10}}} = \xi_\perp e^{\Phi}\sqrt{\frac{\mu_{x0}}{\mu_{10}}}, \tag{9.13.12}$$

where in the last equality we have used our expression (9.12.9) for $\xi_\perp$.

Note that the ratio of ξ_x to $\xi_\perp$ implied by (9.13.12) depends only on g_{10}, α_0, and $\mu_{x0,10}$, and not at all on the exact choice of the functional dependence rescaling exponent $\zeta(\ell)$ on ℓ; *any* choice that leads to $\mu_x(\ell^*) = \mu_1(\ell^*)$ gives the same answer.

A similar argument can be applied to the time scale τ. We start by solving the recursion relation (9.13.6) for $\mu_1(\ell^*)$:

$$\ln\left(\frac{\mu_{\mathrm{ref}}}{\mu_{10}}\right) = \int_0^{\ell^*} z(\ell)\mathrm{d}\ell - 2\ell^* + \Phi_{\mu_1}(g_{10}, \alpha_0), \tag{9.13.13}$$

where we've defined

$$\Phi_{\mu_1}(g_{10}, \alpha_0) \equiv \int_0^{\ell^*} Y_1(g_1(\ell), \alpha(\ell))\mathrm{d}\ell. \tag{9.13.14}$$

Note that, like Φ, Φ_{μ_1} is completely determined by the bare values g_{10} and α_0, and is *independent* of the arbitrary choice of the rescaling exponents $\chi(\ell)$, $\zeta(\ell)$, and

$z(\ell)$ for the same reasons as before: Both integrand and the limits of integration in (9.13.14) depend only on g_{10} and α_0. Solving (9.13.13) for $\int_0^{\ell^*} z(\ell)d\ell$ gives

$$\int_0^{\ell^*} z(\ell)d\ell = \ln\left(\frac{\mu_{\mathrm{ref}}}{\mu_{10}}\right) + 2\ell^* - \Phi_{\mu_1}(g_{10}, \alpha_0)\,. \tag{9.13.15}$$

Inserting this result into (9.13.4), and using (9.12.9) and (9.12.12), gives

$$\tau = \frac{\xi_\perp^2}{\mu_{10}} e^{-\Phi_{\mu_1}}\,. \tag{9.13.16}$$

If the bare parameter g_{10} is small, then, up to factors of $\mathcal{O}(1)$, we can take Φ and Φ_{μ_1} to be zero, which reduces (9.13.12) to

$$\xi_x = \xi_\perp \sqrt{\frac{\mu_{x0}}{\mu_{10}}}\,, \tag{9.13.17}$$

and (9.13.16) to

$$\tau = \frac{\xi_\perp^2}{\mu_{10}}\,. \tag{9.13.18}$$

We can also determine $\xi_\perp$ in this limit by noting that, for small g_1, the recursion relation (9.16.51) for g_1 becomes simply

$$\frac{dg_1}{d\ell} = \epsilon g_1\,, \tag{9.13.19}$$

which is easily solved to give

$$g_1(\ell) = g_{10} e^{\epsilon \ell}\,. \tag{9.13.20}$$

Setting $g_1(\ell^*) = 1$ and solving for e^{ℓ^*}, gives

$$e^{\ell^*} = (g_{10})^{-\frac{1}{\epsilon}}\,. \tag{9.13.21}$$

Using our expression (9.8.5) for g_1, evaluated with the bare parameters, this gives

$$e^{\ell^*} = \Lambda \left(\frac{\mu_{x0}\mu_{10}^5}{D_0^2 \lambda_0^4}\right)^{\frac{1}{2\epsilon}}\,. \tag{9.13.22}$$

Using this in turn in (9.12.9) gives

$$\xi_\perp = \left(\frac{\mu_x^0 \mu_{10}^5}{D_0^2 \lambda_0^4}\right)^{\frac{1}{2\epsilon}}\,. \tag{9.13.23}$$

Note that $\xi_\perp$ is independent of the ultraviolet cutoff Λ in this case, which is to be expected, since the divergent renormalization of the parameters is an infrared phenomenon. Of course, there is still implicit dependence of this result on short-scale physics, since it is such physics that determines the bare parameters μ_x^0, μ_{10}, D_0, and λ_0.

In $d = 3$, where $\epsilon = 1$, this becomes

$$\xi_\perp = \frac{\sqrt{\mu_{x0}\mu_{10}^5}}{D_0\lambda_0^2}, \quad d = 3. \tag{9.13.24}$$

Using this in (9.13.17) and (9.13.18) gives respectively

$$\xi_x = \frac{\mu_{x0}\mu_{10}^2}{D_0\lambda_0^2}, \quad \tau = \frac{\mu_{x0}\mu_{10}^4}{D_0^2\lambda_0^4}, \quad d = 3. \tag{9.13.25}$$

We now turn to the last remaining concern about the scaling form of the correlation function C_u. We will now show that this is also independent of the arbitrary rescaling choices, and we'll also calculate it for small bare coupling g_{10}.

To do this, we see from (9.12.6) that we need to calculate the value of the integral $\int_0^{\ell^*} 2\chi(\ell)d\ell$. We can obtain this by integrating the recursion relation (9.16.39) for the noise strength D from $\ell = 0$ to ℓ^*, and using the fact that $D(\ell^*) = D_{\text{ref}}$. This gives

$$\ln\left(\frac{D_{\text{ref}}}{D_0}\right) = \int_0^{\ell^*} [z(\ell) - 2\chi(\ell) - \zeta(\ell)]d\ell + (1-d)\ell^* + \Phi_D(g_{10}, \alpha_0), \tag{9.13.26}$$

where we've defined

$$\Phi_D(g_{10}, \alpha_0) \equiv \int_0^{\ell^*} g_1(\ell)G_D(\alpha(\ell))d\ell, \tag{9.13.27}$$

which again, like all of our Φ's, depends *only* on the bare coupling g_{10} and the bare ratio α_0.

Solving (9.13.26) for the integral $\int_0^{\ell^*} 2\chi(\ell)d\ell$, gives

$$\int_0^{\ell^*} 2\chi(\ell)d\ell = \int_0^{\ell^*} [z(\ell) - \zeta(\ell)]d\ell + \ln\left(\frac{D_0}{D_{\text{ref}}}\right) + (1-d)\ell^* + \Phi_D(g_{10}, \alpha_0). \tag{9.13.28}$$

Using our results (9.13.15) and (9.13.11) for $\int_0^{\ell^*} z(\ell)d\ell$ and $\int_0^{\ell^*} \zeta(\ell)d\ell$ respectively, this becomes

$$\int_0^{\ell^*} 2\chi(\ell)d\ell = \ln\left(\frac{D_0\mu_{\text{ref}}}{\sqrt{\mu_{10}\mu_{x0}}D_{\text{ref}}}\right) + (2-d)\ell^* + \Phi_A(g_{10}, \alpha_0), \tag{9.13.29}$$

where we've defined

$$\Phi_A \equiv \Phi_D - (\Phi_{\mu_1} + \Phi). \tag{9.13.30}$$

Using this and our expression (9.12.9) for $\xi_\perp$ in equation (9.12.8) for A gives

$$A = a^{-2\xi_{FP}} \left(\frac{\xi_\perp}{a}\right)^{(2-d-2\chi_{\rm FP})} \frac{\mu_{\rm ref} D_0}{\sqrt{\mu_{10}\mu_{x0}} D_{\rm ref}} e^{\Phi_A} . \tag{9.13.31}$$

It is clear from this expression that, as claimed earlier, the value of A depends only on the parameters of the bare model, not on the arbitrary choice of rescaling exponents.

For a model with the bare coupling $g_{10} \ll 1$, we can set $\Phi_A = 0$ in (9.13.31), and obtain an explicit expression for A in terms of the bare parameters:

$$A = a^{-2\xi_{FP}} \left(\frac{\xi_\perp}{a}\right)^{(2-d-2\chi_{\rm FP})} \frac{\mu_{\rm ref} D_0}{\sqrt{\mu_{10}\mu_{x0}} D_{\rm ref}} . \tag{9.13.32}$$

Arguments virtually identical to those just presented can be used to show the scaling form of the density correlation function given by (9.17.11).

The lengths $\xi_\perp$, and ξ_x, and the time τ have significance beyond their appearance in the scaling laws (9.17.10) and (9.17.11): they are also the nonlinear lengths and time. By this, we mean that, if all of the distances $r_\perp$, $|x - \gamma t|$, and the time t, are much smaller than the corresponding length or time – that is, if $r_\perp \ll \xi_\perp$, $|x - \gamma t| \ll \xi_x$, and $t \ll \tau$, the linear theory results of Section 9.2 will apply. This can be seen either by a renormalization group argument, or perturbation theory.

The renormalization group argument starts with the general trajectory integral matching expression (9.12.2). We then note that, if all the conditions $r_\perp \ll \xi_\perp$, $|x - \gamma t| \ll \xi_x$, and $t \ll \tau$ are satisfied, we can always choose to evaluate the right-hand side at a value of $\ell < \ell^*$ at which all the arguments $r_\perp e^{-\ell}$, $(x - \gamma t) \exp\left(-\int_0^\ell \zeta(\ell') d\ell'\right)$, and $t \exp\left(-\int_0^\ell z(\ell')\right)$ are microscopic. The C_u on the right-hand side of (9.12.2) can then simply be treated as a finite constant since it is evaluated at short distances and times, and so will be unaffected by any infrared divergences.

We'll now illustrate this for the special case $r_\perp \ll \xi_\perp$, $x = 0$, $t = 0$. Again we choose $\ell = \ln(r_\perp/a)$. We will also choose χ, z, and ζ to keep $\mu_{1,2,x}$ and D fixed at their initial values. Since $\ell \ll \ell^*$ and $r_\perp \ll \xi_\perp$, $g_1(\ell)$ is small if g_{10} is small. Therefore, the graphical corrections in (9.16.39), (9.16.42), and (9.16.43) are negligible. Then our special choices of the scaling exponents become

$$\chi = \frac{2-d}{2}, \quad z = 2, \quad \zeta = 1 . \tag{9.13.33}$$

Inserting these exponents and $\ell = \ln(r_\perp/a)$ into (9.12.2), we obtain

$$\begin{aligned} C_u\Big(r_\perp, 0, 0; \{D_0, \mu_{x0}, \mu_{10}, \mu_{20}\}\Big) &= \left(\frac{r_\perp}{a}\right)^{2-d} C_u\left(a, 0, 0; \{D_0, \mu_{x0}, \mu_{10}, \mu_{20}\}\right) \\ &\propto r_\perp^{2-d} , \end{aligned} \tag{9.13.34}$$

which agrees with the linear theory (9.2.34). This argument can be easily extended to correlation functions with more-general spatial and temporal separations, and to the density–density correlation function. Therefore, the conclusion that the linear theory results of Section 9.2, in particular equations (9.2.34) and (9.2.43) for the velocity–velocity and density–density correlation functions, hold if all distances and times are short compared with the corresponding nonlinear lengths or times; that is, if the conditions $r_\perp \ll \xi_\perp$, $|x - \gamma t| \ll \xi_x$, and $t \ll \tau$ are satisfied.

9.13.2 Perturbation Theory Approach

In this subsection, we present the perturbation theory approach to the calculation of the nonlinear lengths and time. We remind the reader that we do this by calculating perturbative corrections to the linear theory and finding the length and time scales on which they become appreciable. These prove to be precisely the lengths $\xi_\perp$, and ξ_x, and the time τ that we have just derived from the RG approach, thereby confirming the validity of that approach.

The perturbation theory calculation, as discussed in Chapter 3, can be represented by graphs. Here we focus on the correction to μ_1 obtained from one particular graph, Figure 9.6.1(a), which we have evaluated in detail in Section 9.6.1 and [22, 23]. Using different one-loop graphs, or considering renormalization of different parameters, will lead to the same estimates of the nonlinear lengths and times, up to factors of $O(1)$. We also simplify our calculation by considering the case $\mu_2 = 0$; taking a nonzero μ_2 only modifies the lengths $\xi_\perp$, and ξ_x, and the time τ by an $O(1)$ multiplicative factor.

Our strategy is to estimate crudely the nonlinear lengths $\xi_\perp$ and ξ_x, and the nonlinear time τ by using their inverses as infrared cutoffs of the infrared divergent integrals that appear in a perturbation theory calculation of the renormalized μ_1. We'll then determine the values of $\xi_\perp$, ξ_x, and τ as the values of these cutoffs for which the correction to μ_1 becomes comparable to its bare value μ_{10}. As mentioned earlier, we would get the same values for $\xi_\perp$, ξ_x, and τ had we chosen to apply this logic to one of the parameters (i.e., D or μ_x) instead.

The graph Figure 9.6.1(a) represents a correction to $\partial_t u_j^\perp$ of the form

$$\Delta(\partial_t u_j^\perp)_{\mu,a} = -\frac{2D_0\lambda_0^2 k_u^\perp u_c^\perp(\tilde{\mathbf{k}})}{(2\pi)^{d+1}} \int_{\tilde{q}} (k_i^\perp - q_i^\perp) C_{iu}(\tilde{\mathbf{q}}) G_{jc}(\tilde{\mathbf{k}} - \tilde{\mathbf{q}})$$
$$\equiv -2D_0\lambda_0^2 k_u^\perp u_c^\perp(\tilde{\mathbf{k}}) \left[(I_1^{\mu,a})_{cju}(\tilde{\mathbf{k}}) - (I_2^{\mu,a})_{cju}(\tilde{\mathbf{k}})\right], \quad (9.13.35)$$

where $\tilde{\mathbf{k}} = (\omega, \mathbf{k})$, $\tilde{\mathbf{q}} = (\Omega, \mathbf{q})$,

$$(I_1^{\mu,a})_{cju}(\tilde{\mathbf{k}}) \equiv \frac{k_i^\perp}{(2\pi)^{d+1}} \int_{\tilde{q}} C_{iu}(\tilde{\mathbf{q}}) G_{jc}(\tilde{\mathbf{k}} - \tilde{\mathbf{q}}), \quad (9.13.36)$$

$$(I_2^{\mu,a})_{cju}(\tilde{\mathbf{k}}) \equiv \frac{1}{(2\pi)^{d+1}} \int_{\tilde{q}} q_i^{\perp} C_{iu}(\tilde{\mathbf{q}}) G_{jc}(\tilde{\mathbf{k}} - \tilde{\mathbf{q}}), \tag{9.13.37}$$

$$G_{jc}(\tilde{\mathbf{q}}) \equiv G_T(\tilde{\mathbf{q}})\delta_{jc}^{\perp} = \frac{\delta_{jc}^{\perp}}{-\mathrm{i}\Omega + \mu_{10}q_{\perp}^2 + \mu_{x0}q_x^2}, \tag{9.13.38}$$

$$C_{iu}(\tilde{\mathbf{q}}) \equiv \mid G_T(\tilde{\mathbf{q}}) \mid^2 \delta_{iu}^{\perp} = \frac{\delta_{iu}^{\perp}}{\Omega^2 + \left(\mu_{10}q_{\perp}^2 + \mu_{x0}q_x^2\right)^2}, \tag{9.13.39}$$

and the superscripts "μ, a" indicate that this correction comes from the renormalization of the μ terms due to Figure 9.6.1(a). Note that we have replaced all of the parameters λ, D, μ_1, and μ_x by their bare values λ_0, D_0, μ_{10}, and μ_{x0}, since we are now doing perturbation theory, rather than the renormalization group.

Inserting (9.13.38) and (9.13.39) into (9.13.36) and (9.13.37), we get

$$(I_1^{\mu,a})_{cju}(\tilde{\mathbf{k}}) = \frac{k_u^{\perp}\delta_{jc}^{\perp}}{(2\pi)^{d+1}} \int_{\tilde{\mathbf{q}}} \mid G_T(\tilde{\mathbf{q}}) \mid^2 G_T(\tilde{\mathbf{k}} - \tilde{\mathbf{q}}), \tag{9.13.40}$$

$$(I_2^{\mu,a})_{cju}(\tilde{\mathbf{k}}) = \frac{\delta_{jc}^{\perp}}{(2\pi)^{d+1}} \int_{\tilde{\mathbf{q}}} q_u^{\perp} \mid G_T(\tilde{\mathbf{q}}) \mid^2 G_T(\tilde{\mathbf{k}} - \tilde{\mathbf{q}}). \tag{9.13.41}$$

Since (9.13.35) has already a factor $k_u^{\perp}$ in front of it, and we are only interested in terms of $O(k^2)$ (since only these will be relevant at small $\mathbf{k}$, that being the order of the μ_1 terms in the equation of motion), to get relevant contributions to the linear terms of the equation of motion we can set the external frequency $\omega = 0$ in $(I_{1,2}^a)_{cju}(\tilde{\mathbf{k}})$, and expand both of them to $O(k)$. This gives

$$\begin{aligned}(I_1^{\mu,a})_{cju}(\tilde{\mathbf{k}}) &= \frac{k_u^{\perp}\delta_{jc}^{\perp}}{(2\pi)^{d+1}} \int_{\tilde{\mathbf{q}}} \mid G_T(\tilde{\mathbf{q}}) \mid^2 G_T(-\tilde{\mathbf{q}}) \\ &= \frac{k_u^{\perp}\delta_{jc}^{\perp}}{(2\pi)^{d+1}} \int_{\tilde{\mathbf{q}}} \frac{1}{(\Omega^2 + \left(\mu_{10}q_{\perp}^2 + \mu_{x0}q_x^2\right)^2)(\mathrm{i}\Omega + \mu_{10}q_{\perp}^2 + \mu_{x0}q_x^2)},\end{aligned} \tag{9.13.42}$$

$$\begin{aligned}(I_2^{\mu,a})_{cju}(\tilde{\mathbf{k}}) &= \frac{2\mu_{10}k_\ell^{\perp}\delta_{jc}^{\perp}}{(2\pi)^{d+1}} \int_{\tilde{\mathbf{q}}} q_u^{\perp}q_\ell^{\perp} \mid G_T(\tilde{\mathbf{q}}) \mid^2 [G_T(-\tilde{\mathbf{q}})]^2 \\ &= \frac{2\mu_{10}k_\ell^{\perp}\delta_{jc}^{\perp}}{(2\pi)^{d+1}} \int_{\tilde{\mathbf{q}}} \frac{q_u^{\perp}q_\ell^{\perp}}{\left(\Omega^2 + \left(\mu_{10}q_{\perp}^2 + \mu_{x0}q_x^2\right)^2\right)^2}.\end{aligned} \tag{9.13.43}$$

To calculate the nonlinear length along $\perp$ directions, $\xi_\perp$, we impose an infrared cutoff $|\mathbf{q}|_{\min} = \xi_\perp^{-1}$ on the $\mathbf{q}_\perp$ integrals in this expression. That is, we define

$$\int_{\tilde{\mathbf{q}}} \equiv \int_{-\infty}^{\infty} \mathrm{d}\Omega \int_{-\infty}^{\infty} \mathrm{d}q_x \int_{\Lambda > |\mathbf{q}_\perp| > \xi_\perp^{-1}} \mathrm{d}^{d-1}q_\perp . \tag{9.13.44}$$

The integrals over Ω and q_x in (9.13.42) and (9.13.43) are straightforward, particularly if done in that order (i.e., integrating first over Ω, then over q_x). The results are:

$$(I_1^{\mu,a})_{cju}(\tilde{\mathbf{k}}) = \frac{k_u^{\perp}\delta_{jc}^{\perp}}{16\sqrt{\mu_{x0}\mu_{10}^3}}\frac{1}{(2\pi)^{d-1}}\int_{\Lambda>|\mathbf{q}_\perp|>\xi_\perp^{-1}}\frac{\mathrm{d}^{d-1}q_\perp}{q_\perp^3}$$
$$= \frac{k_u^{\perp}\delta_{jc}^{\perp}}{16\sqrt{\mu_{x0}\mu_{10}^3}}\frac{S_{d-1}}{(2\pi)^{d-1}}\frac{1}{4-d}\left(\xi_\perp^{4-d} - \Lambda^{d-4}\right)$$
$$\approx \frac{k_u^{\perp}\delta_{jc}^{\perp}}{16\sqrt{\mu_{x0}\mu_{10}^3}}\frac{S_{d-1}}{(2\pi)^{d-1}}\frac{\xi_\perp^{4-d}}{4-d}, \tag{9.13.45}$$

and

$$(I_2^{\mu,a})_{cju}(\tilde{\mathbf{k}}) = \frac{3k_\ell^{\perp}\delta_{jc}^{\perp}}{32\sqrt{\mu_{x0}\mu_{10}^3}}\frac{1}{(2\pi)^{d-1}}\int_{\Lambda>|\mathbf{q}_\perp|>\xi_\perp^{-1}}\frac{q_u^{\perp}q_\ell^{\perp}\mathrm{d}^{d-1}q_\perp}{q_\perp^5}$$
$$= \frac{3k_u^{\perp}\delta_{jc}^{\perp}}{32(d-1)\sqrt{\mu_{x0}\mu_{10}^3}}\frac{1}{(2\pi)^{d-1}}\int_{\Lambda>|\mathbf{q}_\perp|>\xi_\perp^{-1}}\frac{\mathrm{d}^{d-1}q_\perp}{q_\perp^3}$$
$$\approx \frac{3k_u^{\perp}\delta_{jc}^{\perp}}{32(d-1)\sqrt{\mu_{x0}\mu_{10}^3}}\frac{S_{d-1}}{(2\pi)^{d-1}}\frac{\xi_\perp^{4-d}}{4-d}, \tag{9.13.46}$$

where, in the penultimate equality, we have used

$$\int \mathrm{d}\Xi_{\mathbf{q}_\perp}\, q_u^{\perp}q_\ell^{\perp} = S_{d-1}\frac{q_\perp^2}{d-1}, \tag{9.13.47}$$

where $\int \mathrm{d}\Xi_{\mathbf{q}_\perp}$ denotes an integral over the $(d-1)$-dimensional solid angle associated with $\mathbf{q}_\perp$. In the ultimate equality, we have used $\xi_\perp^{-1} \ll \Lambda$.

Inserting (9.13.45) and (9.13.46) into (9.13.35), we obtain a correction to $\partial_t u_j$ given by

$$\Delta(\partial_t u_j)_{\mu,a} = -\frac{D_0\lambda_0^2}{\sqrt{\mu_{x0}\mu_{10}^3}}\frac{S_{d-1}}{(2\pi)^{d-1}}\frac{\xi_\perp^{4-d}}{4-d}\left[\frac{1}{8} - \frac{3}{16(d-1)}\right]k_\perp^2 u_j\,. \tag{9.13.48}$$

From the form of this correction (i.e., the fact that it is proportional to $k_\perp^2 u_c$), we recognize this as a perturbative correction to μ_1:

$$(\Delta\mu_1)_{\mu,a} = \frac{D_0\lambda_0^2}{\sqrt{\mu_{x0}\mu_{10}^3}}\frac{S_{d-1}}{(2\pi)^{d-1}}\frac{\xi_\perp^{4-d}}{4-d}\left[\frac{1}{8} - \frac{3}{16(d-1)}\right]. \tag{9.13.49}$$

Equating this correction to the bare μ_{10} gives, ignoring factors of $O(1)$,

$$\xi_\perp \propto \left(\frac{\mu_{x0}^0 \mu_{10}^5}{D_0^2 \lambda_0^4}\right)^{\frac{1}{2(4-d)}} . \tag{9.13.50}$$

This agrees with our earlier RG result (9.13.23).

To calculate the nonlinear length ξ_x along x, we now introduce ξ_x^{-1} as an infrared cutoff on the integrals over q_x, and allow $\mathbf{q}_\perp$ and Ω to run free. That is, we set the limits on our integrals as follows:

$$\int_{\tilde{\mathbf{q}}} \equiv 2\int_{-\infty}^{\infty} \mathrm{d}\Omega \int_{|\mathbf{q}_\perp|<\infty} \mathrm{d}^{d-1} q_\perp \int_{\xi_x^{-1}}^{\infty} \mathrm{d}q_x \,, \tag{9.13.51}$$

where the factor of 2 takes into account the fact that the Brillouin zone in q_x, with this infrared cutoff, consists of two disjoint sections, one running from ξ^{-1} to ∞, the other running from $-\infty$ to $-\xi^{-1}$. These two regions make exactly equal contributions; hence the factor of 2 above.

Then we obtain for the integrals in (9.13.42) and (9.13.43),

$$\begin{aligned}
(I_1^{\mu,a})_{cju}(\tilde{\mathbf{k}}) &= \frac{k_u^\perp \delta_{jc}^\perp}{2(2\pi)^d} \int_{\xi_x^{-1}}^{\infty} \mathrm{d}q_x \int \frac{\mathrm{d}^{d-1} q_\perp}{\left(\mu_{x0} q_x^2 + \mu_{10} q_\perp^2\right)^2} \\
&= \frac{k_u^\perp \delta_{jc}^\perp S_{d-1}}{(2\pi)^d} \frac{H_1(d)}{2} \mu_{x0}^{\frac{d-5}{2}} \mu_{10}^{\frac{1-d}{2}} \int_{\xi_x^{-1}}^{\infty} q_x^{d-5} \mathrm{d}q_x \\
&= k_u^\perp \delta_{jc}^\perp \frac{S_{d-1}}{(2\pi)^d} \frac{H_1(d)}{2} \mu_{x0}^{\frac{d-5}{2}} \mu_{10}^{\frac{1-d}{2}} \frac{\xi_x^{4-d}}{4-d}
\end{aligned} \tag{9.13.52}$$

and

$$\begin{aligned}
(I_2^{\mu,a})_{cju}(\tilde{\mathbf{k}}) &= \frac{\mu_{10} k_\ell^\perp \delta_{jc}^\perp}{2(2\pi)^d} \int_{\xi_x^{-1}}^{\infty} \mathrm{d}q_x \int_{|\mathbf{q}_\perp|<\infty} \frac{q_u^\perp q_\ell^\perp \mathrm{d}^{d-1} q_\perp}{\left(\mu_{x0} q_x^2 + \mu_{10} q_\perp^2\right)^3} \\
&= \frac{\mu_{10} k_u^\perp \delta_{jc}^\perp}{2(d-1)(2\pi)^d} \int_{\xi_x^{-1}}^{\infty} \mathrm{d}q_x \int_{|\mathbf{q}_\perp|<\infty} \frac{q_\perp^2 \mathrm{d}^{d-1} q_\perp}{\left(\mu_{x0} q_x^2 + \mu_{10} q_\perp^2\right)^3} \\
&= \frac{S_{d-1} \mu_{10} k_u^\perp \delta_{jc}^\perp}{(2\pi)^d} \frac{H_2(d)}{2(d-1)} \mu_{x0}^{\frac{d-5}{2}} \mu_{10}^{\frac{1-d}{2}} \int_{\xi_x^{-1}}^{\infty} q_x^{d-5} \mathrm{d}q_x \\
&= \frac{S_{d-1} \mu_{10} k_u^\perp \delta_{jc}^\perp}{(2\pi)^d} \frac{H_2(d)}{2(d-1)} \mu_{x0}^{\frac{d-5}{2}} \mu_{10}^{\frac{1-d}{2}} \frac{\xi_x^{4-d}}{4-d} ,
\end{aligned} \tag{9.13.53}$$

where $H_{1,2}(d)$ are finite, $O(1)$ constants given by

$$H_1(d) \equiv \int_0^{\infty} \frac{y^{d-2} \mathrm{d}y}{\left(1+y^2\right)^2} = \frac{\pi}{4}(d-3) \sec\left(\frac{\pi d}{2}\right) \tag{9.13.54}$$

and

$$H_2(d) \equiv \int_0^\infty \frac{y^d \mathrm{d}y}{\left(1+y^2\right)^3} = \frac{\pi}{16}(d-3)(1-d)\sec\left(\frac{\pi d}{2}\right). \tag{9.13.55}$$

Note that, despite appearances to the contrary, neither $H_1(d=3)$ nor $H_2(d=3)$ vanishes; instead $H_1(3)=1/2$ and $H_2(3)=1/4$, as can be verified either by taking the singular limit $d \to 3$ in (9.13.54) and (9.13.55), or by evaluating the corresponding integrals in exactly $d=3$.

Inserting (9.13.52) and (9.13.53) into (9.13.35), we obtain the perturbative correction to μ_1:

$$(\Delta\mu_1)_{\mu,a} = D_0\lambda_0^2 \mu_{x0}^{\frac{d-5}{2}} \mu_{10}^{\frac{1-d}{2}} \frac{S_{d-1}}{2(2\pi)^d} \frac{\xi_x^{4-d}}{4-d}\left[H_1(d) - \frac{H_2(d)}{(d-1)}\right]. \tag{9.13.56}$$

Equating this correction to μ_1 (9.13.56) to its bare value μ_{10}, gives

$$\xi_x = \left(\frac{\mu_{x0}\mu_{10}^5}{D_0^2\lambda_0^4}\right)^{\frac{1}{2(4-d)}} \sqrt{\frac{\mu_{x0}}{\mu_{10}}} \times O(1) = \xi_\perp \sqrt{\frac{\mu_{x0}}{\mu_{10}}} \times O(1), \tag{9.13.57}$$

which agrees with our earlier RG result (9.13.17).

Now we turn to the nonlinear time scale τ. We impose a lower limit $1/\tau$ on the frequency integral in (9.13.35) and let the wave vectors be completely free:

$$\int_{\tilde{\mathbf{q}}} \equiv 2\int_{\tau^{-1}}^\infty \mathrm{d}\Omega \int_{|\mathbf{q}_\perp|<\infty} \mathrm{d}^{d-1}q_\perp \int_{-\infty}^\infty \mathrm{d}q_x, \tag{9.13.58}$$

where, much as in our treatment of the integral over q_x earlier, the factor of 2 takes into account the fact that the region of integration over Ω, with this infrared cutoff, consists of two disjoint sections, one running from τ^{-1} to ∞, the other running from $-\infty$ to $-\tau^{-1}$. These two regions also make exactly equal contributions; hence the factor of 2 above.

In this case we do the integral over wave vectors first. We obtain

$$\begin{aligned}(I_1^{\mu,a})_{cju}(\tilde{\mathbf{k}}) &= \frac{2H_3(d)}{(2\pi)^{d+1}} k_u^\perp \delta_{jc}^\perp \mu_{x0}^{-\frac{1}{2}} \mu_{10}^{\frac{1-d}{2}} \int_{\frac{1}{\tau}}^\infty \omega^{\frac{d}{2}-3}\mathrm{d}\omega \\ &= \frac{4H_3(d)}{(2\pi)^{d+1}} k_u^\perp \delta_{jc}^\perp \mu_{x0}^{-\frac{1}{2}} \mu_{10}^{\frac{1-d}{2}} \frac{\tau^{\frac{4-d}{2}}}{4-d}\end{aligned} \tag{9.13.59}$$

and

$$\begin{aligned}(I_2^{\mu,a})_{cju}(\tilde{\mathbf{k}}) &= \frac{4H_3(d)}{(2\pi)^{d+1}d} k_u^\perp \delta_{jc}^\perp \mu_{x0}^{-\frac{1}{2}} \mu_{10}^{\frac{1-d}{2}} \int_{\frac{1}{\tau}}^\infty \omega^{\frac{d}{2}-3}\mathrm{d}\omega \\ &= \frac{8H_3(d)}{(2\pi)^{d+1}d} k_u^\perp \delta_{jc}^\perp \mu_{x0}^{-\frac{1}{2}} \mu_{10}^{\frac{1-d}{2}} \frac{\tau^{\frac{4-d}{2}}}{4-d},\end{aligned} \tag{9.13.60}$$

where $H_3(d)$ is a finite, $O(1)$ constant given by

$$H_3(d) \equiv \int_{|\mathbf{Q}|<\infty} \frac{Q^2 d^d Q}{\left(1+Q^4\right)^2} = \frac{(2-d)}{16} S_d \pi \sec\left(\frac{\pi d}{4}\right). \tag{9.13.61}$$

Inserting (9.13.59) and (9.13.60) into (9.13.35), we obtain the perturbative correction to μ_1:

$$(\Delta\mu_1)_{\mu,a} = \frac{8D_0\lambda_0^2 \mu_{x0}^{-\frac{1}{2}} \mu_{10}^{\frac{1-d}{2}}}{(2\pi)^{d+1}} \frac{\tau^{\frac{4-d}{2}}}{4-d} \left(\frac{d-2}{d}\right) H_3(d). \tag{9.13.62}$$

Equating this to μ_{10} gives

$$\tau = \left(\frac{\mu_{x0}\mu_{10}^5}{D_0^2\lambda_0^4}\right)^{\frac{1}{(4-d)}} \frac{1}{\mu_{10}} \times O(1) = \frac{\xi_\perp^2}{\mu_{10}} \times O(1), \tag{9.13.63}$$

which agrees with (9.13.18). This completes our calculation of the crossover length and time scales between linear and nonlinear theories using perturbation theory.

9.14 Universal Amplitude Ratio

The fact that there is a fixed point value of α, even if it's not zero at higher-loop orders, implies an experimentally observable universal amplitude ratio. Specifically, it is the ratio of the damping of the transverse and the longitudinal modes, obtained as follows. For the longitudinal mode, we look at the equal-time correlation

$$C_L(t=0,\mathbf{k}) = \frac{D(\mathbf{k})}{\mu_L(\mathbf{k})k_\perp^2 + \mu_x(\mathbf{k})k_x^2}, \tag{9.14.1}$$

where we have explicitly shown the dependencies of the coefficients on the wavevector $\mathbf{k}$. A similar expression can be obtained for C_T.

If one considers a generic direction of $\mathbf{k}$ (i.e., any direction for which $k_\perp^\zeta \geq k_x$), we can ignore the second term in the above denominator and the ratio of $C_T(t=0,\mathbf{k})/C_L(t=0,\mathbf{k})$ is thus

$$\lim_{\mathbf{k}_\perp\to 0} \frac{\mu_L(\mathbf{k})}{\mu_1(\mathbf{k})} = 1+\alpha^* = 1+O(\epsilon), \tag{9.14.2}$$

which is a universal number.

9.15 Separatrix Between Positive and Negative Density Correlations

In Section 9.2.3 we have shown by using linear theory that the sign of the equal-time density correlation function depends on the spatial difference between the two correlating points **r**. Specifically, in **r**-space the positive and the negative regions of the density correlations are separated by a cone-shaped locus given by (9.2.41), which, up to a $O(1)$ factor, can be rewritten in terms of the nonlinear lengths as

$$\frac{|x|}{\xi_x} = \frac{r_\perp}{\xi_\perp}. \tag{9.15.1}$$

For $x/\xi_x \gg r_\perp/\xi_\perp$ the density correlations are positive, while for $x/\xi_x \ll r_\perp/\xi_\perp$ they become negative. This result only holds for small distances (i.e., $x \ll \xi_x$ and $r_\perp \ll \xi_\perp$) since linear theory is only valid at short length scales.

At large distances we expect a similar separatrix in **r**-space which separates the regions with different signs of the density correlations. The scaling form of these correlations (9.17.11) shows that the sign of the equal-time correlations is determined by the ratio $\frac{|x|/\xi_x}{(r_\perp/\xi_\perp)^\zeta}$ instead of $\frac{|x|/\xi_x}{(r_\perp/\xi_\perp)}$. This implies at large distances (i.e., $x \gg \xi_x$ or $r_\perp \gg \xi_\perp$) the positive and the negative density correlations are separated by a locus given by

$$\frac{|x|}{\xi_x} = \left(\frac{r_\perp}{\xi_\perp}\right)^\zeta. \tag{9.15.2}$$

Note that (9.15.1) and (9.15.2) connect right at $|x| = \xi_x$.

The regions in **r**-space with different signs of the density correlations are illustrated in Figure 9.17.1.

9.16 One-Loop Graphical Corrections with Nonzero μ_2

We now turn to the calculation of the graphical corrections to the parameters for the full model with $\mu_2 \neq 0$. The reasoning of this section is exactly the same as that of the previous section; the only difference is that the algebra is complicated (boy, is it *ever* complicated!) by the nonzero value of μ_2.

Fortunately, it turns out that this complication vanishes upon renormalization (at least to one-loop order) because the $\mu_2 = 0$ fixed point proves to be stable to one-loop order. Furthermore, it proves to be the only stable fixed point in the problem, at least to one-loop order.

The origin of this complication lies in the more complicated form of the propagators and correlation functions. Instead of the simple, diagonal expressions (9.3.14) and (9.3.15) that we have when $\mu_2 = 0$, we now, as shown in Section 9.2, have, for the propagators,

$$G_{ij}(\tilde{\mathbf{k}}) \equiv L_{ij}^\perp(\mathbf{k}_\perp)G_L(\tilde{\mathbf{k}}) + P_{ij}^\perp(\mathbf{k}_\perp)G_T(\tilde{\mathbf{k}}), \tag{9.16.1}$$

with

$$G_L(\tilde{\mathbf{k}}) = \frac{1}{-i\omega + \mu_L k_\perp^2 + \mu_x k_x^2}, \tag{9.16.2}$$

$$G_T(\tilde{\mathbf{k}}) = \frac{1}{-\mathrm{i}\omega + \mu_1 k_\perp^2 + \mu_x k_x^2}\,, \tag{9.16.3}$$

and the longitudinal and transverse projection operators $L_{ij}^{\perp}(\mathbf{k}_\perp)$ and $P_{ij}^{\perp}(\mathbf{k}_\perp)$, respectively, were defined earlier.

We now also have similar decompositions for the correlation functions:

$$C_{ij}(\tilde{\mathbf{k}}) \equiv L_{ij}^{\perp}(\mathbf{k})|G_L(\tilde{\mathbf{k}})|^2 + P_{ij}^{\perp}(\mathbf{k})|G_T(\tilde{\mathbf{k}})|^2\,. \tag{9.16.4}$$

The expansions in powers of external momentum $\mathbf{k}$ that we have to perform now become much more complicated. Expanding the propagator factors $G_{L,T}(\tilde{\mathbf{q}} - \tilde{\mathbf{k}})$ is as straightforward as before:

$$\begin{aligned}
G_L(\tilde{\mathbf{k}} - \tilde{\mathbf{q}}) &= \frac{1}{\mathrm{i}\omega + \mu_L(\mathbf{k}_\perp - \mathbf{q}_\perp)^2 + \mu_x(k_x - q_x)^2} \\
&= \frac{1}{\mathrm{i}(\omega - \Omega) + \Gamma_L(\mathbf{q})}\left(1 + \left(\frac{2\mu_L \mathbf{q}_\perp \cdot \mathbf{k}_\perp + 2\mu_x q_x k_x - \mu_L k_\perp^2 - \mu_x k_x^2}{\mathrm{i}(\omega - \Omega) + \Gamma_L(\mathbf{q})}\right)\right. \\
&\quad \left. + \frac{4\mu_L^2(\mathbf{q}_\perp \cdot \mathbf{k}_\perp)^2 + 4\mu_x^2 q_x^2 k_x^2 + 4\mu_x\mu_L(\mathbf{q}_\perp \cdot \mathbf{k}_\perp)q_x k_x}{(\mathrm{i}(\omega - \Omega) + \Gamma_L(\mathbf{q}))^2} + \mathcal{O}(k^3)\right) \\
&= G_L(-\tilde{\mathbf{q}}) + (2\mu_L \mathbf{q}_\perp \cdot \mathbf{k}_\perp + 2\mu_x q_x k_x)\, G_L(-\tilde{\mathbf{q}})^2 - \left(\mu_L k_\perp^2 + \mu_x k_x^2\right) G_L(-\tilde{\mathbf{q}})^2 \\
&\quad + \left[(4\mu_L^2(\mathbf{q}_\perp \cdot \mathbf{k}_\perp)^2 + 4\mu_x^2 q_x^2 k_x^2 + 4\mu_x\mu_L(\mathbf{q}_\perp \cdot \mathbf{k}_\perp)q_x k_x\right] G_L(-\tilde{\mathbf{q}})^3 + \mathcal{O}(k^3),
\end{aligned} \tag{9.16.5}$$

and

$$\begin{aligned}
G_T(\tilde{\mathbf{k}} - \tilde{\mathbf{q}}) &= \frac{1}{\mathrm{i}\omega + \mu_1(\mathbf{k}_\perp - \mathbf{q}_\perp)^2 + \mu_x(k_x - q_x)^2} \\
&= \frac{1}{\mathrm{i}(\omega - \Omega) + \Gamma_L(\mathbf{q})}\left(1 + \left(\frac{2\mu_1 \mathbf{q}_\perp \cdot \mathbf{k}_\perp + 2\mu_x q_x k_x - \mu_1 k_\perp^2 - \mu_x k_x^2}{\mathrm{i}(\omega - \Omega) + \Gamma_L(\mathbf{q})}\right)\right. \\
&\quad \left. + \frac{4\mu_1^2(\mathbf{q}_\perp \cdot \mathbf{k}_\perp)^2 + 4\mu_x^2 q_x^2 k_x^2 + 4\mu_x\mu_1(\mathbf{q}_\perp \cdot \mathbf{k}_\perp)q_x k_x}{(\mathrm{i}(\omega - \Omega) + \Gamma_L(\mathbf{q}))^2} + \mathcal{O}(k^3)\right) \\
&= G_L(-\tilde{\mathbf{q}}) + (2\mu_1 \mathbf{q}_\perp \cdot \mathbf{k}_\perp + 2\mu_x q_x k_x)\, G_L(-\tilde{\mathbf{q}})^2 - \left(\mu_1 k_\perp^2 + \mu_x k_x^2\right) G_L(-\tilde{\mathbf{q}})^2 \\
&\quad + \left[(4\mu_1^2(\mathbf{q}_\perp \cdot \mathbf{k}_\perp)^2 + 4\mu_x^2 q_x^2 k_x^2 + 4\mu_x\mu_1(\mathbf{q}_\perp \cdot \mathbf{k}_\perp)q_x k_x\right] G_L(-\tilde{\mathbf{q}})^3 + \mathcal{O}(k^3).
\end{aligned} \tag{9.16.6}$$

However, now we also have to expand the projection operators:

$$\begin{aligned}
L_{i\ell}(\mathbf{k} - \mathbf{q}) &= \frac{(k_i - q_i)(k_b - q_b)}{|\mathbf{k} - \mathbf{q}|^2} = \frac{(k_i - q_i)(k_b - q_b)}{q^2} \\
&\left(1 + \frac{2\mathbf{q} \cdot \mathbf{k} - k^2}{q^2} + \frac{4(\mathbf{q} \cdot \mathbf{k})^2}{q^4} + \mathcal{O}(k^3)\right)
\end{aligned}$$

$$= L_{i\ell}(\mathbf{q}) + \left(\frac{2L_{i\ell}(\mathbf{q})\mathbf{q}\cdot\mathbf{k} - k_i q_b - k_b q_i}{q^2}\right) + L_{i\ell}(\mathbf{q})\left(-\frac{k^2}{q^2} + \frac{4(\mathbf{q}\cdot\mathbf{k})^2}{q^4}\right) + \frac{k_i k_\ell}{q^2} - \frac{2\mathbf{q}\cdot\mathbf{k}(k_i q_b + k_b q_i)}{q^4} + \mathcal{O}(k^3). \tag{9.16.7}$$

$$P_{i\ell}(\mathbf{k}-\mathbf{q}) = P_{i\ell}(\mathbf{q}) - \left(\frac{2L_{i\ell}(\mathbf{q})\mathbf{q}\cdot\mathbf{k} - k_i q_b - k_b q_i}{q^2}\right) - L_{i\ell}(\mathbf{q})\left(-\frac{k^2}{q^2} + \frac{4(\mathbf{q}\cdot\mathbf{k})^2}{q^4}\right) - \frac{k_i k_\ell}{q^2} + \frac{2\mathbf{q}\cdot\mathbf{k}(k_i q_b + k_b q_i)}{q^4} + \mathcal{O}(k^3). \tag{9.16.8}$$

Ultimately, we have to expand *products* of the propagator factors $G_{L,T}(\tilde{\mathbf{q}} - \tilde{\mathbf{k}})$; these, obviously, will get to be fairly complicated. Multiplying out the previous expansions (9.16.5), (9.16.6), and (9.16.8), we get, after more algebra than we'd like,

$$\begin{aligned}
&G_L(\tilde{\mathbf{k}} - \tilde{\mathbf{q}})L^{\perp}_{i\ell}(\mathbf{k}-\mathbf{q}) \\
&= G_L(-\tilde{\mathbf{q}})L^{\perp}_{i\ell}(\mathbf{q}) + (2\mu_L \mathbf{q}_\perp\cdot\mathbf{k}_\perp + 2\mu_x q_x k_x)\, G_L(-\tilde{\mathbf{q}})^2 L^{\perp}_{i\ell}(\mathbf{q}) \\
&\quad + \frac{2L^{\perp}_{i\ell}(\mathbf{q})\mathbf{q}_\perp\cdot\mathbf{k}_\perp - k^{\perp}_i q^{\perp}_b - k^{\perp}_b q^{\perp}_i}{q^2_\perp} G_L(-\tilde{\mathbf{q}}) - \left(\mu_L k^2_\perp + \mu_x k^2_x\right) G_L(-\tilde{\mathbf{q}})^2 L^{\perp}_{i\ell}(\mathbf{q}) \\
&\quad + \left[(4\mu^2_L(\mathbf{q}_\perp\cdot\mathbf{k}_\perp)^2 + 4\mu^2_x q^2_x k^2_x + 4\mu_x\mu_L(\mathbf{q}_\perp\cdot\mathbf{k}_\perp)q_x k_x\right] G_L(-\tilde{\mathbf{q}})^3 L^{\perp}_{i\ell}(\mathbf{q}) \\
&\quad + \left[L^{\perp}_{i\ell}(\mathbf{q})\left(-\frac{k^2_\perp}{q^2_\perp} + \frac{4(\mathbf{q}_\perp\cdot\mathbf{k}_\perp)^2}{q^4_\perp}\right) + \frac{k^{\perp}_i k^{\perp}_\ell}{q^2_\perp} - \frac{2\mathbf{q}_\perp\cdot\mathbf{k}_\perp(k^{\perp}_i q^{\perp}_b + k^{\perp}_b q^{\perp}_i)}{q^4_\perp}\right] G_L(-\tilde{\mathbf{q}}) \\
&\quad + (2\mu_L \mathbf{q}_\perp\cdot\mathbf{k}_\perp + 2\mu_x q_x k_x)\, G_L(-\tilde{\mathbf{q}})^2 \left(\frac{2L^{\perp}_{i\ell}(\mathbf{q})\mathbf{q}_\perp\cdot\mathbf{k}_\perp - k^{\perp}_i q^{\perp}_b - k^{\perp}_b q^{\perp}_i}{q^2_\perp}\right).
\end{aligned} \tag{9.16.9}$$

$$\begin{aligned}
&G_T(\tilde{\mathbf{k}} - \tilde{\mathbf{q}})P^{\perp}_{i\ell}(\mathbf{k}-\mathbf{q}) \\
&= G_T(-\tilde{\mathbf{q}})P^{\perp}_{i\ell}(\mathbf{q}) + (2\mu_1 \mathbf{q}_\perp\cdot\mathbf{k}_\perp + 2\mu_x q_x k_x)\, G_T(-\tilde{\mathbf{q}})^2 P^{\perp}_{i\ell}(\mathbf{q}) \\
&\quad - \frac{2L^{\perp}_{i\ell}(\mathbf{q})\mathbf{q}_\perp\cdot\mathbf{k}_\perp - k^{\perp}_i q^{\perp}_b - k^{\perp}_b q^{\perp}_i}{q^2_\perp} G_T(-\tilde{\mathbf{q}}) - \left(\mu_1 k^2_\perp + \mu_x k^2_x\right) G_T(-\tilde{\mathbf{q}})^2 P^{\perp}_{i\ell}(\mathbf{q}) \\
&\quad - \left[(4\mu^2_1(\mathbf{q}_\perp\cdot\mathbf{k}_\perp)^2 + 4\mu^2_x q^2_x k^2_x + 4\mu_x\mu_1(\mathbf{q}_\perp\cdot\mathbf{k}_\perp)q_x k_x\right] G_T(-\tilde{\mathbf{q}})^3 P^{\perp}_{i\ell}(\mathbf{q}) \\
&\quad - \left[L^{\perp}_{i\ell}(\mathbf{q})\left(-\frac{k^2_\perp}{q^2_\perp} + \frac{4(\mathbf{q}_\perp\cdot\mathbf{k}_\perp)^2}{q^4_\perp}\right) + \frac{k^{\perp}_i k^{\perp}_\ell}{q^2_\perp} - \frac{2\mathbf{q}_\perp\cdot\mathbf{k}_\perp(k^{\perp}_i q^{\perp}_b + k^{\perp}_b q^{\perp}_i)}{q^4_\perp}\right] G_T(-\tilde{\mathbf{q}}) \\
&\quad - (2\mu_1 \mathbf{q}_\perp\cdot\mathbf{k}_\perp + 2\mu_x q_x k_x)\, G_T(-\tilde{\mathbf{q}})^2 \left(\frac{2L^{\perp}_{i\ell}(\mathbf{q})\mathbf{q}_\perp\cdot\mathbf{k}_\perp - k^{\perp}_i q^{\perp}_b - k^{\perp}_b q^{\perp}_i}{q^2_\perp}\right).
\end{aligned} \tag{9.16.10}$$

Note that although this algebra is painful (and easy to get wrong!), it is perfectly logically straightforward.

Now, of course, when we evaluate graphs, we'll have to take *products* of the hideous expansions (9.16.9) and (9.16.10), which will, naturally, be even *more* hideous! Once again, of course, the *logic* of the calculation is perfectly straightforward; it's just the algebra that's nasty. Indeed, it was nasty enough that it drove your humble author, who has long thought of himself as the John Henry [82] of theoretical physics, because he does all of his calculations by hand, to occasional use of Mathematica. (Chiu Fan and Leiming have been cagey about whether they were driven to the same extreme!)

In the process, there are many integrals of products of various powers of the propagator factors $G_{L,T}(\tilde{\mathbf{q}})$ that wind up needing to be evaluated. All of these can be done either by simple complex contour techniques, or by even simpler trigonometric substitutions. I will therefore simply quote the results for all of these.

In the following expressions, I'll frequently use our earlier notation for the wavevector dependent dampings:

$$\Gamma_L \equiv \frac{1}{\mu_L q_\perp^2 + \mu_x q_x^2}, \qquad \Gamma_T \equiv \frac{1}{\mu_T q_\perp^2 + \mu_x q_x^2}, \tag{9.16.11}$$

where I remind the reader of our definition $\mu_L \equiv \mu_1 + \mu_2$.

We begin with integrations over Ω and q_x:

$$\begin{aligned}\frac{1}{(2\pi)^{d+1}}\int_{-\infty}^{\infty} dq_x \int_{-\infty}^{\infty} d\Omega \; G_L(\tilde{\mathbf{q}})G_L(-\tilde{\mathbf{q}}) &= \frac{1}{(2\pi)^{d+1}}\int_{-\infty}^{\infty} dq_x \int_{-\infty}^{\infty} \frac{d\Omega}{\Omega^2 + \Gamma_L^2(\mathbf{q})} \\ &= \frac{1}{(2\pi)^{d}}\int_{-\infty}^{\infty} \frac{dq_x}{2\Gamma_L(\mathbf{q})} \\ &= \frac{1}{(2\pi)^{d-1}}\frac{1}{4\sqrt{\mu_x\mu_L}}\frac{1}{q_\perp},\end{aligned} \tag{9.16.12}$$

$$\begin{aligned}&\frac{1}{(2\pi)^{d+1}}\int_{-\infty}^{\infty} dq_x \int_{-\infty}^{\infty} d\Omega \; G_L(\tilde{\mathbf{q}})G_L(-\tilde{\mathbf{q}})^2 \\ &\quad = \frac{1}{(2\pi)^{d}}\int_{-\infty}^{\infty} dq_x \frac{1}{4\Gamma_L(\mathbf{q})^2} = \frac{1}{16\sqrt{\mu_x\mu_L^3}}\frac{1}{(2\pi)^{d-1}}\frac{1}{q_\perp^3}.\end{aligned} \tag{9.16.13}$$

$$\begin{aligned}&\frac{1}{(2\pi)^{d+1}}\int_{-\infty}^{\infty} dq_x \int_{-\infty}^{\infty} d\Omega \; G_L(\tilde{\mathbf{q}})G_L(-\tilde{\mathbf{q}})^3 \\ &\quad = \frac{1}{(2\pi)^{d}}\int_{-\infty}^{\infty} dq_x \frac{1}{8\Gamma_L(\mathbf{q})^3} = \frac{3}{128\sqrt{\mu_x\mu_L^5}}\frac{1}{(2\pi)^{d-1}}\frac{1}{q_\perp^5}.\end{aligned} \tag{9.16.14}$$

$$\frac{1}{(2\pi)^{d+1}}\int_{-\infty}^{\infty} \mathrm{d}q_x \int_{-\infty}^{\infty} \mathrm{d}\Omega \;\; G_L(\tilde{\mathbf{q}})G_L(-\tilde{\mathbf{q}})^4$$
$$= \frac{1}{(2\pi)^d}\int_{-\infty}^{\infty} \mathrm{d}q_x \;\; \frac{1}{16\Gamma_L(\mathbf{q})^4} = \frac{5}{512\sqrt{\mu_x\mu_L^7}}\frac{1}{(2\pi)^{d-1}}\frac{1}{q_\perp^7}\,. \tag{9.16.15}$$

$$\frac{1}{(2\pi)^{d+1}}\int_{-\infty}^{\infty} \mathrm{d}q_x \int_{-\infty}^{\infty} \mathrm{d}\Omega \;\; q_x^2 G_L(\tilde{\mathbf{q}})G_L(-\tilde{\mathbf{q}})^4$$
$$= \frac{1}{(2\pi)^d}\int_{-\infty}^{\infty} \mathrm{d}q_x \;\; \frac{q_x^2}{16\Gamma_L(\mathbf{q})^4} = \frac{1}{512\sqrt{\mu_x^3\mu_L^5}}\frac{1}{(2\pi)^{d-1}}\frac{1}{q_\perp^5}\,. \tag{9.16.16}$$

$$\frac{1}{(2\pi)^{d+1}}\int_{-\infty}^{\infty} \mathrm{d}q_x \int_{-\infty}^{\infty} \mathrm{d}\Omega \;\; G_L(\tilde{\mathbf{q}})^2 G_L(-\tilde{\mathbf{q}})^2$$
$$= \frac{1}{(2\pi)^d}\int_{-\infty}^{\infty} \mathrm{d}q_x \;\; \frac{1}{4\Gamma_L(\mathbf{q})^3} = \frac{3}{64\sqrt{\mu_x\mu_L^5}}\frac{1}{(2\pi)^{d-1}}\frac{1}{q_\perp^5}\,. \tag{9.16.17}$$

$$\frac{1}{(2\pi)^{d+1}}\int_{-\infty}^{\infty} \mathrm{d}q_x \int_{-\infty}^{\infty} \mathrm{d}\Omega \;\; G_T(\tilde{\mathbf{q}})G_T(-\tilde{\mathbf{q}})^2$$
$$= \frac{1}{(2\pi)^d}\int_{-\infty}^{\infty} \mathrm{d}q_x \;\; \frac{1}{4\Gamma_T(\mathbf{q})^2} = \frac{1}{16\sqrt{\mu_x\mu_1^3}}\frac{1}{(2\pi)^{d-1}}\frac{1}{q_\perp^3}\,. \tag{9.16.18}$$

$$\frac{1}{(2\pi)^{d+1}}\int_{-\infty}^{\infty} \mathrm{d}q_x \int_{-\infty}^{\infty} \mathrm{d}\Omega \;\; G_T(\tilde{\mathbf{q}})G_T(-\tilde{\mathbf{q}})^3$$
$$= \frac{1}{(2\pi)^d}\int_{-\infty}^{\infty} \mathrm{d}q_x \;\; \frac{1}{8\Gamma_T(\mathbf{q})^3} = \frac{3}{128\sqrt{\mu_x\mu_1^5}}\frac{1}{(2\pi)^{d-1}}\frac{1}{q_\perp^5}\,. \tag{9.16.19}$$

$$\frac{1}{(2\pi)^{d+1}}\int_{-\infty}^{\infty} \mathrm{d}q_x \int_{-\infty}^{\infty} \mathrm{d}\Omega \;\; G_T(\tilde{\mathbf{q}})G_T(-\tilde{\mathbf{q}})^4$$
$$= \frac{1}{(2\pi)^d}\int_{-\infty}^{\infty} \mathrm{d}q_x \;\; \frac{1}{16\Gamma_T(\mathbf{q})^4} = \frac{5}{512\sqrt{\mu_x\mu_1^7}}\frac{1}{(2\pi)^{d-1}}\frac{1}{q_\perp^7}\,. \tag{9.16.20}$$

$$\frac{1}{(2\pi)^{d+1}}\int_{-\infty}^{\infty} \mathrm{d}q_x \int_{-\infty}^{\infty} \mathrm{d}\Omega \;\; q_x^2 G_T(\tilde{\mathbf{q}})G_T(-\tilde{\mathbf{q}})^4$$
$$= \frac{1}{(2\pi)^d}\int_{-\infty}^{\infty} \mathrm{d}q_x \;\; \frac{q_x^2}{16\Gamma_T(\mathbf{q})^4} = \frac{1}{512\sqrt{\mu_x^3\mu_1^5}}\frac{1}{(2\pi)^{d-1}}\frac{1}{q_\perp^5}\,. \tag{9.16.21}$$

$$\frac{1}{(2\pi)^{d+1}}\int_{-\infty}^{\infty} \mathrm{d}q_x \int_{-\infty}^{\infty} \mathrm{d}\Omega \;\; G_T(\tilde{\mathbf{q}})^2 G_T(-\tilde{\mathbf{q}})^2$$
$$= \frac{1}{(2\pi)^d}\int_{-\infty}^{\infty} \mathrm{d}q_x \;\; \frac{1}{4\Gamma_T(\mathbf{q})^3} = \frac{3}{64\sqrt{\mu_x\mu_1^5}}\frac{1}{(2\pi)^{d-1}}\frac{1}{q_\perp^5}\,. \tag{9.16.22}$$

$$\begin{aligned}
&\frac{1}{(2\pi)^{d+1}}\int_{-\infty}^{\infty}\mathrm{d}q_x\int_{-\infty}^{\infty}\mathrm{d}\Omega\;\; G_L(\tilde{\mathbf{q}})G_L(-\tilde{\mathbf{q}})G_T(\tilde{\mathbf{q}})G_T(-\tilde{\mathbf{q}})\\
&\quad=\frac{1}{(2\pi)^d}\int_{-\infty}^{\infty}\mathrm{d}q_x\;\frac{1}{2\Gamma_L(\mathbf{q})\Gamma_T(\mathbf{q})(\Gamma_L(\mathbf{q})+\Gamma_T(\mathbf{q}))}\\
&\quad=\frac{1}{2\sqrt{\mu_x}(\mu_L-\mu_1)^2}\left(\frac{1}{\sqrt{\mu_L}}+\frac{1}{\sqrt{\mu_1}}-\frac{2\sqrt{2}}{\sqrt{\mu_L+\mu_1}}\right)\frac{1}{(2\pi)^{d-1}}\frac{1}{q_\perp^5}\,.
\end{aligned}\tag{9.16.23}$$

$$\begin{aligned}
&\frac{1}{(2\pi)^{d+1}}\int_{-\infty}^{\infty}\mathrm{d}q_x\int_{-\infty}^{\infty}\mathrm{d}\Omega\;\; G_L(\tilde{\mathbf{q}})G_L(-\tilde{\mathbf{q}})G_T(-\tilde{\mathbf{q}})\\
&\quad=\frac{1}{(2\pi)^d}\int_{-\infty}^{\infty}\mathrm{d}q_x\;\frac{1}{4\Gamma_L(\mathbf{q})(\Gamma_L(\mathbf{q})+\Gamma_T(\mathbf{q}))}\\
&\quad=\frac{1}{4\sqrt{\mu_x}(\mu_L-\mu_1)}\left(\frac{\sqrt{2}}{\sqrt{\mu_L+\mu_1}}-\frac{1}{\sqrt{\mu_L}}\right)\frac{1}{(2\pi)^{d-1}}\frac{1}{q_\perp^3}\\
&\quad\equiv A(\mu_L,\mu_1)\frac{1}{(2\pi)^{d-1}}\frac{1}{q_\perp^3}\,.
\end{aligned}\tag{9.16.24}$$

$$\begin{aligned}
&\frac{1}{(2\pi)^{d+1}}\int_{-\infty}^{\infty}\mathrm{d}q_x\int_{-\infty}^{\infty}\mathrm{d}\Omega\;\; G_L(\tilde{\mathbf{q}})G_L(-\tilde{\mathbf{q}})G_T(-\tilde{\mathbf{q}})^2\\
&\quad=\frac{1}{(2\pi)^d}\int_{-\infty}^{\infty}\mathrm{d}q_x\;\frac{1}{2\Gamma_L(\mathbf{q})(\Gamma_L(\mathbf{q})+\Gamma_T(\mathbf{q}))^2}\\
&\quad=\frac{2(\mu_L+\mu_1)^{3/2}-\sqrt{2\mu_L}(\mu_L+3\mu_1)}{8\sqrt{\mu_x\mu_L}(\mu_L+\mu_1)^{3/2}(\mu_L-\mu_1)^2}\frac{1}{(2\pi)^{d-1}}\frac{1}{q_\perp^5}\\
&\quad\equiv B(\mu_L,\mu_1)\frac{1}{(2\pi)^{d-1}}\frac{1}{q_\perp^5}\,.
\end{aligned}\tag{9.16.25}$$

Now I turn to integrals over $\mathbf{q}_\perp$:

$$\int_{\tilde{\mathbf{q}}}^{>}\equiv\int_{\Lambda>|\mathbf{q}_\perp|>\Lambda e^{-\mathrm{d}\ell}}\mathrm{d}^{d-1}q_\perp\int_{-\infty}^{\infty}\mathrm{d}\Omega\int_{-\infty}^{\infty}\mathrm{d}q_x\,. \tag{9.16.26}$$

The simplest such integral that arises is

$$\frac{1}{(2\pi)^{d-1}}\int_{\Lambda>|\mathbf{q}_\perp|>\Lambda e^{-\mathrm{d}\ell}}\frac{\mathrm{d}^{d-1}q_\perp}{q_\perp^3}=\frac{1}{(2\pi)^{d-1}}\int_{\Lambda e^{-\mathrm{d}\ell}}^{\Lambda}\mathrm{d}q_\perp\int\mathrm{d}\Xi_{\mathbf{q}_\perp}q_\perp^{d-5}\,, \tag{9.16.27}$$

where $\int\mathrm{d}\Xi_{\mathbf{q}_\perp}$ denotes an integral over the $d-1$-dimensional solid angle associated with $\mathbf{q}_\perp$. Since the integrand is independent of the direction of $\mathbf{q}_\perp$, we can do this angular integral trivially; it simply gives a multiplicative factor of S_{d-1}, the surface area of a unit $d-1$-dimensional sphere. Thus we have

$$\frac{1}{(2\pi)^{d-1}}\int_{\Lambda>|\mathbf{q}_\perp|>\Lambda e^{-d\ell}}\frac{d^{d-1}q_\perp}{q_\perp^3}=\frac{S_{d-1}}{(2\pi)^{d-1}}\int_{\Lambda e^{-d\ell}}^{\Lambda}dq_\perp q_\perp^{d-5}=\frac{S_{d-1}}{(2\pi)^{d-1}}\Lambda^{d-4}d\ell, \tag{9.16.28}$$

where in the last step we have used the fact that $d\ell$ is infinitesimal to write $1-e^{-d\ell}=d\ell$.

A slightly harder integral that arises in our calculations is

$$I_{iu}=\frac{1}{(2\pi)^{d-1}}\int_{\mathbf{q}_\perp}\frac{q_i^\perp q_u^\perp}{q_\perp^5}. \tag{9.16.29}$$

We did almost exactly this integral using the angular average trick in Chapter 3. The result is:

$$\frac{1}{(2\pi)^{d-1}}\int_{\mathbf{q}_\perp}\frac{q_i^\perp q_u^\perp}{q_\perp^5}=\frac{\delta_{iu}^\perp}{(d-1)}\frac{S_{d-1}}{(2\pi)^{d-1}}\Lambda^{d-4}d\ell. \tag{9.16.30}$$

We also encounter the somewhat harder integral

$$\frac{1}{(2\pi)^{d-1}}\int_{\mathbf{q}_\perp}\frac{q_i^\perp q_u^\perp q_j^\perp q_\ell^\perp}{q_\perp^7} \tag{9.16.31}$$

in our struggles. This can also be done using symmetry arguments, and leveraging our result (9.16.30).

To do so, we first note that, since $\perp$-space is isotropic, the tensor structure of the integral (9.16.31) must be made entirely out of Kronecker deltas in the perp space (anything else would pick out a special direction in that space). This is, of course, the same reasoning we used to evaluate the simpler integral (9.16.30). Furthermore, the integral (9.16.31) is obviously completely symmetric upon interchanging any pair of indices.

These considerations imply that the integral (9.16.31) must take the form

$$\frac{1}{(2\pi)^{d-1}}\int_{\mathbf{q}_\perp}\frac{q_i^\perp q_u^\perp q_j^\perp q_c^\perp}{q_\perp^7}=C\Pi_{iujc}, \tag{9.16.32}$$

where we've defined

$$\Pi_{iujc}\equiv\delta_{iu}^\perp\delta_{jc}^\perp+\delta_{ij}^\perp\delta_{uc}^\perp+\delta_{ic}^\perp\delta_{ju}^\perp \tag{9.16.33}$$

and C is a constant yet to be determined. We can easily determine that constant by taking the trace of (9.16.32) over any pair of indices. Choosing the pair jc, this gives

$$\frac{1}{(2\pi)^{d-1}}\int_{\mathbf{q}_\perp}\frac{q_i^\perp q_u^\perp}{q_\perp^5}=C\Pi_{iujj}, \tag{9.16.34}$$

where we've used the fact that $q_j^\perp q_j^\perp=q_\perp^2$.

But the integral on the left-hand side of this equation is just (9.16.30). The trace on the right-hand side can be evaluated essentially by counting on your fingers:

$$\Pi_{iujj} = \delta^{\perp}_{iu}\delta^{\perp}_{jj} + \delta^{\perp}_{ij}\delta^{\perp}_{uj} + \delta^{\perp}_{ij}\delta^{\perp}_{ju} = \delta^{\perp}_{iu}(d-1) + \delta^{\perp}_{iu} + \delta^{\perp}_{iu} = (d+1)\delta^{\perp}_{iu}\,. \quad (9.16.35)$$

Using this and (9.16.30) in (9.16.34) gives

$$\frac{\delta^{\perp}_{iu}}{(d-1)}\frac{S_{d-1}}{(2\pi)^{d-1}}\Lambda^{d-4}\mathrm{d}\ell = C(d+1)\delta^{\perp}_{iu}\,, \quad (9.16.36)$$

which, reassuringly, both works (in the sense that both sides are proportional to $\delta^{\perp}_{iu}$), and allows us to solve for C:

$$C = \frac{\delta^{\perp}_{iu}}{(d-1)(d+1)}\frac{S_{d-1}}{(2\pi)^{d-1}}\Lambda^{d-4}\mathrm{d}\ell\,. \quad (9.16.37)$$

Thus we have

$$\frac{1}{(2\pi)^{d-1}}\int_{\mathbf{q}_\perp}\frac{q^{\perp}_i q^{\perp}_u q^{\perp}_j q^{\perp}_c}{q^7_\perp} = \frac{\Pi_{iujc}}{(d-1)(d+1)}\frac{S_{d-1}}{(2\pi)^{d-1}}\Lambda^{d-4}\mathrm{d}\ell\,, \quad (9.16.38)$$

with Π_{iujc} given by (9.16.33).

With these in hand, and a sufficiently strong stomach for hideous algebra, you too can calculate the full recursion relations for a Malthusian flock with $\mu_2 \neq 0$. The remaining gory details are in references [22, 23]; the result is the following recursion relations to one-loop order:

$$\frac{1}{D}\frac{\mathrm{d}D}{\mathrm{d}\ell} = z - 2\chi - d + 1 - \zeta + g_1 G_D(\alpha)\,, \quad (9.16.39)$$

$$\frac{1}{\lambda}\frac{\mathrm{d}\lambda}{\mathrm{d}\ell} = z + \chi - 1\,, \quad (9.16.40)$$

$$\frac{1}{\mu_x}\frac{\mathrm{d}\mu_x}{\mathrm{d}\ell} = z - 2\zeta\,, \quad (9.16.41)$$

$$\frac{1}{\mu_1}\frac{\mathrm{d}\mu_1}{\mathrm{d}\ell} = z - 2 + g_1 G_{\mu_1}(\alpha)\,, \quad (9.16.42)$$

$$\frac{1}{\mu_2}\frac{\mathrm{d}\mu_2}{\mathrm{d}\ell} = z - 2 + g_1 G_{\mu_2}(\alpha), \quad (9.16.43)$$

and G_{D,μ_1,μ_2} are all hideously complicated functions of the dimensionless ratio α. They are

$$G_D(\alpha) \equiv \frac{(d-2)}{2(d-1)}\frac{1}{\alpha^2}\left[1+\frac{1}{\sqrt{\alpha+1}}-\frac{2\sqrt{2}}{\sqrt{\alpha+2}}\right]$$

$$= \frac{1}{\alpha^2}\left[\frac{1}{3}+\frac{1}{3\sqrt{\alpha+1}}-\frac{2\sqrt{2}}{3\sqrt{\alpha+2}}\right], \quad (d=4) \tag{9.16.44}$$

$$G_{\mu_1}(\alpha) \equiv \frac{2}{d^2-1}\left(\frac{(2d^2-6d+3)}{32}+\frac{(d+3)\sqrt{2}}{\alpha^2(\alpha+2)^{3/2}}-\frac{1}{\alpha^2}-\frac{d+1}{2\alpha^2\sqrt{\alpha+1}}+\frac{d-3}{2\alpha}\right.$$
$$\left.+\frac{d+15}{2\sqrt{2}\alpha(\alpha+2)^{3/2}}+\frac{3}{\sqrt{2}(\alpha+2)^{3/2}}-\frac{d+1}{4\alpha\sqrt{\alpha+1}}+\frac{3-d}{2\sqrt{2}\alpha\sqrt{\alpha+2}}\right) \tag{9.16.45}$$

$$= \frac{2}{15}\left(\frac{11}{32}+\frac{7\sqrt{2}}{\alpha^2(\alpha+2)^{3/2}}-\frac{1}{\alpha^2}-\frac{5}{2\alpha^2\sqrt{\alpha+1}}+\frac{1}{2\alpha}+\frac{19}{2\sqrt{2}\alpha(\alpha+2)^{3/2}}\right.$$
$$\left.+\frac{3}{\sqrt{2}(\alpha+2)^{3/2}}-\frac{5}{4\alpha\sqrt{\alpha+1}}-\frac{1}{2\sqrt{2}\alpha\sqrt{\alpha+2}}\right), \quad (d=4) \tag{9.16.46}$$

$$G_{\mu_2}(\alpha) \equiv \frac{2}{(d^2-1)\alpha}\left(-\frac{(3d-1)\sqrt{2}}{\alpha^2(\alpha+2)^{3/2}}+\frac{(d-1)}{\alpha^2}+\frac{(d+1)}{2\alpha^2\sqrt{\alpha+1}}\right.$$
$$+\frac{(d^2-4d+3)\sqrt{2}}{4\alpha\sqrt{\alpha+2}}-\frac{(d^2-7d+8)}{4\alpha}+\frac{d+1}{64(\alpha+1)^{3/2}}+\frac{(13-15d)}{2\sqrt{2}\alpha(\alpha+2)^{3/2}}$$
$$\left.-\frac{3(d-1)}{\sqrt{2}(\alpha+2)^{3/2}}+\frac{(d+1)}{4\alpha\sqrt{\alpha+1}}+\frac{2d^2-9d+11}{32}\right) \tag{9.16.47}$$

$$= \frac{2}{15\alpha}\left(-\frac{11\sqrt{2}}{\alpha^2(\alpha+2)^{3/2}}+\frac{3}{\alpha^2}+\frac{5}{2\alpha^2\sqrt{\alpha+1}}+\frac{3\sqrt{2}}{4\alpha\sqrt{\alpha+2}}+\frac{1}{\alpha}+\frac{5}{64(\alpha+1)^{3/2}}\right.$$
$$\left.-\frac{47}{2\sqrt{2}\alpha(\alpha+2)^{3/2}}-\frac{9}{\sqrt{2}(\alpha+2)^{3/2}}+\frac{5}{4\alpha\sqrt{\alpha+1}}+\frac{7}{32}\right). \quad (d=4) \tag{9.16.48}$$

As you can see, I wasn't exaggerating when I said "hideous"!

Note that in this full, $\mu_2 \neq 0$ calculation, there continue to be no graphical corrections to μ_x. As discussed in Section 9.11, this is almost certainly an artifact of the one-loop approximation.

Note that the appearance of negative powers of α in the expressions (9.16.44)–(9.16.48) is somewhat misleading: Despite those negative powers, none of these functions diverges at $\alpha = 0$; in fact, these singularities all cancel, and G_D, G_{μ_1}, and G_{μ_2} are all smooth, analytic, and finite for all finite α (*including* $\alpha = 0$) that satisfy the stability constraint $\alpha > -1$.

Indeed, difficult as it may be to believe, with a little help from the Marquis de l'Hôpital (OK, with a *lot* of help from the old Marquis!), one can show that these recursion relations reduce to those found earlier in the limit $\mu_2 \to 0$ (that is, for $\alpha \to 0$). This means among other things that they must have the fixed point we found earlier in that $\mu_2 = 0$ limit.

We'll now show that fixed point is the *only* stable fixed point in this problem.

First, from equations (9.16.39)–(9.16.43), we obtain the closed–flow equations for g_1 and α.

We can derive a recursion relation for g_1 as we have done for dimensionless couplings in earlier chapters. We start by using the definition (9.8.5) of g_1, which implies

$$\ln g_1 = 2\ln\lambda + \ln D - \frac{1}{2}\ln\mu_x - \frac{5}{2}\ln\mu_1 + \text{constant}, \tag{9.16.49}$$

where the "constant" involves genuine constants like S_d (the surface area of a unit d-dimensional sphere), π, and so forth, as well as the ultaviolet cutoff Λ. All of these "constants" are independent of RG time ℓ.

Therefore, the recursion relation for $\ln g_1$ follows simply by differentiating (9.16.49) with respect to ℓ:

$$\begin{aligned}\frac{\mathrm{d}\ln g_1}{d\ell} &= 2\frac{\mathrm{d}\ln\lambda}{d\ell} + \frac{\mathrm{d}\ln D}{d\ell} - \frac{1}{2}\frac{\mathrm{d}\ln\mu_x}{d\ell} - \frac{5}{2}\frac{\mathrm{d}\ln\mu_1}{d\ell} \\ &= \frac{2}{\lambda}\frac{\mathrm{d}\lambda}{\mathrm{d}\ell} + \frac{1}{D}\frac{\mathrm{d}D}{\mathrm{d}\ell} - \frac{1}{2\mu_x}\frac{\mathrm{d}\mu_x}{\mathrm{d}\ell} - \frac{5}{2\mu_1}\frac{\mathrm{d}\mu_1}{\mathrm{d}\ell}.\end{aligned} \tag{9.16.50}$$

Using the recursion relations (9.16.39), (9.16.40), (9.16.41), (9.16.42), and (9.16.43) in this, and multiplying both sides by g_1, gives

$$\frac{\mathrm{d}g_1}{\mathrm{d}\ell} = g_1\left[\epsilon + g_1(G_D - \frac{5}{2}G_{\mu_1})\right] \equiv g_1\left[\epsilon + g_1 G_{g_1}(\alpha)\right], \tag{9.16.51}$$

where

$$\begin{aligned}G_{g_1}(\alpha) = G_D - \frac{5}{2}G_{\mu_1} &= \frac{(-10d^2+30d-15)}{32(d^2-1)} + \frac{(d^2-d+8)}{2(d^2-1)\alpha^2} - \frac{(2d^2+3d+11)\sqrt{2}}{(d^2-1)\alpha^2(\alpha+2)^{3/2}} \\ &+ \frac{(d+3)}{2(d-1)\alpha^2\sqrt{\alpha+1}} + \frac{(15-5d)}{2(d^2-1)\alpha} - \frac{\sqrt{2}(4d^2-9d+97)}{4(d^2-1)\alpha(\alpha+2)^{3/2}} \\ &+ \frac{(5d-45)}{2\sqrt{2}(d^2-1)(\alpha+2)^{3/2}} + \frac{5}{4(d-1)\alpha\sqrt{\alpha+1}}\end{aligned} \tag{9.16.52}$$

$$\begin{aligned}&= -\frac{11}{96} + \frac{2}{3\alpha^2} - \frac{11\sqrt{2}}{3\alpha^2(\alpha+2)^{3/2}} + \frac{7}{6\alpha^2\sqrt{\alpha+1}} - \frac{1}{6\alpha} - \frac{25}{6\sqrt{2}\alpha(\alpha+2)^{3/2}} \\ &- \frac{5}{6\sqrt{2}(\alpha+2)^{3/2}} + \frac{5}{12\alpha\sqrt{\alpha+1}}. \quad (d=4)\end{aligned} \tag{9.16.53}$$

A similar, but much simpler, application of the same "logarithmic derivative" trick gives the recursion relation for the ratio $\alpha = \frac{\mu_2}{\mu_1}$:

$$\frac{d\alpha}{d\ell} = g_1\alpha(G_{\mu_2} - G_{\mu_1}) \equiv g_1\alpha G_\alpha(\alpha), \tag{9.16.54}$$

where

$$\begin{aligned} G_\alpha(\alpha) = &\left(\frac{2}{d^2-1}\right)\left(\frac{(1-3d)\sqrt{2}}{\alpha^3(\alpha+2)^{3/2}} + \frac{(d-1)}{\alpha^3} + \frac{(d+1)}{2\alpha^3\sqrt{\alpha+1}} + \frac{(1-19d)}{2\sqrt{2}\alpha^2(\alpha+2)^{3/2}}\right. \\ &+ \frac{(d^2-4d+3)}{2\sqrt{2}\alpha^2\sqrt{\alpha+2}} - \frac{(d^2-7d+4)}{4\alpha^2} + \frac{3(d+1)}{4\alpha^2\sqrt{\alpha+1}} - \frac{3\sqrt{2}}{2(\alpha+2)^{3/2}} \\ &- \frac{(9+7d)}{2\sqrt{2}\alpha(\alpha+2)^{3/2}} + \frac{(2d^2-25d+59)}{32\alpha} + \frac{d+1}{64\alpha(\alpha+1)^{3/2}} + \frac{(d+1)}{4\alpha\sqrt{\alpha+1}} \\ &\left. + \frac{(d-3)}{2\sqrt{2}\alpha\sqrt{\alpha+2}} - \frac{((2d^2-6d+3))}{32}\right) \end{aligned} \tag{9.16.55}$$

$$\begin{aligned} = &-\frac{22\sqrt{2}}{15\alpha^3(\alpha+2)^{3/2}} + \frac{2}{5\alpha^3} + \frac{1}{3\alpha^3\sqrt{\alpha+1}} - \frac{5}{\sqrt{2}\alpha^2(\alpha+2)^{3/2}} \\ &+ \frac{1}{5\sqrt{2}\alpha^2\sqrt{\alpha+2}} + \frac{4}{15\alpha^2} + \frac{1}{2\alpha^2\sqrt{\alpha+1}} - \frac{\sqrt{2}}{5(\alpha+2)^{3/2}} \\ &- \frac{37}{15\sqrt{2}\alpha(\alpha+2)^{3/2}} - \frac{3}{80\alpha} + \frac{1}{96\alpha(\alpha+1)^{3/2}} + \frac{1}{6\alpha\sqrt{\alpha+1}} \\ &+ \frac{1}{15\sqrt{2}\alpha\sqrt{\alpha+2}} - \frac{11}{240}. \quad (d=4) \end{aligned} \tag{9.16.56}$$

A plot of $\frac{d\alpha}{d\ell}$ versus α is given in Figure 9.16.1, and clearly shows the only fixed point for α is at $\alpha = 0$.

To show that this fixed point is stable, we'll use the recursion relation (9.8.16) in arbitrary spatial dimension d, and setting $\frac{dg_1}{d\ell} = 0$, we get the fixed point value $g_1^* = \frac{64(4-d)(d-1)}{14d-23}$ of g_1 found earlier in equation (9.10.1).

To prove the stability of the fixed point, we'll need the values of G_{g_1} and G_α at $\alpha = 0$ for general d, which with a bit more assistance from le Marquis de l'Hôpital, are

$$G_{g_1}^* = G_{g_1}(\alpha = 0) = \frac{23-14d}{64(d-1)}, \tag{9.16.57}$$

$$G_\alpha^* = G_\alpha(\alpha = 0) = \frac{5(4-d-d^2)}{64(d^2-1)}. \tag{9.16.58}$$

To demonstrate the stability of this fixed point, we show that small departures from it decay to zero upon renormalization (see also Figure 9.16.2). Specifically,

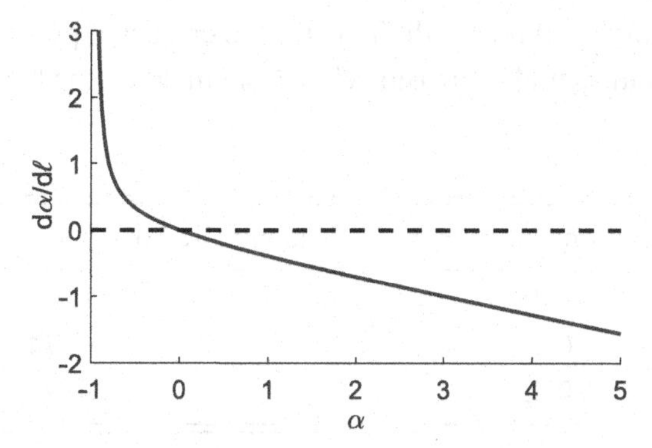

Fig. 9.16.1 Plot of $\frac{d\alpha}{d\ell}$ versus α for $\epsilon = 1$ and g_1 fixed at 64/11, showing that $\frac{d\alpha}{d\ell}$ vanishes only at $\alpha = 0$. The plot only changes by a constant multiplicative factor if we change the value of g_1, so for all values of g_1, $\frac{d\alpha}{d\ell}$ vanishes only at $\alpha = 0$. Reproduced from [23].

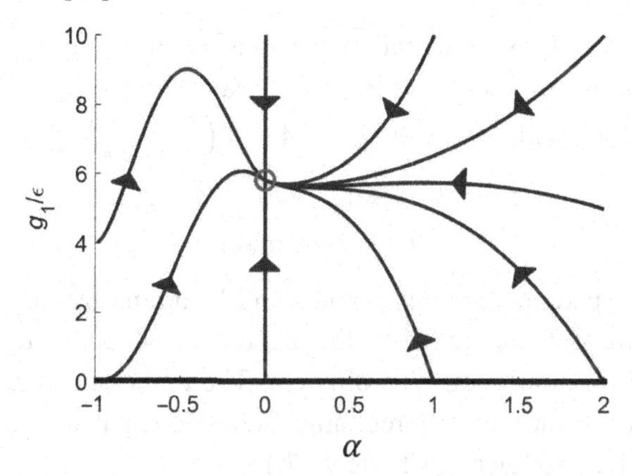

Fig. 9.16.2 The RG flow of the coupling constant g_1 and ratio $\alpha \equiv \frac{\mu_2}{\mu_1}$ at $\epsilon = 1$. The stable fixed point (circle) is at $g_1^* = 64\epsilon/11$ and $\alpha^* = 0$, while the entire α axis (i.e., the locus $g_1 = 0$) is an unstable Gaussian fixed line and is indicated by a thick black line. Reproduced from [22] and [23].

we linearize the recursion relations (9.16.51) and (9.16.54) around the fixed point, writing

$$g_1(\ell) = g_1^* + \delta g_1(\ell) \tag{9.16.59}$$

and expanding the recursion relations (9.16.51) and (9.16.54) to linear order in $\delta g_1(\ell)$ and $\alpha(\ell)$. This leads to the recursion relations:

$$\begin{aligned}\frac{d\delta g_1}{d\ell} &= (d-4)\delta g_1(\ell) + \left(g_1^*\right)^2 G'_{g_1}(\alpha = 0)\,\alpha \\ &= (d-4)\delta g_1(\ell) - \frac{160(d-1)\left(3d^2 - 8d - 1\right)}{(14d-23)^2(d+1)}(4-d)^2\alpha\,,\end{aligned} \tag{9.16.60}$$

Table 9.17.1 Summary of the predictions for the critical exponents for Malthusian flocks in spatial dimension $d = 3$ obtained by the two different approaches.

Exponents	ϵ-Expansion, $d = 3$	Uncontrolled one-loop expansion, $d = 3$
z	1.45	1.47
ζ	0.73	0.74
χ	−0.45	−0.47

$$\frac{d\alpha}{d\ell} = G_\alpha(\alpha = 0)g_1^*\alpha = \left(\frac{5(4-d-d^2)(4-d)}{(d+1)(14d-23)}\right)\alpha. \tag{9.16.61}$$

Because $d^2+d-4 > 0$ for all spatial dimensions d in the range of interest $2 \leq d \leq 4$, it is obvious from (9.16.60) and (9.16.61) that the fixed point at $g_1^* = 64\epsilon/11$ and $\alpha^* = 0$ is stable, with eigenvalues $d - 4$ and $\left(\frac{5(4-d-d^2)(4-d)}{(d+1)(14d-23)}\right)$ respectively.

9.17 Summary

There was a lot to take in from this chapter, so I'll summarize the results here.

First and foremost, there are the scaling exponents which tell us essentially everything we need to know about this problem. The numerical values predicted for these in $d = 3$ from the two different approaches (ϵ-expansion, and uncontrolled one-loop expansion) are given in Table 9.17.1.

The expressions for these exponents in the $d = 4 - \epsilon$ expansion, to leading order in ϵ, are

$$z = 2 - \frac{6\epsilon}{11} + \mathcal{O}(\epsilon^2), \tag{9.17.1}$$

$$\zeta = 1 - \frac{3\epsilon}{11} + \mathcal{O}(\epsilon^2), \tag{9.17.2}$$

$$\chi = -1 + \frac{6\epsilon}{11} + \mathcal{O}(\epsilon^2). \tag{9.17.3}$$

Setting $\epsilon = 4 - d = 1$ in these expressions gives the values in three dimensions listed in the first column of Table 9.17.1.

In the one-loop order (i.e., lowest order in perturbation theory) we derived perturbative renormalization group recursion relations in arbitrary spatial dimensions. This approach, although strictly speaking an uncontrolled approximation, can easily be shown to give exponents for the $\mathcal{O}(n)$ model critical point in $d = 3$ that are at least as accurate as the first order in $d = 4 - \epsilon$ expansion with ϵ set to 1. Therefore, we expect this approach should provide a very effective interpolation formula for d between (and including) 3 and 4.

Using this approach, we find

$$z = 2 - \frac{2(4-d)(4d-7)}{14d-23}, \tag{9.17.4}$$

$$\zeta = 1 - \frac{(4-d)(4d-7)}{14d-23}, \tag{9.17.5}$$

$$\chi = -1 + \frac{2(4-d)(4d-7)}{14d-23}, \tag{9.17.6}$$

which indeed recover our ϵ expansion results near $d = 4$, as the readers can verify for themselves.

In the physically interesting case $d = 3$, these give

$$z = \frac{28}{19} \approx 1.47, \tag{9.17.7}$$

$$\zeta = \frac{14}{19} \approx 0.74, \tag{9.17.8}$$

$$\chi = -\frac{9}{19} \approx -0.47. \tag{9.17.9}$$

These are our best numerical estimates of the values of these exponents in $d = 3$. We suspect that they are accurate to ± 1 percent, an error estimate based on the difference between the ϵ-expansion and the uncontrolled one-loop results.

These exponents can be directly measured experimentally, because they govern the scaling behavior of the experimentally measurable velocity correlation function:

$$C_u(\mathbf{r}, t) \equiv \langle \mathbf{u}_\perp(\mathbf{r}, t) \cdot \mathbf{u}_\perp(\mathbf{0}, 0) \rangle = r_\perp^{2\chi} F_u \left(\frac{(|x - \gamma t|/\xi_x)}{(r_\perp/\xi_\perp)^\zeta}, \frac{(t/\tau)}{(r_\perp/\xi_\perp)^z} \right)$$

$$\propto \begin{cases} r_\perp^{2\chi}, & (r_\perp/\xi_\perp)^\zeta \gg |x - \gamma t|/\xi_x, \quad (r_\perp/\xi_\perp)^z \gg (|t|/\tau), \\ |x - \gamma t|^{\frac{2\chi}{\zeta}}, & |x - \gamma t|/\xi_x \gg (r_\perp/\xi_\perp)^\zeta, \quad |x - \gamma t|/\xi_x \gg (|t|/\tau)^{\frac{\zeta}{z}}, \\ |t|^{\frac{2\chi}{z}}, & (|t|/\tau) \gg (r_\perp/\xi_\perp)^z, \quad (|t|/\tau) \gg (|x - \gamma t|/\xi_x)^{\frac{z}{\zeta}}, \end{cases} \tag{9.17.10}$$

where F_u is a universal scaling function (i.e., the same for all Malthusian flocks), γ is a nonuniversal (i.e., system-dependent) speed, $\xi_{\perp,x}$ are nonuniversal lengths, and τ is a nonuniversal time. We also note that fluctuations of the velocity field $\mathbf{u}_\perp$ are always positively correlated, i.e., C_u is always positive.

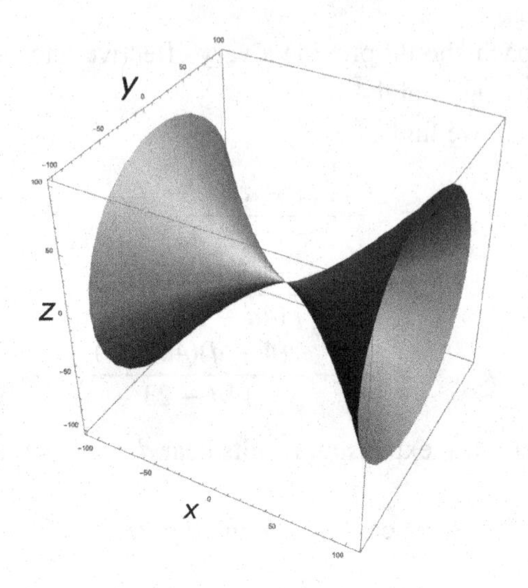

Fig. 9.17.1 In $d = 3$, the equal time density correlations are positive inside the trumpet-shaped surface defined by (9.17.12), and negative outside it. Reproduced from [23].

Density correlations also obey a scaling law involving the same universal exponents z, ζ, and χ, and nonuniversal lengths $\xi_{\perp,x}$ and time τ:

$$C_\rho(\mathbf{r},t) \equiv \langle \delta\rho(\mathbf{r},t)\delta\rho(\mathbf{0},0)\rangle = r_\perp^{2(\chi-1)} F_\rho\left(\frac{(|x-\gamma t|/\xi_x)}{(r_\perp/\xi_\perp)^\zeta}, \frac{(t/\tau)}{(r_\perp/\xi_\perp)^z}\right)$$

$$\propto \begin{cases} r_\perp^{2(\chi-1)}, & (r_\perp/\xi_\perp)^\zeta \gg |x-\gamma t|/\xi_x, \quad (r_\perp/\xi_\perp)^z \gg (|t|/\tau), \\ |x-\gamma t|^{\frac{2(\chi-1)}{\zeta}}, & |x-\gamma t|/\xi_x \gg (r_\perp/\xi_\perp)^\zeta, \quad |x-\gamma t|/\xi_x \gg (|t|/\tau)^{\frac{\zeta}{z}}, \\ |t|^{\frac{2(\chi-1)}{z}}, & (|t|/\tau) \gg (r_\perp/\xi_\perp)^z, \quad (|t|/\tau) \gg (|x-\gamma t|/\xi_x)^{\frac{z}{\zeta}}, \end{cases} \tag{9.17.11}$$

where we've defined $\delta\rho(\mathbf{r},t) \equiv \rho(\mathbf{r},t) - \rho_0$, with ρ_0 the mean density (see [37, 83, 84, 85, 86]; see also [87] and references therein).

In contrast to the velocity–velocity correlation function, C_ρ can be positive or negative. Indeed, the equal-time density correlation is positive when $|x|/\xi_x > r_\perp/\xi_\perp$ for $|x| < \xi_x$ or $|x|/\xi_x > (r_\perp/\xi_\perp)^\zeta$ for $|x| > \xi_x$, and negative otherwise. Therefore, there is a separatrix on which the equal-time density correlations are exactly zero (Figure 9.17.1). This separatrix is given by

$$\frac{x}{\xi_x} = \begin{cases} \frac{r_\perp}{\xi_\perp}, & x < \xi_x, \\ \left(\frac{r_\perp}{\xi_\perp}\right)^\zeta, & x > \xi_x. \end{cases} \tag{9.17.12}$$

Bibliography

[1] S. Ramaswamy, The mechanics and statics of active matter. *Ann. Rev. Condens. Matt. Phys.* **1**, 323–345 (2010).

[2] M. C. Marchetti, J. F. Joanny, S. Ramaswamy, et al., Hydrodynamics of soft active matter. *Rev. Mod. Phys.* **85**, 1143 (2015).

[3] S. Ramaswamy and R. A. Simha, Hydrodynamic fluctuations and instabilities in ordered suspensions of self-propelled particles. *Phys. Rev. Lett.* **89**, 058101 (2002).

[4] R. A. Simha and S. Ramaswamy, Statistical hydrodynamics of ordered suspensions of self-propelled particles: waves, giant number fluctuations and instabilities. *Physica A* **306**, 262 (2002).

[5] R. A. Simha, Ph.D. thesis, Indian Institute of Science (2003).

[6] Y. Hatwalne, S. Ramaswamy, M. Rao and R. A. Simha, Rheology of active-particle suspensions. *Phys. Rev. Lett.* **92**, 118101 (2004).

[7] J. Toner, Why walking is easier than pointing: Hydrodynamics of dry active matter, in *Active Matter and Non-Equilibrium Statistical Physics: Lecture Notes of the Les Houches Summer School, September 2018* **112**, eds. G. Gompper, M. C. Marchetti, J. Tailleur, J. M. Yeomans and C. Salomon (Oxford University Press, Oxford, 2022), p. 52–101.

[8] C. Reynolds, Flocks, herds and schools: A distributed behavioral model. *Comput. Graph.* **21**, 25 (1987).

[9] J. L. Deneubourg and S. Goss, Collective patterns and decisions making. *Ethol. Ecol. Evol.* **1**, 295 (1989).

[10] A. Huth and C. Wissel, in *Biological Motion*, eds. W. Alt and E. Hoffmann (Springer Verlag, Berlin/Heidelberg, 1990), p. 577–590.

[11] B. L. Partridge, The structure and function of fish schools. *Sci. Amer.* **246**, 6, 114–123 (June 1982).

[12] A. Bricard, J.-B. Caussin, N. Desreumaux, O. Dauchot, and D. Bartolo, Emergence of macroscopic directed motion in populations of motile colloids. *Nature* **503**, 95–104 (2013).
[13] D. Geyer, A. Morin, and D. Bartolo, *Nature Mater.* **17**, 789 (2018).
[14] W. Loomis, *The Development of Dictyostelium Discoideum* (Academic, New York, 1982).
[15] J. T. Bonner, *The Cellular Slime Molds* (Princeton University Press, Princeton, NJ, 1967).
[16] W. J. Rappel, A. Nicol, A. Sarkissian, H. Levine, and W. F. Loomis, Self-organized vortex state in two-dimensional dictyostelium dynamics. *Phys. Rev. Lett.* **83,** 1247 (1999).
[17] H. Levine, W. J. Rappel, and I. Cohen, *Phys. Rev. E* **63**, 17101 (2001).
[18] T. Vicsek, A. Czirok, E. Ben-Jacob, I. Cohen, and O. Shochet, Novel type of phase transition in a system of self-driven particles. *Phys. Rev. Lett.* **75,** 1226 (1995).
[19] A. Czirok, H. E. Stanley, and T. Vicsek, Spontaneous ordered motion of self-propelled particles. *J. Phys. A* **30,** 1375 (1997).
[20] P. M. Chaikin and T. C. Lubensky, *Principles of Condensed Matter Physics* (Cambridge University Press, Cambridge, 1995).
[21] J. Toner, Birth, death, and flight: A theory of Malthusian flocks. *Phys. Rev. Lett.* **108**, 088102 (2012).
[22] L. Chen, C. F. Lee, and J. Toner, Moving, reproducing, and dying beyond Flatland: Malthusian flocks in dimensions $d > 2$. *Phys. Rev. Lett.* **125**, 098003 (2020).
[23] L. Chen, C. F. Lee, and J. Toner, Universality class for a non-equilibrium state of matter: a $d = 4 - \epsilon$ expansion study of Malthusian flocks. *Phys. Rev. E* **102**, 022610 (2020).
[24] P. G. deGennes and J. Prost, *The Physics of Liquid Crystals*, 2nd ed. (Clarendon Press, Oxford, 1993).
[25] N. D. Mermin and H. Wagner, Absence of ferromagnetism or antiferromagnetism in one- or two-dimensional isotropic Heisenberg models. *Phys. Rev. Lett.* **17**, 1133 (1966).
[26] P. C. Hohenberg, Existence of long-range order in one and two dimensions. *Phys. Rev.* **158**, 383 (1967).
[27] N. D. Mermin, Absence of ordering in certain classical systems. *J. Math. Phys.* **8**, 1061–1064 (1967).
[28] S. F. Edwards and D. R. Wilkinson, The surface statistics of a granular aggregate. *Proc. R. Soc. Lond. Ser. A* **381**, 1780 (1982).
[29] D. Forster, D. R. Nelson, and M. J. Stephen, Large-distance and long-time properties of a randomly stirred fluid. *Phys. Rev. A* **16**, 732 (1977).

[30] S.-K. Ma, *Modern Theory of Critical Phenomena* (Westview Press, Boulder, CO, 2000).

[31] M. Kardar, G. Parisi, and Y.-C. Zhang, Dynamic scaling of growing interfaces. *Phys. Rev. Lett.* **56**, 889 (1986).

[32] D. R. Nelson, Crossover scaling functions and renormalization-group trajectory integrals. *Phys. Rev. B* **11**, 3504 (1975).

[33] J. Toner and Y. Tu, Long-range order in a two-dimensional dynamical XY model: How birds fly together. *Phys. Rev. Lett.* **75**, 4326 (1995).

[34] Y. Tu, M. Ulm, and J. Toner, Sound waves and the absence of Galilean invariance in flocks. *Phys. Rev. Lett.* **80**, 4819 (1998).

[35] J. Toner and Y. Tu, Flocks, herds, and schools: A quantitative theory of flocking. *Phys. Rev. E* **58,** 4828 (1998).

[36] J. Toner, Y. Tu, and S. Ramaswamy, Hydrodynamics and phases of flocks. *Ann. Phys.* **318,** 170 (2005).

[37] J. Toner, A reanalysis of the hydrodynamic theory of fluid, polar-ordered flocks. *Phys. Rev. E* **86**, 031918 (2012).

[38] J. Toner, N. Guttenberg, and Y. Tu, Swarming in the dirt: Ordered flocks with quenched disorder. *Phys. Rev. Lett.* **121**, 248002 (2018).

[39] J. Toner, N. Guttenberg, and Y. Tu, Hydrodynamic theory of flocking in the presence of quenched disorder. *Phys. Rev. E* **98**, 062604 (2018).

[40] L. Chen, C. F. Lee, A. Maitra, and J. Toner, Packed swarms on dirt: two dimensional incompressible flocks with quenched and annealed disorder. *Phys. Rev. Lett.* **129**, 188004 (2022).

[41] L. Chen, C. F. Lee, A. Maitra, and J. Toner, Incompressible polar active fluids with quenched disorder in two dimensions. *Phys. Rev. E.* **106**, 044608 (2022).

[42] L. Chen, C.F. Lee, A. Maitra, and J. Toner, Incompressible polar active fluids with quenched disorder in dimensions $d > 2$. *Phys. Rev. Lett.* **129**, 198001 (2022).

[43] S. Ramaswamy, R. A. Simha, and J. Toner, Active nematics on a substrate: Giant number fluctuations and long-time tails. *Europhys. Lett.* **62**, 196 (2003).

[44] H. Chaté, F. Ginelli, G. Grégoire and F. Raynaud. Collective motion of self-propelled particles interacting without cohesion. *Phys Rev E*, **77**, 046113 (2008).

[45] F. Ginelli, The physics of the Vicsek model. *Eur. Phys. J. Special Topics* **225**, 2099–2117 (2016).

[46] S. Shankar, S. Ramaswamy, and M. C. Marchetti, Low-noise phase of a two-dimensional active nematic system. *Phys. Rev. E* **97**, 012707 (2018).

[47] J. Toner, Giant number fluctuations in dry active polar fluids: A shocking analogy with lightning rods. *J. Chem. Phys.* **150**, 154120 (2019).

[48] S. Henkes, Y. Fily, and M. C. Marchetti, Active jamming: self-propelled soft particles at high density. *Phys. Rev. E* **84,** 040301(R) (2011).

[49] L. Chen, C. F. Lee, and J. Toner, Incompressible polar active fluids in the moving phase in dimensions $d > 2$. *New J. Phys.* **20**, 113035 (2018).

[50] L. Chen, C. F. Lee, and J. Toner, Mapping two-dimensional polar active fluids to two-dimensional soap and one-dimensional sandblasting. *Nat. Commun.* **7**, 12215 (2016).

[51] R. Ramaswamy, G. Bourantas, F. Jülicher, and I. F. Sbalzarini, A hybrid particle-mesh method for incompressible active polar viscous gels. *J. Comput. Phys.* **291**, 334 (2015).

[52] H. H. Wensink, J. Dunkel, S. Heidenrich, et al., Meso-scale turbulence in living fluids. *Proc. Nat. Acad. Sci. USA* **109**, 14308 (2012).

[53] D. J. G. Pearce, A. M. Miller, G. Rowlands, and M. Turner, Role of projection in the control of bird flocks *Proc. Nat. Acad. Sci. USA* **111**, 10422 (2014).

[54] E. Bertin, M. Droz, and G. Grégoire, Boltzmann and hydrodynamic description for self-propelled particles *Phys. Rev. E* **74**, 022101 (2006).

[55] E. Bertin, M. Droz, and G. Grégoire, Hydrodynamic equations for self-propelled particles: microscopic derivation and stability analysis *J. Phys. A: Math. Theor.* **42**, 445001 (2009).

[56] S. Mishra, A. Baskarin, and M. C. Marchetti, Fluctuations and pattern formation in self-propelled particles. *Phys. Rev. E* **81**, 061916 (2010).

[57] F. D. C. Farrell, M. C. Marchetti, D. Marenduzzo, and J. Tailleur, Pattern formation in self-propelled particles with density dependent motility. *Phys. Rev. Lett.* **108**, 248101 (2012).

[58] S. Yamanaka and T. Ohta, Formation and collision of traveling bands in interacting deformable self-propelled particles. *Phys. Rev. E* **89**, 012918 (2014).

[59] J. Bialké H. Löwen, and T. Speck, Microscopic theory for the phase separation of self-propelled repulsive disks. *EPL* **103**, 30008 (2013).

[60] T. Ihle, 2013 Invasion-wave-induced first-order phase transition in systems of active particles. *Phys. Rev. E.* **88**, 040303 (2013).

[61] P. M. Chaikin and T. C. Lubensky, *Principles of Condensed Matter Physics* (Cambridge University Press, Cambridge, 1995).

[62] A. Caille, Remarques sur la diffusion des rayons X dans les smectiques. *C. R. Acad. Sci., Ser. B* **274**, 891 (1972).

[63] P. G. de Gennes, Theory of the smectic state of liquid crystals. *J. Phys. (Paris)* **30**, 9 (1969).

[64] G. Grinstein and R. A. Pelcovits, Anharmonic effects in bulk smectic liquid crystals and other "one-dimensional solids." *Phys. Rev. Lett.* **47,** 856 (1981).
[65] G. F. Mazenko, S. Ramaswamy, and J. Toner, Viscosities diverge as $1/\omega$ in smectic-A liquid crystals. *Phys. Rev. Lett.* **49**, 51 (1982).
[66] G. F. Mazenko, S. Ramaswamy, and J. Toner, Breakdown of conventional hydrodynamics for smectic-A, hexatic-B, and cholesteric liquid crystals. *Phys. Rev. A* **28**, 1618 (1983).
[67] A. Kashuba, Exact scaling of spin-wave correlations in the 2D XY ferromagnet with dipolar forces. *Phys. Rev. Lett.* **73,** 2264 (1994).
[68] L. Golubović and Z.-G. Wang, Anharmonic elasticity of smectic A and the Kardar–Parisi–Zhang model. *Phys. Rev. Lett.* **69,** 2535 (1992).
[69] L. Golubović and Z.-G. Wang, Kardar–Parisi–Zhang model and anomalous elasticity of two- and three-dimensional smectic A liquid crystals. *Phys. Rev. E* **49,** 2567 (1994).
[70] L. Chen, C. F. Lee, A. Maitra, and J. Toner, Dynamics of packed swarms: time-displaced correlators of two-dimensional incompressible flocks. *Phys. Rev. E* **109**, L012601 (2024).
[71] T. R. Malthus, *An Essay on the Principle of Population*, ed. J. Johnson (St. Paul's Churchyard, London, 1798).
[72] J. Ranft, M. Basan, J. Elgeti, et al., Fluidization of tissues by cell division and apoptosis. *Proc. Nat. Acad. Sci. USA* **107**, 20863 (2010).
[73] J. Ranft, J. Prost, F. Jülicher, and J.-F. Joanny, Tissue dynamics with permeation. *Eur. Phys. J. E* **35**, 46 (2012).
[74] M. E. Cates, D. Marenduzzo, I. Pagonabarraga, and J. Tailleur, Arrested phase separation in reproducing bacteria creates a generic route to pattern formation. *Proc. Nat. Acad. Sci. USA* **107**, 26, 11715 (2010).
[75] D. Dell'Arciprete, M. L. Blow, et al. A growing bacterial colony in two dimensions as an active nematic, *Nature Comm.* **9**, 4190 (2018).
[76] Z. You, D. J. G. Pearce, A. Sengupta, and L. Giomi, Geometry and mechanics of microdomains in growing bacterial colonies. *Phys. Rev. X* **8**, 031065 (2018).
[77] A. Doostmohammadi, S. P. Thampi, and J. M. Yeomans, Defect-mediated morphologies in growing cell colonies. *Phys. Rev. Lett.* **117**, 048102 (2016).
[78] K. Kruse, J.-F. Joanny, F. Jülicher, J. Prost, and K. Sekimoto, Generic theory of active polar gels: a paradigm for cytoskeleton dynamics. *Eur. Phys. J. E* **16**, 5 (2005).
[79] J. Brugués and D. Needleman, Physical basis of spindle self-organization. *Proc. Nat. Acad. Sci. USA* **111**, 52, 18496 (2014).
[80] L. P. Dadhichi, J. Kethapelli, R. Chajwa, S. Ramaswamy, and A. Maitra, Nonmutual torques and the unimportance of motility for long-range order in two-dimensional flocks. *Phys. Rev. E* **101**, 052601 (2020).

[81] G. Grinstein, D. H. Lee, and S. Sachdev, Conservation laws, anisotropy, and "self-organized criticality" in noisy non-equilibrium systems. *Phys. Rev. Lett.* **64**, 1927 (1990).

[82] W. Nikola-Lisa, John Henry: Then and Now. *African American Rev.* **32**, 51 (1998).

[83] G. Grégoire, H. Chaté, and Y. Tu, Comment on "Particle diffusion in a quasi-two-dimensional bacterial bath." *Phys. Rev. Lett.* **86**, 556 (2001).

[84] G. Grégoire, H. Chaté, and Y. Tu, Active and passive particles: Modeling beads in a bacterial bath. *Phys. Rev. E* **64**, 11902 (2001).

[85] G. Grégoire, H. Chaté, and Y. Tu, Moving and staying together without a leader. *Physica D* **181**, 157–171 (2003).

[86] G. Grégoire and H. Chaté, Onset of collective and cohesive motion. *Phys. Rev. Lett.* **92**(2), 025702 (2004).

[87] F. Family and D. P. Landau, *Kinetics of Aggregation and Gelation* (North-Holland, Amsterdam, 1984).

[88] F. Ginelli and H. Chaté, Relevance of metric-free interactions in flocking phenomena. *Phys. Rev. Lett.* **105**, 168103 (2010).

[89] A. Peshkov, S. Ngo, E. Bertin, H. Chaté, and F. Ginelli, Continuous theory of active matter systems with metric-free interactions. *Phys. Rev. Lett.* **109**, 098101 (2012).

[90] B. Mahault, X.-c. Jiang, E. Bertin, et al., Self-propelled particles with velocity reversals and ferromagnetic alignment: Active matter class with second-order transition to quasi-long-range polar order. *Phys. Rev. Lett.* **120**, 258002 (2018).

[91] D. Nesbitt, G. Pruessner, and C. F. Lee, Uncovering novel phase transitions in dense dry polar active fluids using a lattice Boltzmann method. Preprint, ArXiv:1902.00530.

[92] A. Aharony, Critical behavior of amorphous magnets. *Phys. Rev. B* **12**, 1038 (1975); in *Phase Transitions and Critical Phenomena VI* (Academic Press, London, 1976).

[93] L. Chen, C. F. Lee, and J. Toner, Squeezed in three dimensions, moving in two: Hydrodynamic theory of three-dimensional incompressible easy-plane polar active fluids. *Phys. Rev. E* **98**, 040602(R) (2018).

[94] A. B. Webb, I. M. Lengyel, D. J. Jörg, et al., Persistence, period and precision of autonomous cellular oscillators from the Zebrafish segmentation clock. *eLife*, 08438 (2016).

[95] F. Jülicher and S. Eaton, Emergence of tissue shape changes from collective cell behaviours. *Sem. Cell Dev. Biol.* **67**, 103 (2017).

[96] C. Reynolds, Flocks, herds, and schools: a distributed behavioral model. *Comp. Graph.* **21**, 25 (1987).

[97] J. L. Deneubourg and S. Goss, Collective patterns and decision-making. *Ethol. Ecol. Evol.* **1,** 295 (1989).

[98] B. L. Partridge, The structure and function of fish schools. *Sci. Amer.*, **246,** 114–123 (June 1982).

[99] R. Voituriez, J. F. Joanny, and J. Prost, Spontaneous flow transition in active polar gels. *Europhys. Lett.* **70,** 404 (2005).

[100] K. Kruse, J. F. Joanny, F. Jülicher, J. Prost, and K. Sekimoto, Generic theory of active polar gels: a paradigm for cytoskeletal dynamics. *Euro. Phys. J.* **16,** 5 (2005).

[101] L. D. Landau and E. M. Lifshitz, *Fluid Mechanics* (Pergamon Press, Oxford, 1959).

[102] L. Chen, J. Toner, and C. F. Lee, Critical phenomenon of the order–disorder transition in incompressible active fluids. *New J. Phys.* **17,** 042002 (2015).

[103] L.-H. Tang, Steady-state scaling function of the (1+1)-dimensional single-step model. *J. Stat. Phys.* **67,** 819 (1992).

[104] E. Frey, U. C. Täuber, and T. Hwa, Mode-coupling and renormalization group results for the noisy Burgers equation. *Phys. Rev. E* **53,** 4424 (1996).

[105] M. E. Fisher and A. Aharony, Dipolar interactions at ferromagnetic critical points. *Phys. Rev. Lett.* **30**, 559–562 (1973).

[106] A. Aharony and M. E. Fisher, Critical behavior of magnets with dipolar interactions. I. Renormalization group near four dimensions. *Phys. Rev. B* **8**, 3323 (1973).

[107] A. Aharony, Critical behavior of magnets with dipolar interactions. II. Feynman graph expansion for ferromagnets near four dimensions. *Phys. Rev. B* **30**, 3342 (1973).

[108] A. D. Bruce and A. Aharony, Critical exponents of ferromagnets with dipolar interactions: Second-order ε expansion. *Phys. Rev. B* **10**, 2078 (1974).

[109] M. Prähofer and H. Spohn Exact scaling functions for one-dimensional stationary KPZ growth. *J. Stat. Phys.* **115**, 255–279 (2004).

[110] S. Ramaswamy, The mechanics and statics of active matter. *Ann. Rev. Condens. Matt. Phys.* **1**, 323–345 (2010).

[111] C. Bechinger, R. Di Leonardo, H. Löwen, et al., Active particles in complex and crowded environments. *Rev. Mod. Phys.* **88**, 045006 (2016).

[112] F. Schweitzer, *Brownian Agents and Active Particles: Collective Dynamics in the Natural and Social Sciences*. Springer Series in Synergetics (Springer, New York, 2003).

[113] H. Chaté, Dry aligning dilute active matter. *Ann. Rev. Condens. Matt. Phys.* **11**, 189 (2020).

[114] B. Mahault, F. Ginelli, and H. Chate, Quantitative assessment of the Toner and Tu theory of polar flocks. *Phys. Rev. Lett.* **123**, 218001 (2019).

Index

Printed in the United States
by Baker & Taylor Publisher Services

Printed in the United States
by Baker & Taylor Publisher Services